New Concepts and Trends of Hybrid Multiple Criteria Decision Making

New Concepts and Trends of Hybrid Multiple Criteria Decision Making

By
Gwo-Hshiung Tzeng and Kao-Yi Shen

CRC Press is an imprint of the
Taylor & Francis Group, an **informa** business

CRC Press
Taylor & Francis Group
6000 Broken Sound Parkway NW, Suite 300
Boca Raton, FL 33487-2742

© 2017 by Taylor & Francis Group, LLC
CRC Press is an imprint of Taylor & Francis Group, an Informa business

No claim to original U.S. Government works

Printed on acid-free paper

International Standard Book Number-13: 978-1-4987-7708-7 (Hardback)

This book contains information obtained from authentic and highly regarded sources. Reasonable efforts have been made to publish reliable data and information, but the author and publisher cannot assume responsibility for the validity of all materials or the consequences of their use. The authors and publishers have attempted to trace the copyright holders of all material reproduced in this publication and apologize to copyright holders if permission to publish in this form has not been obtained. If any copyright material has not been acknowledged please write and let us know so we may rectify in any future reprint.

Except as permitted under U.S. Copyright Law, no part of this book may be reprinted, reproduced, transmitted, or utilized in any form by any electronic, mechanical, or other means, now known or hereafter invented, including photocopying, microfilming, and recording, or in any information storage or retrieval system, without written permission from the publishers.

For permission to photocopy or use material electronically from this work, please access www.copyright.com (http://www.copyright.com) or contact the Copyright Clearance Center, Inc. (CCC), 222 Rosewood Drive, Danvers, MA 01923, 978-750-8400. CCC is a not-for-profit organization that provides licenses and registration for a variety of users. For organizations that have been granted a photocopy license by the CCC, a separate system of payment has been arranged.

Trademark Notice: Product or corporate names may be trademarks or registered trademarks, and are used only for identification and explanation without intent to infringe.

Library of Congress Cataloging-in-Publication Data

Names: Tzeng, Gwo-Hshiung, author.
Title: New concepts and trends of hybrid multiple criteria decision making /
Gwo-Hshiung Tzeng and Kao-Yi Shen.
Description: New York : CRC Press, 2017.
Identifiers: LCCN 2017000939 | ISBN 9781498777087 (hbk : alk. paper)
Subjects: LCSH: Multiple criteria decision making. | Decision making. |
Problem solving.
Classification: LCC T57.95 .T94 2017 | DDC 658.4/03--dc23
LC record available at https://lccn.loc.gov/2017000939

Visit the Taylor & Francis Web site at
http://www.taylorandfrancis.com

and the CRC Press Web site at
http://www.crcpress.com

Contents

Preface xi

Authors xiii

1 Introduction 1
  1.1 Overview of Traditional MCDM Techniques and Methods............ 1
  1.2 Statistics versus MCDM Approach... 4
  1.3 History of MADM.. 5
  1.4 History of MODM.. 8
  1.5 Developments in Computational Intelligence, Machine
       Learning, and Soft Computing for Decision Aids 10
       1.5.1 Basic Concepts of Fuzzy Sets 13
       1.5.2 Basic Notions of Rough Sets 14
  1.6 Emerging Trend in Multiple Rule-Based Decision Making............16
  1.7 Outline of the Book...19

SECTION I CONCEPTS AND THEORY

2 New Concepts and Trends in MCDM 23
  2.1 Problem Solving in Traditional MCDM...25
  2.2 Why New Hybrid MCDM Approaches Are Needed 26
  2.3 Framework of New Hybrid MCDM Models for Tomorrow 28

3 Basic Concepts of DEMATEL and Its Revision 35
  3.1 Background and Basic Notions of DEMATEL 36
  3.2 Operational Steps of the Original DEMATEL............................... 36
  3.3 Infeasibility of the Original DEMATEL Technique...................... 39
  3.4 Revised DEMATEL 40
  3.5 Two Numerical Examples 41
  3.6 Conclusion 47

4 DEMATEL Technique for Forming INRM and DANP Weights 49
  4.1 Methodology for Assessing Real-World Problems.......................... 49
  4.2 Constructing an Influential Network Relations Map 50
  4.3 Determining Influential Weights Using DANP 55

v

vi ■ *Contents*

4.4 Problem Solving for Ranking or Selection Decision by INRM and DANP 57

4.5 Conclusion 58

5 Traditional MADM and New Hybrid MADM for Problem Solving 61

5.1 Traditional MADM for Ranking and Selection 61

5.1.1 AHP and ANP.. 62

5.2 New Hybrid Modified MADM... 63

5.2.1 DEMATEL-Based ANP Instead of AHP and ANP......... 64

5.2.2 Modified VIKOR for Measuring Performance Gaps..........65

5.3 Additive and Nonadditive Types of Aggregators............................. 68

5.3.1 Additive-Type Aggregators 68

5.3.2 Nonadditive-Type Aggregators (Fuzzy Integrals).............. 68

5.4 Conclusion 72

6 MODM with De Novo and Changeable Spaces 73

6.1 Basic Concepts and Trends of MODM 74

6.2 De Novo Programming.. 78

6.3 MOP with Changeable Parameters.. 83

6.4 Discussion 89

6.5 Conclusion 91

7 Multiple Rules-Based Decision Making for Solving Data-Centric Problems...93

7.1 Variable-Consistency Dominance-Based Rough Set Approach 94

7.2 Basic Notions of the Reference Point-Based MRDM Approach 96

7.3 Core Attribute-Based MRDM Approach 100

7.4 Hybrid Bipolar MRDM Approach 100

7.4.1 Dominance-Based Rough Set Approach 101

7.4.2 Evaluations for an Aggregated Bipolar Decision Model......104

7.5 Conclusion 105

SECTION II   APPLICATIONS OF MCDM

8 The Case of DEMATEL for Assessing Information Risk 109

8.1 Background of the Case and the Research Framework.................110

8.2 DEMATEL Analysis with INRM...113

8.3 DANP Influential Weights for Criteria... 114

8.4 Discussion and Conclusion..117

9 E-Store Business Evaluation and Improvement Using a Hybrid MADM Model...119

9.1 Background of the Case and the Research Framework 120

9.2 DANP for Finding Influential Weights 122

Contents ■ **vii**

9.3 Performance Measures and Modified VIKOR for Evaluations 126
9.4 Discussion 126
9.5 Conclusion 132

10 Improving the Performance of Green Suppliers in the TFT-LCD Industry 133
10.1 Background of the Case... 134
10.2 Research Framework and the Selected Criteria 135
10.3 DANP for Finding Influential Weights of Criteria 136
10.4 Modified PROMETHEE 137
10.5 Discussion 147
10.6 Conclusion 149

11 Exploring Smartphone Improvements Based on a Hybrid MADM Model 151
11.1 Background of the Case... 152
11.2 Research Framework 152
11.3 DEMATEL Analysis and DANP Influential Weights for Criteria... 156
11.4 Modified VIKOR for Performance Gap Aggregation 161
11.5 Discussion on the Improvements of Smartphone Vendors 165
11.6 Conclusion 166

12 Evaluating the Development of Business-to-Business M-Commerce of SMEs 169
12.1 Research Background 169
12.2 Research Framework 170
12.2.1 Technological Environment Aspect...............................170
12.2.2 Organizational Environment Aspect.............................171
12.2.3 External Environment Aspect 171
12.3 DEMATEL Analysis and DANP Influential Weights for Criteria... 172
12.4 Modified VIKOR for Performance Gap Aggregation 174
12.5 Discussion 177
12.6 Conclusion 181

13 Evaluation and Selection of Glamour Stocks by a Hybrid MADM Model 183
13.1 Research Background and Investment Strategy 184
13.2 Research Framework for the G-Score and Hybrid MADM Models...184
13.3 DEMATEL Analysis and DANP Influential Weights of the G-Score Model 186

viii ◾ *Contents*

13.4 Modified VIKOR for Performance Gap Aggregation and Evaluation 192
13.5 Discussion and Examination of Stock Returns 195
13.6 Conclusion 198

14 Nonadditive Hybrid MADM Model for Selecting and Improving Suppliers 201
14.1 Research Background and Literature Review.............................. 202
14.1.1 Multiple Attribute Decision Making............................. 203
14.1.2 Mathematical Programming Models 203
14.1.3 Combined and Integrated Hybrid Approaches.............. 203
14.3 Hybrid MADM Model Using Nonadditive-Type Aggregators..... 204
14.3.1 Determining Gap Values Based on the New Concepts of Modified VIKOR Method.. 205
14.3.2 Applying λ Fuzzy Measures for Fuzzy Integrals.............. 206
14.4 Numerical Example.. 207
14.4.1 Measuring the Influential Relations and Weights by DEMATEL and DANP... 208
14.4.2 Integrated Weighted Gaps Using the Fuzzy Integral Technique 208
14.5 Discussion on Improving toward the Aspired Levels 216
14.6 Conclusion 217

15 New Perspectives on Modeling Strategic Alliances by De Novo Programming 219
15.1 Introduction to Strategic Alliances and Literature Review............219
15.1.1 Transaction Cost Theory.. 220
15.1.2 Resource-Dependent Theory... 220
15.1.3 Strategic Behavior and Organizational Learning Perspectives 221
15.2 Resource Allocation and De Novo Programming Perspectives 221
15.3 Numerical Example.. 224
15.4 Discussion 226
15.5 Conclusion 227

16 Automated Factory Planning Using the New Idea of Changeable Spaces 229
16.1 Background of the Case... 230
16.2 Research Framework and the Evolution of Optimization 231
16.2.1 Conventional Pareto Solution 232
16.2.2 De Novo Programming for Optimization...................... 232
16.2.3 New Ideas of Changeable Spaces...................................233
16.3 Discussion 235
16.4 Conclusion 236

Contents ■ ix

17 Fuzzy Inference-Supported MRDM for Technical Analysis: A Case of Stock Investment 239

17.1 Background of Technical Analysis and Computational Intelligence Techniques 240

17.2 Hybrid Investment Support System Based on Fuzzy and Rough Set Techniques 242

17.3 Numerical Experiments 244

    17.3.1 Data Preprocessing 244

    17.3.2 Discretization of Fuzzy TA Signals by Fuzzy Inference Systems 249

    17.3.3 VC-DRSA Model 250

17.4 Simulated Investment Performance and Discussion 252

17.5 Conclusion 254

18 Financial Improvements of Commercial Banks Using a Hybrid MRDM Approach 255

18.1 Research Background 255

18.2 Core Attributes–Based MRDM Approach for Financial Performance Improvement 257

18.3 An Empirical Case of Five Commercial Banks 259

18.4 Analytical Results by the Modified VIKOR Method 264

18.5 Discussion 268

18.6 Conclusion 270

19 FCA-Based DANP Model Using the Rough Set Approach: A Case Study of Semiconductor Companies 273

19.1 Research Background 273

19.2 Reviews of MCDM and Soft Computing Methods in Financial Applications 275

19.3 Framework of the Hybrid MRDM Model 278

19.4 Case Study of Semiconductor Companies 280

19.5 FCA-Based DANP Model for Ranking Improvement Plans 284

19.6 Discussion on Improvement Planning 286

19.7 Conclusion 288

20 Hybrid Bipolar MRDM Model For Business Analytics 289

20.1 Background of Business Analytics 290

20.2 Hybrid MRDM Model Using the Bipolar Approach 291

20.3 A Case from the Semiconductor Industry 298

    20.3.1 Data 298

    20.3.2 Bipolar Weighting System 299

    20.3.3 Aggregate Fuzzy Performance Evaluations Using Modified VIKOR 302

x ■ *Contents*

20.4 Results of the Evaluations 305
20.5 Discussion on Contextual Improvement Planning 306
20.6 Conclusion 308

References **309**

Index **343**

Preface

This book introduces new hybrid multiple criteria decision-making (MCDM) methods for solving real-world problems with improvement planning. Traditional MCDM ignores some vital issues, such as certain unrealistic assumptions and constraints from statistics or economics. This book proposes several new concepts and demonstrates how current trends in hybrid MCDM models can be adopted to solve practical problems. In this book, three interrelated parts are introduced: (1) hybrid MRDM (multiple rule/rough-based decision-making), (2) hybrid MADM (multiple attribute decision-making), and (3) hybrid MODM (multiple objective decision-making). The features and contributions of this book are as follows:

In Hybrid MRDM (Multiple Rule/Rough-Based Decision-Making)

First, how to select the minimum and critical criteria to forming a decision model is difficult while facing a complex decision problem. One of the plausible approaches to resolving this issue is extracting the core attributes based on the essential concepts of rough set theory (RST). The core attributes can then be applied to construct a hybrid MRDM model. The MRDM approach can generate decision rules in the form of "if ... then ..." logic. Furthermore, this book suggests a new approach by using dominance-based rough set approach (DRSA) or variable-consistency dominance-based rough set approach (VC-DRSA) to build a directional flow graph (DFG). It combines decision-making trial and evaluation laboratory (DEMATEL) analytics with the extracted rules, to form a hybrid MRDM model, which help decision makers (DMs) unravel the hidden or vague patterns/logics of a complex problem.

In Hybrid MADM (Multiple Attribute Decision-Making)

Second, traditional MADM studies have focused on obtaining the final ranking results or forming a utility function for modeling preferences. However, in most real-world problems, the relationships between criteria or aspects are usually interdependent. Therefore, the DEMATEL technique is proposed to resolve this issue by identifying the interdependent relations among dimensions (criteria). To construct a decision model that allows the existence of interrelationship among criteria helps avoid some unrealistic assumptions in statistics and economics, which is usually overlooked in the mainstream

xi

social science research. Furthermore, the analytical result from DEMATEL can form an influential network relation map (INRM), which supports to adjusting and identifying the influential weights in DEMATEL-based ANP (DANP) method.

Third, in this book, we shift the focus from selecting a relatively favorable one from a group of candidates to pursuing the aspired levels on all aspects. By searching for aspiration levels at each stage, the decision process continuously improves and enhances a system or model. For example, the classical VIKOR method was modified, where the traditional settings adopt maximum–minimum values as the positive and negative ideal points. In the modified VIKOR method, the aspiration level and the least desirable value of each criterion replace the tradition settings, which can avoid decisions that pick the best apple among a barrel of rotten ones, so to speak.

Fourth, the new hybrid MADM analytical tools put more emphases on performance improvement. The focus in the field has shifted from ranking and selection to performance improvement. In practice, a systematic approach for problem-solving is much more valuable than merely addressing the systematic structure of the problem.

Fifth, Kahneman and Tversky (Kahneman received the Nobel Prize in Economics in 2002) used additive-type utility aggregation models in their studies during the 1960s and found that consumers' product selection behaviors were not consistent with the traditional concept of *multivariate utility* based on value function aggregation. In 1974, Michio Sugeno completed his doctoral thesis, "Theory of fuzzy integrals and its applications," in the Tokyo Institute of Technology. In that work, the fuzzy integral technique was proposed to aggregate the values of multiple attributes in a nonadditive approach. The mathematical modeling of nonadditive aggregation can adjust the problematic assumption in traditional additive type models. This finding inspired Kahneman to propose the *Prospect theory* in 1978 by adopting the basic concept of nonadditive aggregation. In this book, this nonadditive aggregation method is incorporated with the other methods to form new hybrid MADM models.

In Hybrid MODM (Multiple Objective Decision-Making)

Sixth, conventional MODM methods in planning and designing are based on a decision space defined by a set of fixed conditions and resources. The objective function is defined in a feasible space with fixed requirements. However, the new concept of changeable spaces programming in MODM may transform or extend those fixed requirements or constraints, to search for results approaching the aspiration levels. This new method can support DMs to achieve win–win solutions and pursue continuous improvements. Therefore, this new concept of changeable spaces would be superior to merely searching the ideal points or Pareto-optimal solutions in conventional MODM approaches.

Finally, we hope that this book provides value to our readers and benefits their research ahead. Also, we are grateful to the financial supports from the Ministry of Science and Technology of Taiwan under the following grant numbers: (1) 104-2410-H-305-052-MY3 and (2) 105-2410-H-034-019-MY2.

Authors

Gwo-Hshiung Tzeng was born in 1943 in Taiwan. In 1977, he received a PhD at the Graduate School of Economics (majoring in management science) from Osaka University, Japan. He has served as a distinguished chair professor at Tatung University, Taiwan, from 1977 to 1978, and at National Taipei University, Taiwan, since 2013. Formerly, he was an associate professor at National Chiao Tung University, Taiwan, from 1978 to 1981; a research associate at Argonne National Laboratory, Lemont, Illinois, from 1981 to 1982; a visiting professor in the Department of Civil Engineering at University of Maryland, College Park, from 1989 to 1990; a visiting professor in the Department of Engineering and Economic Systems, Energy Modeling Forum, at Stanford University, California, from 1997 to 1998, a professor at National Chiao Tung University from 1981 to 2003 and a lifelong chair professor there from 2003, and a president from 2004 to 2005 and a distinguished chair professor from 2004 to 2013 at Kainan University, Taiwan. His current research interests include statistics, multivariate analysis, econometrics, network routing and scheduling, multiple criteria decision making, fuzzy theory, technology management, energy conservation, environment management, transportation systems and management, transportation planning and investment, logistics, locations, urban planning, tourism management, and electronic commerce. He received the MCDM Edgeworth-Pareto Award from the International Society on Multiple Criteria Decision Making in June 2009, the Pinnacle of Achievement Award in 2005, the lifelong research project funding from the National Science Council (now the Ministry of Science and Technology Affairs, Taiwan) after obtaining three outstanding research awards (1985, 1993, 1995) and two special investigator awards (1997–2000, 2000–2003), and the National Distinguished Chair Professor Award (the highest honor offered) from the Ministry of Education Affairs, Taiwan, in 2000.

Kao-Yi Shen received his bachelor and master degrees in industrial engineering from Tunghai University, Taiwan, in 1992 and 1998, respectively. In 2002, he received a PhD degree in business administration from Chengchi University, Taiwan. He has worked as a senior analyst in the venture capital industry and as a marketing manager and head of project management in a Taiwan-based IT company. He has developed more than ten IT products, and he has also held several patents in TV signal-related applications in Taiwan, the United States, and China. He was an assistant professor from August 2009 to February 2015 and is now an associate professor in the Department of Banking and Finance at Chinese Culture University (School of Continuing Education), Taiwan. He has published several papers on solving financial problems using soft computing and MCDM methods in journals such as *Soft Computing*, *Applied Soft Computing*, *Knowledge-Based Systems*, *International Journal of Fuzzy Systems*, and *Information Sciences*. He has also served as an associate editor for the *Journal of International Fuzzy Systems* since 2016.

Chapter 1

Introduction

This book introduces new concepts and trends in hybrid multiple criteria decision making (MCDM) methods for solving real-world problems with improvement planning. According to Zavadskas et al. (2016), hybrid MCDM (or HMCDM) denotes an approach that combines or integrates more than two methods to solve the same multiple criteria problem. Traditional MCDM ignores some vital issues such as certain unrealistic assumptions and constraints. This book proposes several new concepts and demonstrates how current trends in hybrid MCDM models can facilitate solving practical problems (Liou and Tzeng 2012; Liou 2013; Peng and Tzeng 2013; Huang and Tzeng 2014). In this book, several new techniques are introduced, including the dominance-based rough set approach (DRSA), the decision rule–based bipolar method, formal concept analysis, the DEMATEL technique for building influential network relation maps and finding the influential weights of a DEMATEL-based analytic network process (ANP), modified evaluation methods (e.g., SAW, VIKOR, grey analysis, PROMETHEE, and ELECTRE) for setting performance improvement strategies, and multiple objective programming (MOP) with changeable spaces for optimizing (aspiration-level) resource allocation (Figure 1.1).

Hybrid MCDM models have already successfully analyzed the characteristics of various real-world problems (Zavadskas and Turskis 2011; Tzeng and Huang 2011; Liou and Tzeng 2012; Peng and Tzeng 2013; Greco et al. 2016). Several empirical cases from the financial and information technology (IT) industries illustrating MCDM methods for researchers and practitioners are provided in Section II.

1.1 Overview of Traditional MCDM Techniques and Methods

The goals of traditional decision making can be sorted into three categories: ranking, selection, and optimization. A typical MCDM problem usually begins with

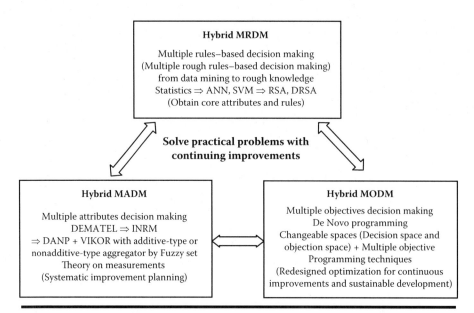

Figure 1.1 New hybrid approach in MCDM based on three pillars.

identifying the goal and the related criteria, attributes, or constraints. Subsequently, to evaluate the available alternative solutions to a problem, researchers must collect the preferences of the decision makers (DMs) regarding the available alternatives in the context of the pertinent criteria. If only a single criterion (or attribute) was considered, it would be intuitive to choose the alternative with the highest preference rating. In the presence of multiple criteria (or attributes), DMs are usually compelled to compromise, even though none of the alternatives outperforms or dominates the others on all the criteria. Therefore, various quantitative methods have been proposed to support DMs who must make decisions that advance predefined goals.

Following Hwang and Yoon (1981), MCDM problems can be further categorized into two subfields: multiple attribute decision making (MADM) and multiple objective decision making (MODM). MADM is concerned with ranking or selecting by evaluating predetermined alternatives, and MODM is aimed at identifying the optimal outcome by searching for an efficient frontier within a solution space under the given constraints. Most conventional MCDM research comprises these two subfields (Köksalan et al. 2011).

Because MADM focuses on evaluating alternatives according to a DM's preference, it is reasonable to adopt the utility theory from economics to depict a DM's preferences for multiple attributes in the form of preference (value or utility) functions; this is performed in multiple attribute utility theory (MAUT). Although MAUT has solid foundations from economic theories, its essential assumption is based on the independence of attributes, which is not realistic in most real-world cases. The utility additive (UTA) method extends the MAUT method and uses regression and linear programming techniques to reach the solution of optimal value (utility) that is consistent with a DM's global preferences. However, just as MAUT assumes the independence of attributes, UTA methods assume the independence of variables (i.e., attributes or criteria).

Another prevalent approach in MADM research is the analytic hierarchy process (AHP), which was introduced by Saaty (1972, 1992). AHP decomposes a problem into hierarchical levels (mainly two: dimensions and criteria), and uses pairwise comparisons to calculate a DM's global preferences regarding the final goal. Certain aggregation methods (additive-type operators), such as SAW, TOPSIS, VIKOR, and PROMETHEE, may integrate with the preference weights obtained by traditional MADM methods in order to support DMs who must make the final ranking or selection. MADM methods are mainly devised for evaluations.

In contrast, MODM is more suitable for design or planning, because it optimizes the allocation of limited resources. Because real-world environments are often full of competing and even conflicting constraints, DMs may rely on MODM methods to achieve multiple goals simultaneously; mathematical programming is mainly adopted to solve those optimization problems. Nevertheless, traditional MODM optimization methods are only concerned with *Pareto-optimal solutions* (also called *noninferior or nondominated solutions* or *effective solutions*) within a system; notions of *essential* optimization are not addressed (Zeleny 1990). When encountering an MODM problem, DMs usually find that the optimization of all objectives in a given system is impossible. Most problems involve trade-offs; one cannot increase the levels of satisfaction for one objective without reducing those for another. Conventional multi-objective programming techniques are often applied to search for noninferior solutions (i.e., Pareto-optimal solutions) or compromise solution, which are transformed into single-objective programs that produce acceptable solutions to DMs. When objective functions conflict with one another, DMs can formalize trade-offs by using conventional methods.

Because trade-offs often prevent the simultaneous optimization of all criteria, Zeleny (1982) offered an MODM method called *De Novo programming* for redesigning systems. Zeleny believed that most resources can be acquired at a reasonable price, and thus, the only constraint is the total budget required to purchase those resources (Babic and Pavic 1996). The main difference between De Novo programming and traditional MODM, then, is that De Novo programming reallocates the resources of a redesigned (or reshaped) system. With this system, the objective functions are subject to existing constraints and can eliminate the trade-offs to achieve a solution that is considered ideal (i.e., an ideal point) in the context of the reconstructed decision space. The aforementioned theories and methods of MADM and MODM can contribute to solving traditional MCDM problems. The latest developments in MCDM are introduced in Chapter 2.

4 ■ *Trends of Hybrid Multiple Criteria Decision Making*

1.2 Statistics versus MCDM Approach

To understand the strengths and limitations of MCDM, one must distinguish the differences between the statistical and MCDM approaches. Researchers who study social science topics usually depend on statistics as a major analytical tool and seek to generalize from sample data collected from a population. The fundamental assumptions of the statistical approach, such as the assumed probabilistic distributions of data sets and the independence of variables, are unrealistic and unsuitable for certain real-world problems with complex and interrelated variables, attributes, and criteria (Liou and Tzeng 2012; Liou 2013; Shen and Tzeng 2015a). Furthermore, MCDM studies are often aimed at solving a predefined problem; therefore, more emphasis is placed on constructing models that may be close to DMs' preferences and yield ideal or satisfactory guidance for decisions. MCDM may select an alternative with the highest preference rating or allocate existing resources efficiently based on the given constraints.

For numerous widely adopted statistical methods and models, such as regression and structural equation modeling, the models' assumptions influence the analytical results. Consider a multiple regression model that explains the performance of suppliers by using several variables (e.g., workforce quality and operational efficiency); the performance of suppliers would be defined as a response variable y, and the explanatory variables x_j (where j is the number of explanatory variables) are assumed to be independent. If n observations are available and each observation's response value is indicated as y_i (for $i = 1,\ldots,n$), then y_i is assumed to be collected from a distribution with theoretical mean μ_i. In addition, y_i would be conceptualized as $\mu_i + \varepsilon_i$ in the regression, where ε_i represents the effect of random errors; again, ε_i is assumed to be generated independently from a normal distribution with zero mean and a specific variation. In the addressed vendor selection model, the assumption for the probabilistic distribution of ε_i is neither identifiable nor examinable (Berk and Freedman 2003); however, it has certain effects on the obtained regression model.

In addition, the assumption that the explanatory variables x_j are independent is questionable in most social science problems. In the assumed case of supplier selection, the quality of a supplier's workforce should have a certain degree of influence on its operational efficiency; nevertheless, statistics requires that those two explanatory variables (i.e., workforce quality and operational efficiency) must be assumed to be independent. As mentioned, a research project based on statistics attempts to generalize its models to support its hypotheses and theories; consequently, such projects must collect data samples that are sufficiently large to be representative of the assumed population, which can only provide averaged numbers (Spronk et al. 2005) from the sample data. Such averaged results can describe or explain the relationships among the explanatory and response variables. By contrast, MCDM studies often address a predefined case in which DMs attempt to select the optimal decision (ranking or resource allocation). In contrast to the statistical approach, the

MCDM approach avoids questionable probabilistic assumptions and seeks to solve problems.

Consider how these two approaches would form the strategy of a large-scale company (e.g., Microsoft or Apple). The statistical approach would tend to collect questionnaires from all available employees or shareholders to determine the average opinion, but the MCDM approach would query the preferences, knowledge, and experience of the managers of the company to devise an optimal strategy. Thus, we may roughly conclude that the statistical approach puts more emphasis on examining the relationships among the variables for theoretical purposes, whereas the MCDM approach focuses on supporting DMs who must solve complicated decision problems in practice.

1.3 History of MADM

When von Neumann and Morgenstern (1947) published a generalized mathematical theory to model the dynamics among the competitors or players in a complex socioeconomic environment, their ideas marked the origin of MADM (Köksalan et al. 2011). This general theory provided various quantitative models with which researchers were able to tackle complicated decision problems. Early advances in MADM developed three major avenues: utility-based approaches, outranking methods, and pairwise comparison approaches. At present, MADM continues to grow rapidly and has incorporated multidisciplinary theories, methodologies, and techniques such as applied mathematics and information science.

Because, as stated in Section 1.1, MADM considers DM preferences regarding multiple attributes, it can benefit from theories of utility that have been developed in economics. The initial attempt was MAUT (Keeney and Raiffa 1972, 1976), which was based on the utility theory of Bernoulli (1738). MAUT attempts to model DM preferences in a hierarchical structure by using various utility functions. Conjoint measurement is a key concern that utility-based methods address by aggregating the multiple values of various attributes with different scales to form a final utility value; numerous pioneering works have addressed this issue (Bouyssou and Pirlot 2005). Although early utility-based methods (e.g., MAUT and UTA) proposed certain techniques to resolve conjoint measurement, the major concern was and is the assumed independence among criteria, which is also termed *preferential independence* (Grabisch 1995; Hillier 2001). In numerous real-world problems, especially social problems, the preferential outcome of one criterion over another is often influenced by the other criteria. Therefore, nonadditive-type aggregators, such as the *Choquet integral* (Choquet 1953) and *fuzzy integral* (Sugeno 1974) aggregations, have been proposed for measuring the interactions among criteria. However, forming an adequate fuzzy measurement for those nonadditive-type aggregators was and remains a challenging task in practice (Tzeng and Huang 2011; Liou et al. 2014). The historical development of MADM is shown in Figure 1.2.

6 ■ *Trends of Hybrid Multiple Criteria Decision Making*

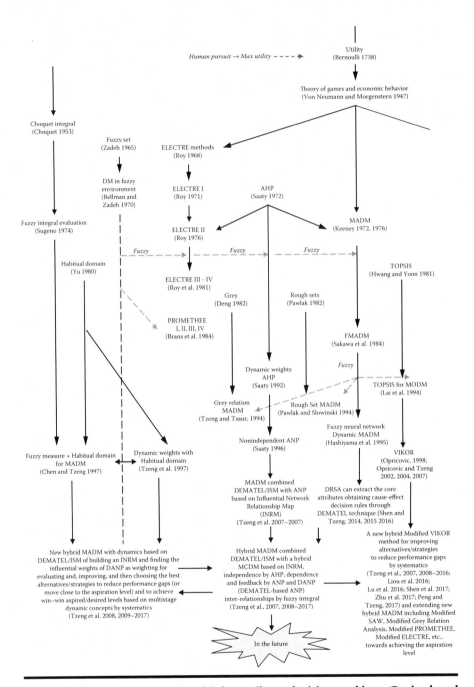

Figure 1.2 Development of multiple attribute decision making. (Revised and reprinted with permission from Liou, J.J.H., and Tzeng, G.H., *Technological and Economic Development of Economy*, 18(4), 672–695, 2012.)

Outranking methods, especially the ELECTRE (Benayoun et al. 1966; Roy 1968) and PROMETHEE families (Brans et al. 1984), compare the preference relations among alternatives to acquire information regarding the optimal alternative. Although empirical outranking methods were proposed for overcoming the practical difficulties experienced with utility functions, outranking methods have been criticized for their lack of axiomatic foundations, particularly for classical aggregation problems, structural problems, and noncompensatory problems (Bouyssou and Vansnick 1986). In addition, the outranking approach requires DMs to pass subjective judgments regarding the performance of each alternative on each criterion. Each weight on each criterion must be assigned before the outranking decision model can be formed. Although each criterion has a different scale, DMs are expected to give consistent and precise judgments on the relative importance of weight. The synthesized utility function must be calculated from a partial set of pairwise comparisons between alternatives, and direct comparisons present difficulties if the number of criteria is large. Unfortunately, practical problems in complicated social or business settings often must account for numerous criteria.

In 1965, fuzzy set theory (Zadeh 1965; Bellman and Zadeh 1970) was proposed to confront the problems of linguistic definitions and uncertain information. Fuzzy sets generalize conventional set theory and have enriched the field of automatic control. Since the 1970s, MADM has applied fuzzy sets to situations of subjective uncertainty.

The third approach forms evaluation models by using pairwise comparisons of attributes or criteria from DMs. This approach has the advantage of requiring each DM to focus on only two criteria or attributes at a time, which streamlines complex analyses. According to the theory that human brains are constrained by bounded rationality (Simon 1972, 1982), pairwise comparisons should assist DMs to provide precise opinions. In contrast to the outranking approach, the pairwise comparison approach can usefully evaluate the relative weights of criteria—even when considering a large number of criteria. Because the third approach also relies heavily on retrieving and transforming experts' or DMs' opinions or judgments to form evaluation models, the impreciseness and uncertainty of those opinions or judgments cannot be overlooked; it is reasonable to incorporate the fuzzy set theory to enhance pairwise comparison methods (Sakawa and Yano 1985; Tzeng and Huang 2011).

The widely adopted AHP (Saaty 1988) was based on the assumption of independent criteria, but the generalized ANP (Saaty 1996) allows for interrelationships among criteria. ANP has been applied to various decision problems in fields such as science, military operations, engineering, and socioeconomics. According to a recent survey from Mardani et al. (2015), except for hybrid fuzzy MADM methods, the AHP and ANP methods might be the most prevalent type of MADM research published over the past 20 years.

8 ■ *Trends of Hybrid Multiple Criteria Decision Making*

1.4 History of MODM

MODM aims to solve optimization problems under specified constraints. In these problems, several (conflicting) objectives must be achieved simultaneously. Traditional MODM adopts mathematical programming to identify the optimal (ideal) solution, but traditional MODM must address two major issues: (1) trade-offs among objectives and (2) scale problems, which are sometimes termed the *curse of dimensionality*.

First, the trade-off problem arises from the conflicts among the objectives; to resolve this issue, the trade-offs between the considered objectives can be identified and multiple objectives can be transformed into a weighted single objective to simplify the original problem. Second, the scaling problem arises when increases in the number of relevant dimensions or criteria cause the computational cost to increase tremendously. The scale problem was considered to be a major challenge to the mainframes of the 1950s and 1960s, but since the late 1980s, personal computers have provided ample computational resources to general researchers. In addition, certain evolutionary techniques, such as genetic algorithms (GAs) (Holland 1975) and particle swarm optimization (PSO) (Coello et al. 2004), have increased the efficiency of searches in solution spaces.

As shown in Figure 1.3, the concept of vector optimization (Kuhn and Tucker 1951) might be considered to be the beginning of MODM. After Yu (1973) proposed the compromise solution method to cope with MODM problems, various applications began to grow rapidly, such as transportation investment, transportation planning, econometrics, development planning, financial planning, business administration, investment portfolio selection, land-use planning, water resource management, public policy, and environmental issues.

Although the main body of traditional MODM methods was extended from simple MOP to multilevel multiobjective programming and multistage multiobjective programming in the 1980s, soft computing, such as fuzzy set theory (Zadeh 1965) and grey theory (Deng 1982), and goal programming, such as data envelopment analysis (DEA) (Charnes et al. 1978), also emerged and began to be incorporated with certain MODM methods for investigating real-world problems.

The uncertainty of real-world environments and the subjective nature of DM judgments have led to soft computing techniques (e.g., fuzzy set theory) playing an influential role in MODM. Objectives and constraints may be provided in the form of natural language terms (e.g., low debt ratios), and fuzzy numbers (variables) are often useful for representing the uncertainty or impreciseness underlying MODM models. Ever since Bellman and Zadeh (1970) proposed various concepts of decision making under fuzzy environments, MADM and MODM researchers have appreciated the effectiveness of fuzzy set theory. Systematic discussions on fuzzy MODM can be found in previous works such as Zimmermann (1978), Hwang and Yoon (1981), Sakawa et al. (1984), Lee and Li (1993), and Tzeng and Huang (2014).

Introduction ■ 9

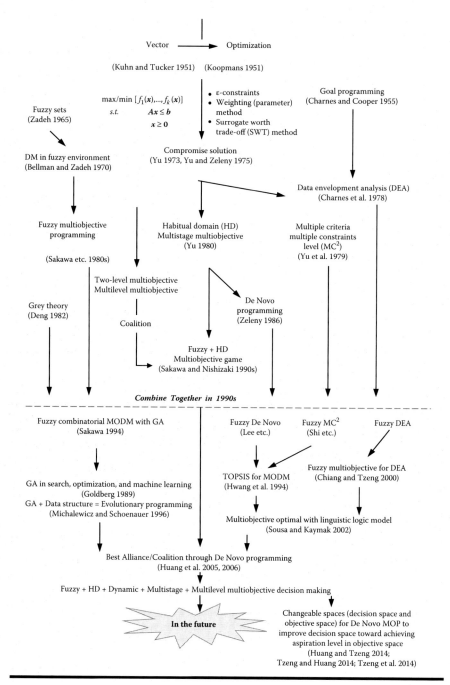

Figure 1.3 Development of multiple objective decision making. (Revised and reprinted with permission from Liou, J.J.H., and Tzeng, G.H., *Technological and Economic Development of Economy*, 18(4), 672–695, 2012.)

DEA might be the most prevalent goal-programming method; it has been widely used to measure the technical efficiency of decision-making units (DMUs). Unlike regression-based methods, DEA is a nonparametric approach and is more suitable for resolving practical problems (Tzeng and Huang 2014) without dubious assumptions concerning parameters in a model.

The first DEA method was the famous Charnes–Cooper–Rhodes (CCR) Model (Charnes et al. 1978), which can evaluate the efficiency and inefficiency of DMUs by converting multiple inputs into multiple outputs through mathematical programming techniques. Subsequently, network DEA (Färe and Grosskopf 1997, 2012) extended classical DEA by forming the network relationships and models that treat subunits of DMUs as basic DEA units. A considerable number of studies (e.g., Tone and Tsutsui 2009, 2014; Cook et al. 2010) in management sciences and economics have evaluated DMUs and have sought to improve those that were underperforming. In the 1990s, DEA incorporated fuzzy set theory to capture the uncertainty and impreciseness in MOP; one example is the bisection method proposed by Sakawa and Yumine (1983), which outperformed nonlinear programming with fuzzy set theory in MOP. Certain fuzzy MOP studies have continued this trend (Chiang and Tzeng 2000a; Tzeng and Huang 2014). The holistic development of MODM is illustrated in Figure 1.3.

Recently, MODM has gradually shifted from win–lose strategies to win–win strategies (Tzeng and Huang 2014). A win–lose strategy means a firm can only optimize its system by using the given resources and restricted capabilities. That is, the firm faces traditional optimization problems. Increasing numbers of firms seek to create added value and achieve new heights of performance through flexible resources and expanding competencies. Those new concepts and developments in MODM are discussed in the first section of this book.

1.5 Developments in Computational Intelligence, Machine Learning, and Soft Computing for Decision Aids

In the 1950s and 1960s, MCDM researchers attempted to develop mathematical models and methods to support DMs who made decisions concerning ranking, selection, or optimization (Figure 1.3). The early digital computers of the 1950s were large and expensive vacuum tube machines that were not easily accessed or used by researchers. Consequently, early MCDM studies did not expect to leverage digital computational resources to resolve complicated decision problems. In the 1970s, as transistors began to replace vacuum tubes, and in the 1980s, when affordable integrated circuits provided billions of transistors on a single chip, low-cost and convenient computational power gave researchers the opportunity to devise complicated algorithms to enhance the quality of decisions. In the present text, we mainly discuss three subfields of computer science that have influenced MCDM research over the past several decades: (1) computer-supported linear programming, (2) machine learning in computational intelligence, and (3) soft computing techniques.

First, computer-supported linear programming techniques have been adopted to solve optimization problems in operational research and MODM since the Second World War. Famous software packages such as Lingo and the Solver module of Excel have been widely used to solve the aforementioned MODM problems in various fields. Consider the following typical MODM problem:

$$\max f_i(\boldsymbol{x}), i = 1, \ldots, n,$$

$$s.t. \boldsymbol{Ax} \leq \boldsymbol{b}, \tag{1.1}$$

$$\boldsymbol{x} \geq \boldsymbol{0}.$$

In Equation 1.1, $f_i(\boldsymbol{x})$ denotes the ith goal that must be maximized, and decision variables x vector is a matrix of multiple constraints, all of which must be satisfied. Suppose the number of goals is 10 (i.e., $n = 10$) and the number of constraints is 100 with 50 variables; it would be nearly impossible to solve this problem without the use of computer-supported linear programming techniques. Because the modern social and business environments have grown greatly in complexity, typical real-world problems must consider numerous goals, constraints, and variables in practice. Since the 1980s, to maximize the efficiency of computational resource utilization, linear programming techniques have been combined with certain evolutionary computing techniques (e.g., GA) to search for optimal results (e.g., Goldberg 1989). The latest developments and newest concepts in this research thread are discussed in Chapter 2.

Second, although the distinction between computational intelligence and artificial intelligence (AI) is not always clear, computational intelligence is often regarded as a complementary field to AI in computer science (Eiben and Smith 2003), and both fields attempt to generate intelligent outputs or intelligent actions based on the calculations of machines. The term *artificial intelligence* originated from a proposal by McCarthy and others in 1955, and a major goal of AI (in the long term) is to create an intelligent machine with the capability to learn and sense with consciousness; this ambitious goal remains controversial in certain respects (e.g., philosophy, psychology, and ethics). AI mainly concerns multidisciplinary knowledge-based approaches from various fields. The discipline of computational intelligence applies computers or specialized hardware to prediction and pattern classification problems; here, we discuss only the machine learning techniques in computational intelligence that have been applied in MCDM research. Interested readers may refer to the work edited by Doumpos and Grigoroundis (2013) for a more comprehensive discussion.

Machine learning techniques have strengths in analyzing the nonlinear relationships in data, and most machine learning techniques (e.g., artificial neural networks [ANNs], decision trees [DTs], and support vector machines [SVMs]) are not required to assume any probabilistic distribution of data. *Classifiers* are machine learning techniques that learn from historical data to provide predictive outcomes for DMs as decision aids. ANN-related techniques (e.g., back-propagation, self-organizing map ANNs, and competitive learning neural networks) may have been the most prevalent techniques for solving prediction problems (e.g., financial market predictions) over the past several decades (Lin et al. 2012). ANN techniques mimic the learning mechanisms of brains, and the learned results are stored in the connections between neurons; because it is difficult to extract comprehensible meaning from the ANN learning process and the learned results, they are often criticized as a "black box" (Shen and Tzeng 2014a). Some other single classifiers, such as SVMs (Wang et al. 2005) and random forests, also suffer from similar drawbacks. To enhance the accuracy of single classifiers, hybrid classifiers combine two or more techniques (e.g., GA–SVM [Huang et al. 2006e] and DT–ANN–SVM [Hung and Chen 2009]). In such a hybrid system, one technique performs the initial classification and the others tune some of the parameters of the hybrid models (GAs or ANNs are often adopted to optimize the parameters during the learning phase). Generally, machine learning techniques focus on increasing the accuracy of classification or prediction.

Some machine learning techniques that learn from decision examples have been proposed to form the preference weights for MADM models. For example, Malakooti and Zhou (1994) used ANNs to formulate and assess utility functions by eliciting information from DMs, and the trained ANN models were adopted to rank alternatives. Several recent studies in MCDM could be categorized as part of this approach (Taha and Rostam 2011), but academic interest in this approach seems to have waned gradually. Aside from machine learning techniques, evolutionary computing methods (e.g., GA and PSO; Section 1.4) have been widely integrated with linear programming techniques to solve complex optimization problems. Evolutionary computing algorithms emulate natural evolutionary processes, which are mainly used in MODM studies (Tzeng and Huang 2014).

Third, soft computing techniques, particularly fuzzy sets (Zadeh 1965) and rough sets (Pawlak 1982), are based on solid mathematical foundations to model the impreciseness or uncertainty in a system, and have been widely applied in engineering (Zimmermann 2001) and social economics (Tzeng and Huang 2011, 2014). One of the key advantages of soft computing techniques is that meaningful logic or rules can be obtained for the addressed problem. Soft computing techniques are often integrated with those of machine learning to maximize accuracy. For example, fuzzy inference was integrated with ANNs to form an adaptive network-based fuzzy inference system (Jang 1993), which was used to support the financial predictions of commercial banks (Shen and Tzeng 2014a). We briefly introduce the essential ideas of these two soft computing techniques in Subsections 1.5.1 and 1.5.2.

1.5.1 Basic Concepts of Fuzzy Sets

Fuzzy set theory (Zadeh 1965) was proposed to denote the degrees to which particular elements can be said to belong to specific sets; therefore, fuzzy set calculations require greater computational resources than crisp sets do. Fuzzy set theory has been an emerging field of applied mathematics since the 1970s, and the extended subfields of fuzzy sets include fuzzy relations, fuzzy graphs, fuzzy clustering, fuzzy inference (fuzzy logic), and fuzzy linguistics. Among the aforementioned subfields, fuzzy linguistics and fuzzy ranking (which is based on fuzzy linguistics) play the most influential roles in MCDM research. This helps collect DMs' imprecise judgments to form an adequate preference model. The fundamental notions of the fuzzy sets could be expressed as follows.

Suppose that a fuzzy subset $\tilde{A}$ of a universe of discourse U^f is defined by its membership function $\mu_{\tilde{A}}(x)$, as in Equation 1.2:

$$\tilde{A} = \{(x, \mu_{\tilde{A}}(x)) \mid x \in U^f\}, \tag{1.2}$$

where $x \in U^f$ denotes the elements belonging to the universe of discourse, and

$$\mu_{\tilde{A}}(x) : U^f \to [0,1]. \tag{1.3}$$

Given a discrete finite set $U^f = \{x_1, x_2, \ldots, x_n\}$, a fuzzy subset $\tilde{A}$ of U^f can also be represented as

$$\tilde{A} = \sum_{i=1}^{n} \mu_{\tilde{A}}(x_i) / x_i. \tag{1.4}$$

For a continuous case, a fuzzy set $\tilde{A}$ of U^f can be represented as

$$\tilde{A} = \int_{U^f} \mu_{\tilde{A}}(x) / x. \tag{1.5}$$

Equation 1.3 indicates that for any $x \in U^f$, the value of $\mu_{\tilde{A}}(x)$ can be defined between 0 and 1 (i.e., $\mu_{\tilde{A}}(x) \in [0,1]$), which denotes the degree to which x belongs to the fuzzy subset $\tilde{A}$. Then, the support of $\tilde{A}$ is a crisp set of U^f, defined as $\text{supp}(\tilde{A}) = \{x \in U^f \mid \mu_{\tilde{A}}(x) > 0\}$; in addition, the α-cut of a fuzzy subset $\tilde{A}$ of U^f can be defined by $\tilde{A}(\alpha) = \{x \in U^f \mid \mu_{\tilde{A}}(x) \geq \alpha\}$, for $\forall \alpha \in [0,1]$. The height of $\tilde{A}$ is the least upper bound (sup) of $\mu_{\tilde{A}}(x)$, which is defined by

$$h(A) = \sup_{x \in U^f} \mu_{\tilde{A}}(x). \tag{1.6}$$

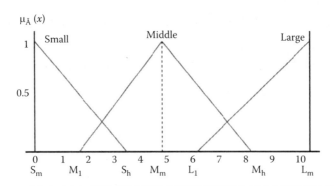

Figure 1.4 Membership function for three levels of linguistic concepts.

When $h(A) = 1$, the fuzzy subset $\tilde{A}$ of a set U^f is said to be normalized. Furthermore, α could be regarded as the confidence level, and the α-cut helps decide a crisp set in $\tilde{A}$ that supports the calculation of fuzzy rankings in MADM models. Consider the broadly adopted triangular fuzzy membership function: it can be used to denote imprecise three-level concepts such as *small*, *medium*, and *large* for a linguistic variable, as shown in Figure 1.4.

In recent trends of MADM research, models that are based on the pairwise comparison approach (such as the aforementioned AHP and ANP) also adopt fuzzy sets to indicate the preference weights of the involved criteria in a decision model. These weights can be synthesized with the performance values based on the linguistic evaluation of an alternative for each criterion.

1.5.2 Basic Notions of Rough Sets

The basic notions of rough sets were introduced by Pawlak in the early 1980s after fuzzy set theory had been developed. Although fuzzy sets and rough sets were all devised to model impreciseness, rough set theory (RST) is more information oriented. It is based on collecting vague patterns or knowledge from available data.

The essential idea of classical RST is the indiscernibility of the objects in the universe of discourse U^r, in the view of the available attributes about those objects. Any subset of all indiscernible objects is called an *elementary set*, which denotes a basic granule of knowledge about the universe. Any union of elementary sets is a crisp set; all other unions are rough sets. In the presence of rough sets in U^r, each rough set has boundary line cases (defined by upper and lower approximations). Because RST assumes that objects can only be discerned through the information that can be gained about them from their observed attributes, RST regards knowledge as a granular structure. The basic notions of the granules of knowledge in a rough set are illustrated in Figure 1.5.

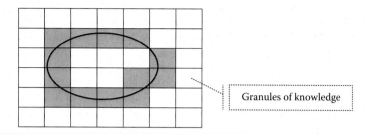

Figure 1.5 Granules of knowledge in a rough set.

A more formal mathematical definition of the indiscernibility relation can be given as follows. Begin with an information system, *IS*, which comprises pairs of approximation relations $APR = (U^r, A)$; given the finite sets of objects in U^r and that A is a set consisting of n attributes (i.e., $a_i \in A$, for $i = 1,\ldots,n$), there exists a value function $f(\cdot)$ that maps each object on the ith attribute to a value V_{a_i} (i.e., $f_{a_i}(x) = V_{a_i}$), and V is the value domain of attribute a. Then, any subset S of A may determine a binary relation $I(S)$ (i.e., indiscernibility relation) on U^r, defined as

$$xI(S)y \text{ iff } f_a(x) = f_a(y), \text{ for every } a \in S. \tag{1.7}$$

On the basis of the approximation relations, we can define the lower and upper approximations of a set. Let X be a subset of U^r, and the lower approximation of A in X is

$$\underline{APR}(A) = \{x \mid x \in U^r, U^r \mid Ind(A) \subset X\}, \tag{1.8}$$

where *Ind*(*A*) denotes the aforementioned indiscernibility relation. The upper approximation of X in A is

$$\overline{APR}(A) = \{x \mid x \in U^r, U^r \mid Ind(A) \cap X \neq \varnothing\}. \tag{1.9}$$

After constructing the upper and lower approximations, the boundary can be represented as

$$Bnd(A) = \overline{APR}(A) - \underline{APR}(A). \tag{1.10}$$

According to the defined upper and lower approximations and the boundary regions, RST may help reduce the required attributes (i.e., CORE attributes; the related discussion is provided in Section 1.6) to discern the *target attribute* (also termed the *decision attribute*), which may support the minimal and representative attributes (criteria) for MADM models.

16 ■ Trends of Hybrid Multiple Criteria Decision Making

1.6 Emerging Trend in Multiple Rule-Based Decision Making

Extended from classical RST, DRSA (Greco et al. 1999, 2002a, 2002b, 2005) was proposed to consider the preferential characteristics in attributes and the *dominance* relationships among alternatives (with respect to certain attributes in the system for classification). The idea of preference regarding criteria is crucial in decision making. For example, in a stock selection problem, the alternative with higher profitability (which is a criterion) should be preferred. Thus, the potential usefulness of DRSA in resolving MCDM problems has been recognized since the late 1990s (Greco et al. 2001, 2005, 2008; Słowiński 2008). A new trend in MCDM has also emerged since then and may be termed *multiple rules–based decision making* (MRDM); one of the key advantages of DRSA is that it can generate a set of decision rules to denote the rough knowledge in data. The basic ideas of DRSA are as follows.

DRSA may begin with organizing data in *IS* in the form of a table, with the attributes and objects (alternatives) arranged in rows and columns, respectively. The DRSA table is a 4-tuple *IS*; that is,

$$IS = (U^{DR}, Q, V, f),$$

where:

U^{DR}	is a finite state of the universe
$Q = \{q_1, q_2, ..., q_p\}$	is a finite set of p attributes (for decision making, Q usually comprises condition attributes C and a decision attribute D in two sets)
Vq	is the value domain of attribute q (V is the union of all value domains of q_i, for $i = 1,..., p$)
f	is a total function, such that $f : U^{DR} \times Q \rightarrow V$, where $f(x,q) \in V_q$ for each $x \in U^{DR}$ and $q \in Q$

In typical MADM applications of DRSA, only a single decision attribute exists in D, and decision classes (DCs) can be denoted as $Cl = \{Cl_t, t = 1,..., n\}$ in a general case.

Define $\succeq_q$ as a weak preference relation on U^{DR} when considering a criterion q (for $q \in Q$). For objects x and y, $x, y \in U$, if $x \succeq_q y$, which denotes that x is at least as good as y, considering attribute q, and the weak preference relation means that x and y are comparable on attribute q. Assume that DCs are all preference ordered (i.e., for all $r, s = 1,..., n$, if $r \succ s$, then Cl_r is preferred to Cl_s). Subsequently, given a set of DCs, the upward and downward union of positive DCs can be defined as follows (the upward union is mainly used for explanations in this text):

$$Cl_s^{\geq} = \bigcup_{r \geq s} Cl_r,$$ (1.11)

$$Cl_s^{\leq} = \bigcup_{r \leq s} Cl_r.$$ (1.12)

The upward and downward unions of DCs support the definitions of dominance relations for $G \subseteq C$, where D_G represents the dominance relation regarding G and xD_Gy denotes that x G-dominates y with regard to any subset criteria in G. Accordingly, the G-dominating set and G-dominated set can be denoted by Equations 1.13 and 1.14:

$$D_G^+(x) = \{y \in U : yD_Gx\},$$ (1.13)

$$D_G^-(x) = \{y \in U : xD_Gy\}.$$ (1.14)

Sets of collected unions of upward DCs can then be used to define G-lower and G-upper approximations, as in Equations 1.15 and 1.16. For brevity, only the upward unions of DCs are discussed hereafter, and the downward unions of DCs can be defined by analogy with Equations 1.15 and 1.16:

$$\underline{G}(Cl_r^{\geq}) = \{x \in U : D_G^+(x) \subseteq Cl_r^{\geq}\},$$ (1.15)

$$\overline{G}(Cl_r^{\geq}) = \{x \in U : D_G^-(x) \cap Cl_r^{\geq} \neq \varnothing\}.$$ (1.16)

Similar to classical RST, which utilizes the indiscernibility relation $I(S)$ for classification, DRSA further enhances indiscernibility with dominance relations to denote the preferential characteristics of attributes. Considering both certain and uncertain classifications, the G-lower approximation represents the certain classification to categorize an alternative in $Cl_r^{\geq}$, the G-upper approximation the uncertain one. Therefore, $\underline{G}(Cl_r^{\geq})$ is a subset of $\overline{G}(Cl_r^{\geq})$; that is, $\underline{G}(Cl_r^{\geq}) \subseteq \overline{G}(Cl_r^{\geq})$. The doubtful region can be defined by Equation 1.18, which is called the G-boundary (doubtful region) regarding the criteria set $G(G \subseteq C)$.

$$Bn_G(Cl_r^{\geq}) = \overline{G}(Cl_r^{\geq}) - \underline{G}(Cl_r^{\geq}).$$ (1.17)

18 ■ *Trends of Hybrid Multiple Criteria Decision Making*

In DRSA, the quality of approximation $\gamma_G(Cl)$ for every $G \subseteq C$ for ordinal DCs with respect to a set of attributes G can be defined by Equation 1.18:

$$\gamma_G(Cl) = \left| U^{DR} - \left(\bigcup_{r \in \{2,\dots,n\}} Bn_G\left(Cl_r^{\geq}\right) \right) \right| \Big/ \left| U^{DR} \right|. \tag{1.18}$$

The quality of approximation defines the ratio of the objects G-consistent with the dominance relationship divided by the total number of objects in U^{DR}, and $| \bullet |$ denotes the cardinality of a set in Equation 1.18. With the dominance-based rough approximation of upward and downward unions of DCs, a generalized description of decision rules can be obtained in terms of "if *antecedent*, then *consequent*." A decision rule $r \equiv$ "if $f_{i_1}(x) \geq r_{i_1} \wedge \dots \wedge f_{ip}(x) \geq r_{ip}$, then $x \in Cl_t^{\geq}$," and an object $y \in U$ supports r if $f_{i_1}(y) \geq r_{i_1} \wedge \dots \wedge f_{ip}(y) \geq r_{ip}$. The total set of y in the *IS* is called the *SUPPORT* of the decision rule r, which indicates the supportive evidence that a rule can provide, and the strength of r can be defined as $n(r) = \text{supp}(r)/|U^{DR}|$, where supp($r$) denotes the number of supports for r. Furthermore, each minimal subset $P \subseteq C$ that satisfies $\gamma_P(Cl) = \gamma_C(Cl)$ is termed a *REDUCT* of Cl, and the intersection of all REDUCTs represents the indispensable attributes for maintaining the same quality of rough approximations by DRSA. The details of DRSA may be found in previous works by Greco et al. (2001, 2002a, 2002b, 2005).

The high potential of DRSA as a decision aid in MCDM, for ranking or selection, was first noticed by Greco et al. (1999, 2001, 2005) and has emerged as a new research field in MCDM. The original idea was based on collecting a certain preference order (i.e., partially preordering the available alternatives in which a DM has confidence) and forming a pairwise comparison table (PCT), as introduced by Greco et al. (1997, 1999). Subsequently, the dominance principle, referred to as *multigraded dominance* by Slowinski et al. (2005) or as the dominance relation D_2 by Szeląg et al. (2013), is defined on pairs of objects. For example: if objects $a,b,c,d \in U$ and (a, b) and (c, d) are defined as pairs over set H, pair (a, b) is regarded as having D_2 dominance over (c, d) with respect to criteria $P \subseteq C$ (where C is the aforementioned condition attribute set in DRSA, and P is a partial set of C), iff $V_i(a) - V_i(b) \geq V_i(c) - V_i(d)$ for each criterion $i \in P$ (here, $V_i(a)$ denotes the value of alternative a on criterion i). Next, on the basis of the essential ideas of DRSA, decision rules were adopted for capturing rough knowledge from the aforementioned PCT. The obtained decision rules are thus applied to the other alternatives (objects) and to certain exploitation methods, such as the net flow score of Greco et al. (1999). This can be adopted to give rankings for the whole set of alternatives. This approach enables ranking by collecting a set of reference objects to denote the preference structure of a DM. The obtained DRSA decision rule–based exploitations are to be calculated for the preference order of all alternatives. It may be regarded as the multidisciplinary integration of diverse fields: soft computing, machine learning, and the outranking approach. The brilliant contributions of Greco et al. (1997, 1999, 2001, 2002a, 2002b, 2005), Slowinski et al. (2005), and Szeląg et al. (2013) to this emerging field have inspired us to propose new hybrid MCDM models to resolve practical problems, especially regarding improvement planning. The proposed framework of the new hybrid MCDM models is discussed in Chapter 2, Section 2.3.

1.7 Outline of the Book

Aside from the background to MCDM research and certain computational intelligence techniques for problem solving, the remainder of this book comprises two sections; Section I introduces the essential concepts and theories of the relevant methods and techniques, and Section II illustrates certain real-world applications, mainly in the financial and IT industries. Although the cases discussed in this book mainly fall in the category of business, the proposed hybrid approach for problem solving is not limited to this field; numerous types of real-world problems, in fields such as economics, business, psychology, social welfare, engineering, transportation planning, new product development, and national policy formation, can be addressed using these methods. In addition, the three pillars—MRDM, MADM, and MODM—may be regarded as three types of methods for solving different types of problems, and the combination or integration of the techniques from those three fields should be based on the problems that are addressed. The details of the new hybrid approach are discussed in the following chapters. We hope that interested readers may find it helpful to apply the new hybrid MCDM models to real-world problems in various fields, thus bridging the gap between academia and practice.

CONCEPTS AND THEORY

1

Chapter 2

New Concepts and Trends in MCDM

Before delving into a detailed discussion on solving problems with MCDM, this chapter provides an overview of the three building blocks of the new hybrid MCDM approach, as proposed in recent years by Professor Gwo-Hshiung Tzeng's research group. As shown in Figure 1.1, the three subfields in this new MCDM are (1) hybrid multiple rules–based decision making (MRDM), (2) hybrid multiple attribute decision making (MADM), and (3) hybrid multiple objective decision making (MODM).

Although the algorithms related to hybrid MRDM might be complicated, the essential ideas are easy to comprehend. Two new concepts from MRDM are as follows:

1. *Critical factors or criteria can be objectively retrieved from historical data.*
   The growing complexity in social problems has impeded the identification of the most crucial variables for modeling by intuition or conventional linear methods (e.g., statistics). MRDM leverages the strengths of computational intelligence by retrieving the critical factors for forming hybrid decision models, which are more reasonable and reliable in practice.
2. *MRDM can induce understandable decision rules in the form of "if…, then…" logic.*
   The manifest benefit of understandable decision rules is that they are close to how human brains learn from experience or examples.

Usually in hybrid MADM, two or more analytical MADM methods (techniques) are combined or integrated for ranking, selection, and improvement planning, in the context of multiple attributes. Four new concepts are highlighted here.

1. *Nearly all the considered attributes in a MADM problem are interrelated.*
   Traditional methods, such as regression and the analytic hierarchy process (AHP), presume that all attributes are independent. This assumption sometimes works in certain engineering applications; however, analysts of most social problems, ranging from business to national policy making, must concede that interrelations do exist between variables in most cases. Adequate methods that can acknowledge and measure those interrelations are required in practice.

2. *Pursuing aspiration levels.*
   Traditional MADM methods often define a goal for each attribute; that goal is often derived from comparisons of a group of readily available alternatives. This might lead decision makers (DMs) to be satisfied with the available solutions. A new concept in MADM is that DMs can pursue aspiration levels to avoid stagnation at the local optima. The importance of this concept is magnified in highly competitive business environments. Consider a country in which the service quality of each airline is rated on a scale of 0–100 points. If only five domestic airlines operated in this country and their service quality scores ranged from 60 to 78, it would be problematic to adopt the highest one as the benchmark or goal (because 78 is far from excellent). The management teams of those operators should set the full score as their aspiration level and plan for continuous improvements; otherwise, those domestic operators will find it difficult to survive under the pressure of global competition.

3. *Systematic improvements for the underperformed attributes.*
   Once the underperforming attributes are identified, systematic improvement planning is required. As the aforementioned interrelations among attributes do exist in most social problems, a plan for systematic improvement should identify the source factors that influence those underperforming attributes. Using medical treatment as an analogy, if a patient is suffering from a headache caused by work-related stress and high blood pressure (hypertension), it would be in vain to treat the patient by merely administering aspirin. The concept of systematic improvements is explained using several empirical cases in the second part of this book.

4. *Nonadditive performance aggregation.*
   The influences of this new concept are twofold. First, when modeling the preference (which economists call the *utility*) of a DM, nonadditive effects are not rare. When consumers evaluate smartphones, for example, although several factors (e.g., function, price, quality, and so on) are considered simultaneously, consumers might be overwhelmingly influenced by appealing designs that are easy to use, which partially explains the leading position of the iPhone. Second, certain complicated social or business phenomena have

the so-called synergy effect $(1 + 1 \neq 2)$, which is commonly observed in strategic alliances or merger and acquisition deals. A case of vendor selection is illustrated in Chapter 14 by using nonadditive performance aggregation.

Finally, in hybrid MODM, a new concept called *changeable spaces* is introduced. Conventional MODM research is confined by the assumption of fixed resources; DMs thus tend to accede to compromised solutions. The concept of changeable spaces attempts to challenge the existing frameworks or constraints by offering a new perspective on redesigning or redefining systems. The details are discussed, with illustrations, in Chapter 6.

As stated in Section 1.2, compared with the statistical approach, MCDM research puts more emphasis on problem solving. In this chapter, the basic concepts of problem solving and certain topics that are often ignored in traditional MCDM research are discussed; in addition, a new framework of hybrid MCDM is proposed to address those underexplored topics.

2.1 Problem Solving in Traditional MCDM

Traditional MCDM research often begins with a predefined problem: single or multiple goals, multiple criteria (in MADM), or multiple constraints (in MODM). Limited alternatives can then be ranked or selected (in MADM), or an efficient frontier can be obtained for the optimal resource allocation (in MODM). However, two underexplored but critical issues must be addressed in traditional MCDM research: (1) the initial problem definition might exert unexpected influence on the obtained results, and (2) decision making should play a more constructive role in problem solving. Merely ranking solutions or optimizing based on existing constraints might not suffice to determine an optimal solution. These two critical issues are discussed in the following paragraphs.

First, the involved or observed criteria (or attributes) in MCDM research are usually obtained from three approaches: (1) subjective judgments by researchers, (2) statistical analysis from historical data, and (3) theoretical support. The subjective judgment approach is constrained by the limited knowledge and experience of researchers. In addition, real-world problems are becoming ever more complex and complicated; it would be unlikely for researchers to choose the minimal and essential criteria (attributes) when considering numerous plausible attributes (e.g., 25 or more) by subjective judgments. This could be explained by *bounded rationality*, a concept proposed by Nobel Prize laureate Herbert A. Simon (1972, 1982) in the context of economics. The statistical approach has certain unrealistic assumptions such as the independence of the considered variables and questionable probabilistic assumptions regarding data, which are discussed in Section 1.2. Theoretical support seems to be ideal at first glance, but a real-world problem is often based on (or rooted in) a specific context. Generalized theories might not suffice for a specific case; therefore, merely using theories to select the minimal and critical criteria (variables) might not be practical for a real case.

Second, MCDM research should play a more proactive or constructive role in problem solving; to select among a group of inferior options would not help DMs achieve satisfactory outcomes. For example, once a performance evaluation model has been constructed by certain MADM methods, such as AHP or the analytic network process (ANP), researchers or practitioners (e.g., the management team of a company) may consider how to improve the proposed alternative. Most previous MCDM research has ignored the idea of improvement planning in problem solving. Traditional MODM studies only make plans regarding their predefined constraints and resources, but in a dynamic and rapidly changing environment, DMs must pursue methods to extend the currently known area of competence to achieve their aspiration levels.

To summarize, traditional MCDM research has ignored the aforementioned two essential questions/issues for problem solving, and the specific reasons for adopting the new hybrid MCDM approach are discussed in Section 2.2.

2.2 Why New Hybrid MCDM Approaches Are Needed

As mentioned in Chapter 1, Section 1.1, and Section 2.1 in this chapter, DMs might find that traditional MCDM techniques are not sufficient to resolve complicated and entwined problems in practice; therefore, we highlight the background and specific motivations of the new concepts and trends in MCDM as follows.

First, as discussed in Section 2.1, traditional MCDM studies have ignored the objective and reliable selection of minimal and representative criteria for forming MCDM models. Complex social or business environments require reasonable approaches to help researchers identify critical attributes. In the so-called Big Data era, human brains (i.e., those of DMs or researchers) inevitably encounter difficulties when they attempt to process the complex and imprecise patterns behind a complex problem (or information system).

Second, most traditional MADM studies have either focused on obtaining final ranking results or forming a utility function to aid DMs. However, traditional MADM studies have typically overlooked the relationships among the involved criteria and have failed to identify the imprecise reasoning embedded in their criteria with respect to the addressed problem.

Third, traditional MCDM models assume that the criteria are independent and hierarchical in structure. However, in most real-world problems, the relationships among criteria or aspects (also called *dimensions*) are usually interdependent with certain feedback effects. To identify the interrelated relationships among variables, the DEMATEL technique can build an influential network relations map (INRM) and find the influential weights of DEMATEL-based ANP (DANP) by using the basic concept of ANP (Saaty 1996), based on the influential relationship matrix of

DEMATEL (Ou Yang et al. 2008, 2013; Peng and Tzeng 2013; Shen et al. 2014; Hu et al. 2014). This can model some interdependent and feedback relationships among criteria. It also helps DMs avoid unrealistic assumptions of statistics and economics when resolving practical problems.

Fourth, the proposed method replaces the relatively favorable solutions of existing alternatives with aspiration levels that can support DMs in the pursuit of excellence. Aspiration levels were proposed by Simon (1955, 1956, 1959) as replacements for classical selections or optimization decisions based on minimum–maximum values. By searching for aspiration levels at each stage, the decision process can become a pursuit for continuous improvements or enhancements. Therefore, the classical VIKOR method was modified (Opricovic and Tzeng 2004, 2007) by correcting the traditional settings that adopt maximum–minimum values as ideal and negative ideal points. In the modified VIKOR method, the aspiration level and the least desirable values for each criterion are used instead to avoid decisions that pick the best apple out of a barrel of rotten ones, so to speak.

Fifth, the new hybrid MADM analytical tools are not only used for ranking or selection; they also be used to improve performance gaps for multiple criteria and the corresponding aspects (or dimensions). Therefore, the emphasis in the research field has shifted from ranking or selection when determining the most preferable approaches to improving the performance of existing methods according to INRM, because a systematic approach for problem solving is required. Instead of addressing the systems of the problem, the sources of the problem must be identified for performance improvement (i.e., improvement planning) to avoid stopgap, piecemeal measures.

Sixth, Daniel Kahneman (who received the Nobel Prize in Economics in 2002) and Amos Tversky used an additive-type utility aggregation model based on their works during the 1960s and found that consumers' product selection behaviors were not in accordance with the traditional concept of *multiattribute utility*, the notion of which is value function aggregation. They opined that traditional multiattribute utility models might not indicate consumers' real behaviors. In 1974, Michio Sugeno completed his doctoral thesis, "Theory of Fuzzy Integrals and Its Applications," at the Tokyo Institute of Technology. In that work, fuzzy integrals were proposed to aggregate the values of multiple attributes in a nonadditive approach. The mathematical modeling of nonadditive aggregation (i.e., value functions) can correct the problematic assumption (i.e., linear independence) found in traditional additive-type models. This fact inspired Kahneman to propose *prospect theory* in 1978 by adopting the basic concept from nonadditive aggregation.

Seventh, conventional MODM methods in planning and design are based on a decision space defined by a set of fixed conditions or resources. The objective function is defined in a feasible space of fixed requirements. (The fixed feasible region is called the *objective space*.) New concepts, especially MODM models with changeable spaces, may transform or extend those fixed requirements or constraints, facilitating searches for results approaching aspiration levels. This new thinking

28 ■ *Trends of Hybrid Multiple Criteria Decision Making*

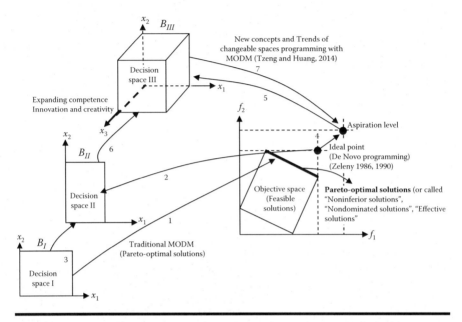

Figure 2.1 Basic concepts of changeable spaces programming (decision space and objective space).

in changeable spaces programming can help DMs achieve win–win planning and design, and also can identify the desired points (i.e., aspiration levels) for continuous improvements. Therefore, these new concepts are expected to be superior to searches for Pareto-optimal solutions or ideal points. The concept of changeable spaces is shown in Figure 2.1; the details are explained and discussed in Chapter 6. The aforementioned seven points echo the new concepts highlighted at the beginning of this chapter.

2.3 Framework of New Hybrid MCDM Models for Tomorrow

On the basis on the two aforementioned issues of traditional MCDM in Section 2.1 and the seven specific motivations in Section 2.2, a framework for the new hybrid MCDM models for problem solving is proposed herein. The overall vision and main ideas are illustrated in Figure 2.2.

Two key points of the proposed conceptual framework are further illustrated in Figure 2.3: (1) identify the core criteria and rough knowledge in a complex environment for appropriate problem settings, and (2) explore the interrelated cause–effect relationships among the core criteria and rough knowledge (e.g., rules, logic, and fuzzy inferences) for systematic improvement planning.

New Concepts and Trends in MCDM ■ 29

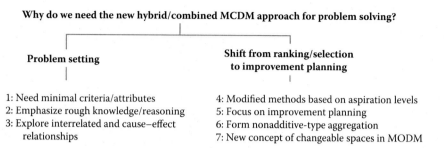

Figure 2.2 Main ideas of problem solving in new hybrid MCDM.

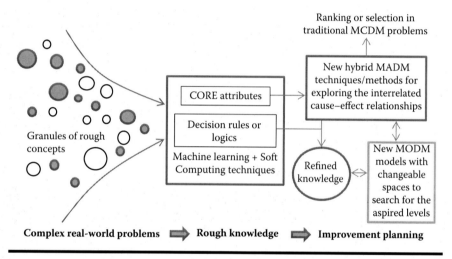

Figure 2.3 Conceptual framework of new hybrid/combined MCDM.

Traditional MCDM research has not adequately addressed the complexity of data, but rapidly growing, information-centric research on practical contemporary problems cannot overlook such complexity. As a result, in this book, we expand past notions of MADM and MODM by further incorporating the field of data-centric analytics, including statistical analyses, machine learning, soft computing, and data-mining techniques (as shown on the left side of Figure 2.4).

Data-centric analytics can be further simplified into MRDM (as shown in Figure 1.1), which is aimed at retrieving critical attributes or criteria to form hybrid MADM or MODM models with understandable rules or logic. The details of data-centric analytics are further illustrated in Figure 2.5.

Therefore, the modeling of new hybrid MCDM may be regarded as a process of transforming data (or information) and knowledge (or experience) from experts (or DMs) to form understandable decision aids for problem solving. This new framework is proposed to leverage the computational capabilities of machines and the

30 ■ *Trends of Hybrid Multiple Criteria Decision Making*

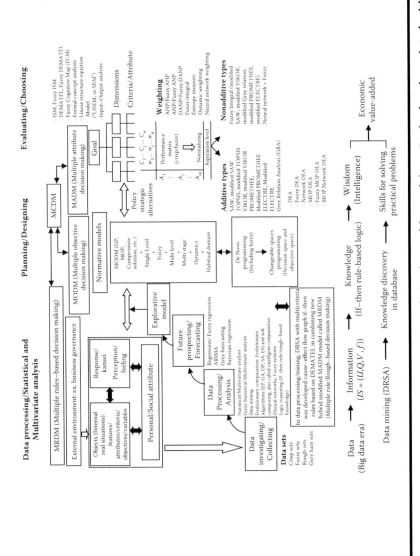

Figure 2.4 New MCDM considering data-centric analytics for problem solving. (Revised and reprinted with permission from Liou, J.J.H., and Tzeng, G.H., *Technological and Economic Development of Economy*, 18(4), 672–695, 2012.)

New Concepts and Trends in MCDM ■ 31

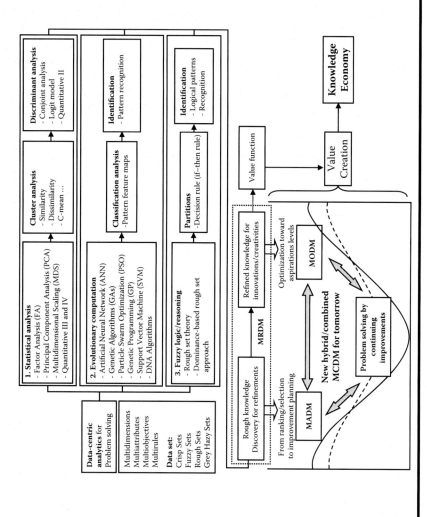

Figure 2.5 New hybrid/combined MCDM based on data-centric analytics. (Revised and reprinted with permission from Liou, J.J.H., and Tzeng, G.H., *Technological and Economic Development of Economy*, 18(4), 672–695, 2012.)

32 ■ *Trends of Hybrid Multiple Criteria Decision Making*

Table 2.1   Overview of the Following Chapters

Part One			Part Two
Subfields	*Methods/ Techniques*	*Chapters*	*Applications (Chapters)*
Hybrid MADM	DEMATEL technique	Chapter 3	Assessing information risk (Chapter 8)
	ANP, INRM, and DANP	Chapter 4	Improving e-store businesses (Chapter 9) Evaluating green suppliers in the TFT-LCD industry (Chapter 10)
	DEMATEL, DANP, modified VIKOR, and fuzzy integrals	Chapter 5	Smartphone improvements (Chapter 11) Evaluating the B2B m-commerce of SMEs (Chapter 12) Selecting glamour stocks (Chapter 13) Performance evaluations of suppliers using a nonadditive-type aggregator (Chapter 14)
Hybrid MODM	Pareto optimality, De Novo programming, and changeable spaces	Chapter 6	Strategic alliance modeling (Chapter 15) Numerical examples of changeable spaces (Chapter 16)
Hybrid MRDM	Dominance-based rough set approach (DRSA)/variable-consistency (VC)-DRSA, formal concept analysis (FCA), and net flow score (NFS)	Chapter 7	Technical analysis for financial investment (Chapter 17) Evaluating commercial banks (Chapter 18) Financial performance improvement planning for IT companies (Chapter 19) Business analytics using a hybrid bipolar decision model (Chapter 20)

knowledge and experience of human beings to enhance the problem-solving capabilities of various fields. Such enhancements can foster the growth of the future knowledge economy. The new concepts, theoretical backgrounds, and applications introduced in the following chapters are summarized in Table 2.1. It is not necessary to read those chapters in sequence; readers who are interested in a specific subfield (i.e., hybrid MRDM, hybrid MADM, or MODM) may find it useful to examine the associated empirical application(s) after reading the theoretical chapter(s).

Chapter 3

Basic Concepts of DEMATEL and Its Revision

The decision-making trial and evaluation laboratory (DEMATEL) technique has been widely applied to various complex social problems, such as setting marketing strategies, evaluating science parks, assessing safety issues, and making group decisions. In addition, it has been integrated with other methods such as the analytical network process (ANP), fuzzy set theory, and rough set theory to enhance the visual illustrations of the analytics. The DEMATEL technique can model the influential relationships among the attributes/variables/factors (i.e., criteria for MCDM problems) in a system by collecting opinions/knowledge/experience from decision makers (DMs) or experts to form an initial direct-relation matrix. The influences of some attributes can ripple transitively to others in a system, which is modeled by raising the powers of the initial direct-relation matrix. Then, the total-influence matrix is computed by summing up the matrices in all powers. The original DEMATEL assumes that the matrix will converge to zero by raising its power to infinity, which cannot be guaranteed. However, the revised DEMATEL can be proved to guarantee the presumption (Lee et al. 2013), and the total-influence matrix can be obtained accordingly. A simplified numerical example will be explained in this chapter.

36 ■ *Trends of Hybrid Multiple Criteria Decision Making*

3.1 Background and Basic Notions of DEMATEL

The DEMATEL technique (Fontela and Gabus 1974, 1976; Warfield 1976) was devised to model the influential relations among a complex system, mainly for solving complicated and interrelated social problems. Various applications have adopted DEMATEL for forming strategies or conducting evaluations in fields such as marketing (Chiu et al. 2006), innovation policy (Huang and Tzeng 2006), and airline safety measurement (Liou et al. 2007).

Considering the complexity of many real-world problems, the original ideas of DEMATEL aimed to model the interrelationships among all the attributes (or variables, factors, and criteria) involved in a system. The initial influence of one attribute on another is represented by a value between 0 and 1; in practice, such a value can be obtained from DMs or experts to denote their understanding/knowledge of the addressed problem/system. A value of 0 means that the attribute exercises no influence, and 1 means that it exercises an absolute influence. A matrix with such entries is used to represent the initial influence between the attributes of a problem/system. In addition, the DEMATEL technique assumes that the influence may ripple transitively in the system. On the modeling side, the transitive influence is calculated by matrix multiplications. The initial matrix represents the direct influence, and the multiplication of the matrix z times ($z \geq 2$) represents the z-indirect influence exercised by the attributes of a system. For example, 2-indirect influence, which is obtained by raising the initial influence matrix to the power of 2, represents the influence exercised by an attribute after a ripple of length 2; moreover, the 1-indirect influence is the same as the initial direct influence. The total influence exercised by an attribute is obtained by summing up the direct initial influence and the indirect influence of all lengths; therefore, the total influence is the sum of an infinite series.

3.2 Operational Steps of the Original DEMATEL

The original DEMATEL technique is summarized in the following four steps.

Step 1: Find the initial average matrix A.

Suppose that we have H (h = 1, 2,…,H) to consider n attributes (i,j belong to $\{1, 2,…,n\}$) for the evaluated problem (otherwise termed a *system*) on hand. Each expert is asked to indicate the degree to which he or she feels that an attribute i affects the attribute j ($i \neq j$). The pairwise comparisons between any two attributes are denoted by a_{ij}, and the expected values/scores are 0, 1, 2, 3, and 4, representing no influence (0), low influence (1), medium influence (2), high influence (3), and very high influence (4), respectively. Experts are requested to provide all of the pairwise comparisons between any two different factors (for any $i \neq j$) as scores, which means $n \times (n-1)$ comparisons should be provided by each expert.

The scores by each expert can h ($h=1,2,\ldots,H$) form an $n \times n$ (the diagonal elements are all set to zero) answer matrix $S^{(h)} = [s_{ij}^{(h)}]$, where $h=1,2,\ldots,H$. Thus, $S^{(1)},\ldots,S^{(k)},\ldots,S^{(m)}$ are the answer matrices from those H experts. Then, the $n \times n$ average matrix A can be computed by averaging the H experts' opinions (scores), as in Equation 3.1:

$$a_{ij} = \frac{1}{H} \sum_{h=1}^{H} s_{ij}^{(h)} \tag{3.1}$$

The average matrix $A = [a_{ij}]_{n \times n}$ is also called the *initial direct-relation matrix*. A shows the initial direct effects that a factor exerts on and receives from the other factors.

Step 2: Normalize A to have the initial direct-relation matrix D.

The normalized initial direct-relation matrix D is obtained by normalizing the average matrix A, as in Equation 3.2:

$$\text{Set } a = \max \left(\max_{1 \le i \le n} \sum_{j=1}^{n} a_{ij}, \max_{1 \le j \le n} \sum_{i=1}^{n} a_{ij} \right). \tag{3.2}$$

Then, A can be normalized by a as the initial direct-relation matrix D (i.e., $D = A/\alpha$).

Since the sum of each row j of matrix A represents the total direct effects that attribute j gives to the other attributes, $\max_{1 \le i \le n} \sum_{j=1}^{n} a_{ij}$ represents the total direct effects of the attribute with the highest direct effects on others. Likewise, since the sum of each column i of matrix A represents the total direct effects received by the attribute i, $\max_{1 \le j \le n} \sum_{i=1}^{n} a_{ij}$ represents the total direct effects received by the attribute that receives the highest direct effects from the others. The positive scalar α takes the higher of the two as the upper bound, and the matrix D is obtained by dividing each element of A by the scalar α. Note that each element d_{ij} of matrix D is between 0 and less than 1.

Step 3: Compute the total relation matrix T.

A continuous decrease of the indirect effects of problems along the powers of matrix D, for example $D^2, D^3, \ldots, D^\infty$ is presumed to reach convergent solutions to the matrix inversion, similar to an absorbing Markov chain matrix. Note that $\lim_{z \to \infty} D^z = [0]_{n \times n}$, where $D = [d_{ij}]_{n \cdot n}, 0 \le d_{ij} < 1$; also, $0 \le \sum_i d_{ij} \le 1$ and $0 \le \sum_j d_{ij} \le 1$. Furthermore, at least one row or one column of summation equals 1, but not all; this will guarantee $\lim_{z \to \infty}(I + D + D^2 + \ldots + D^z) = (I - D)^{-1}$, where

38 ■ *Trends of Hybrid Multiple Criteria Decision Making*

$[0]_{n \times n}$ is an $n \times n$ null matrix and I is an $n \times n$ identity matrix. The total relation matrix T is an $n \times n$ matrix defined as follows: $T = [t_{ij}]$, where $i, j = 1,2,\ldots,n$; thus,

$$T = D + D^1 + D^2 + \ldots + D^\infty = \lim_{z \to \infty}(D + D^2 + \ldots + D^z) = D(I - D)^{-1}. \quad (3.3)$$

Explanation

$$T = D + D^2 + \cdots + D^z,$$

$$= D(I + D + D^2 + \cdots + D^{z-1})(I - D)(I - D)^{-1},$$

$$= D(I - D^z)(I - D)^{-1}.$$

When

$$z \to \infty, \quad D^z = [0]_{n \times n},$$

then

$$T = D(I - D)^{-1}.$$

Next, r and c can be expressed as two $n \times 1$ vectors representing the sum of rows and sum of columns of the total relation matrix T.

$$r = \left[\sum_{j=1}^{n} t_{ij}\right]_{n \times 1} = [t_{i\cdot}]_{n \times 1} = (r_1, \ldots, r_i, \ldots, r_n)' \quad (3.4)$$

$$c = \left[\sum_{i=1}^{n} t_{ij}\right]'_{1 \times n} = [t_{\cdot j}]_{n \times 1} = (c_1, \ldots, c_j, \ldots, c_n)' \quad (3.5)$$

where prime (') denotes the transpose operation for a vector.

According to Equation 3.4, r_i indicates the sum of the ith row in T, which shows the total effects, both direct and indirect, given by the attribute i to the other attributes. Similarly, in Equation 3.5, c_j indicates the sum of the jth column in T, where c_j shows the total effects, both direct and indirect, received by the attribute j from the other attributes. Therefore, while $i = j$, the sum $r_i = c_i$ (or $r_i = c_i$) denotes the total effects both given and received by the attribute i (or j) of the system, which shows the degree of influence (i.e., the total sum of effects given and received) that the attribute i holds in the system. In addition, the difference $r_i - c_i$ shows the net effect that the attribute i contributes to the system, which can categorized an attribute as the cause or effect group. When $r_i - c_i > 0$, the attribute i belongs to the cause group; when $r_i - c_i < 0$, it belongs to the effect group (Tzeng et al. 2007; Tamura et al. 2002).

Step 4: Set a threshold value and obtain the influential network relationship map (INRM).

To comprehend the structural relations among the attributes while keeping the complexity of the system to a manageable level, it is necessary to set a threshold value θ to filter out some negligible effects in matrix T. While each attribute of matrix T provides information on how one attribute affects another, DMs must set a threshold value in order to reduce the complexity of the structural relation model implicitly in matrix T. Only the attributes whose effects in matrix T are greater than the threshold value should be chosen and shown in an INRM (Tzeng et al. 2007).

3.3 Infeasibility of the Original DEMATEL Technique

To demonstrate the infeasibility of the original DEMATEL, let us consider the following example. Assume that the answer matrices from two experts a and b are as follows:

$$S^{(a)} = \begin{bmatrix} 0 & 4 & 1 & 0 \\ 4 & 0 & 0 & 1 \\ 2 & 0 & 0 & 3 \\ 0 & 2 & 4 & 0 \end{bmatrix} \text{ and } S^{(b)} = \begin{bmatrix} 0 & 3 & 1 & 1 \\ 4 & 0 & 1 & 0 \\ 0 & 0 & 0 & 5 \\ 0 & 1 & 3 & 0 \end{bmatrix}.$$

The initial direct relation matrix, which is obtained by averaging the answer matrices, is as follows:

$$A = \begin{bmatrix} 0 & 3.5 & 1 & 0.5 \\ 4 & 0 & 0.5 & 0.5 \\ 1 & 0 & 0 & 4 \\ 0 & 1.5 & 3.5 & 0 \end{bmatrix}.$$

By finding the maximum of the row sums, which is 5, and the maximum of the column sums, which is also 5, the normalized initial direct-relation matrix D can then be calculated as

$$D = \begin{bmatrix} 0 & 0.7 & 0.2 & 0.1 \\ 0.8 & 0 & 0.1 & 0.1 \\ 0.2 & 0 & 0 & 0.8 \\ 0 & 0.3 & 0.7 & 0 \end{bmatrix},$$

40 ■ *Trends of Hybrid Multiple Criteria Decision Making*

where each row and each column of summation are all equal to 1. This outcome is different from Step 3 (i.e., at least one row or one column of summation equals 1, *but not all*, when $0 \le \Sigma_i d_{ij} \le 1$ and $0 \le \Sigma_j d_{ij} \le 1$) in the previous section.

Therefore, if we compute the total relation matrix T by the original DEMATEL technique for this case, D^∞ must be derived as a null matrix for the total relation matrix to converge. However,

$$D^\infty = \begin{bmatrix} 0.169447 & 0.169447 & 0.169447 & 0.169447 \\ 0.075505 & 0.075505 & 0.075505 & 0.075505 \\ 0.410887 & 0.410887 & 0.410887 & 0.410887 \\ 0.344162 & 0.344162 & 0.344162 & 0.344162 \end{bmatrix},$$

which is not as expected as a null matrix $[0]_{n \times n}$. Thus, here, the total relation matrix T cannot be obtained because D^z does not converge to a zero matrix. The infeasibility of DEMATEL is summarized in the following theorem:

Theorem 3.1: In certain cases, $\lim D^z$ may not converge to null matrix $[0]_{n \times n}$; therefore, $T = D + D^2 + \ldots + D^z$ might not converge in all cases.

3.4 Revised DEMATEL

Lee et al. (2013) proposed the revised DEMATEL technique, which guaranteed the convergence of the total relation matrix T by modifying Step 2 in the original DEMATEL. The revised DEMATEL technique is rooted from Theorem 3.2:

Theorem 3.2: Let W be a matrix in which at least one row or one column of summation equals 1, but not all; then we can guarantee $\lim_{k \to \infty} W^k = [0]_{n \times n}$.

Based on Theorem 3.2, the required steps are as follows:

Step 1: The same as the original DEMATEL.
Step 2: Calculate the initial influence matrix X.

In the revised DEMATEL, the initial influence matrix X is divided by α^+, as in Equation 3.6:

$$X = \frac{A}{\alpha^+}, \tag{3.6}$$

where:

$$\alpha^+ = \max(\max_{1 \le i \le n} \sum_{j=1}^{n} a_{ij}, \varepsilon + \max_{1 \le j \le n} \sum_{i=1}^{n} a_{ij}).$$

In Equation 3.6, a very small positive value ε (such as 10^{-5}) can be added with either $\max_{1 \le j \le n} \sum_{i=1}^{n} a_{ij}$ or $\max_{1 \le i \le n} \sum_{j=1}^{n} a_{ij}$; as a result, even while the summations of each row and column in X are all equal, the newly added ε can satisfy the requirement in Theorem 3.2.

Step 3: Derive the new total-influence matrix T^*

Since all indirect-influence matrices are X^2, X^3,...,X^k,...,X^∞, the new total influence matrix will be $T^* = X + X^2 + ... + X^k + ... + X^\infty$, which can be rewritten as Equation 3.7:

$$T^* = \sum_{b=1}^{\infty} X^b = X(I - X)^{-1}, \tag{3.7}$$

where I is an identity matrix. The original DEMATEL only assumes that the infinite power of the initial influence matrix becomes a null matrix; however, the newly revised DEMATEL can guarantee that $\lim_{N \to \infty} X^N = [0]_{n \times n}$. In other words, the new total-influence matrix T^* will converge to a stable state as N approaches infinity.

Step 4: The same as the original DEMATEL.

3.5 Two Numerical Examples

To illustrate the revised DEMATEL technique, a case from previous research is used to show the original DEMATEL in the aforementioned four steps. Also, a simple numerical case using the revised DEMATEL is explained in this section.

First, the empirical case for analyzing the financial performance of the life insurance industry in Taiwan is used to show the original DEMATEL, which is from a previous work (Shen & Tzeng 2015b). The raw financial summary report for the life insurance companies comprises 19 attributes, and the CORE attributes (retrieved by DRSA or VC-DRSA; the definition of CORE attributes can be found in Section 7.1) were reduced to 13 in five dimensions (definitions are shown in Table 3.1).

Step 1: The initial average matrix A came from five domain experts, as shown in Table 3.2.

Step 2: Normalize A to have the initial direct-relation matrix D; in this case, $\alpha = 35.20$. Therefore, $D = \dfrac{A}{a} = \dfrac{A}{35.20}$, as shown in Table 3.3.

42 ■ *Trends of Hybrid Multiple Criteria Decision Making*

Table 3.1 Descriptions of CORE Attributes

Dimensions	Criteria		Definitions/Descriptions
Capital structure (D_1)	Debt	C_1	Total debt/total assets
	ΔProvision	C_2	Change rate of provision for life insurance reserve
Payback (D_2)	1st Y-Premium	C_3	First-year premium ratio
	RY-Premium	C_4	Renewable premium ratio
Operational efficiency (D_3)	N-Cost	C_5	New contract cost/new contract revenue
	ΔEquity	C_6	Change rate of shareholders' equity
	ΔNetProfit	C_7	Change rate of net profit
	CapInvest	C_8	Total invested capital/total assets
Earning quality (D_4)	Persistency	C_9	Persistency of the valid contracts in the 25th month
Capital efficiency (D_5)	NetProfit	C_{10}	Net profitability of capital utilization
	ROI	C_{11}	Return on investment ratio
	O-Profit	C_{12}	Operational profits/operational incomes
	RealEstate	C_{13}	Investment and loan on real estate/total assets

Source: Data from Shen, K.Y., and Tzeng, G.H., *Rough Sets, Fuzzy Sets, Data Mining, and Granular Computing*, Springer, 233–244, 2015b.

Step 3: Compute the total relation matrix T, and find out r and c. Since $T = D(I - D)^{-1}$, as in Equation 3.3, it can be calculated as shown in Table 3.4. Furthermore, r and c can thus be obtained; $r_i - c_i$ divides the attributes into the cause group ($r_i - c_i > 0$) and the effect group ($r_i - c_i < 0$), as indicated in Table 3.5.

Step 4: Set a threshold value and obtain the INRM. The construction of the INRM will support DMs to comprehend the addressed problem by visual illustration. A more detailed example will be provided in the next chapter.

Table 3.2 Initial Average Matrix A

	C_1	C_2	C_3	C_4	C_5	C_6	C_7	C_8	C_9	C_{10}	C_{11}	C_{12}	C_{13}
C_1	0.00	3.00	1.20	1.40	1.00	3.20	2.20	3.20	1.00	2.40	2.20	1.00	1.60
C_2	1.60	0.00	1.60	1.40	0.80	2.20	1.80	1.40	1.40	1.20	0.20	0.20	1.40
C_3	0.60	1.80	0.00	3.20	3.20	1.40	2.60	2.00	3.00	0.80	0.20	3.00	1.40
C_4	0.80	1.40	2.00	0.00	3.00	1.80	3.20	1.80	3.80	0.60	0.40	3.60	1.00
C_5	1.60	1.60	2.80	3.60	0.00	1.60	3.20	1.40	3.20	0.80	0.80	3.80	0.80
C_6	2.00	0.80	0.20	0.20	0.40	0.00	2.00	1.40	0.60	1.20	1.80	1.20	1.20
C_7	2.20	0.80	2.00	1.80	2.00	1.60	0.00	1.80	1.60	2.80	2.00	2.40	1.40
C_8	2.00	1.60	1.20	0.80	2.20	1.60	3.60	0.00	1.40	3.80	3.40	3.40	3.00
C_9	1.20	2.40	1.20	3.20	2.00	1.60	3.20	2.00	0.00	1.60	1.20	3.80	1.20
C_{10}	2.40	1.00	1.20	1.20	2.20	1.60	3.80	2.80	1.40	0.00	3.60	3.80	3.20
C_{11}	2.80	0.80	1.40	1.20	2.00	1.80	4.00	3.80	1.60	2.60	0.00	3.80	3.80
C_{12}	2.80	0.80	2.20	2.60	3.00	1.80	3.20	3.40	2.80	2.00	1.60	0.00	0.80
C_{13}	1.00	0.20	0.60	0.20	0.60	1.60	2.40	3.40	0.60	3.00	2.60	2.40	0.00

Source: Data from Shen, K.Y., and Tzeng, G.H., Rough Sets, Fuzzy Sets, *Data Mining, and Granular Computing*, Springer, 233–244, 2015b.

Table 3.3 Direct-Relation Matrix D

	C_1	C_2	C_3	C_4	C_5	C_6	C_7	C_8	C_9	C_{10}	C_{11}	C_{12}	C_{13}
C_1	0.00	0.09	0.03	0.04	0.03	0.09	0.06	0.09	0.03	0.07	0.06	0.03	0.05
C_2	0.05	0.00	0.05	0.04	0.02	0.06	0.05	0.04	0.04	0.03	0.01	0.01	0.04
C_3	0.02	0.05	0.00	0.09	0.09	0.04	0.07	0.06	0.09	0.02	0.01	0.09	0.04
C_4	0.02	0.04	0.06	0.00	0.09	0.05	0.09	0.05	0.11	0.02	0.01	0.10	0.03
C_5	0.05	0.05	0.08	0.10	0.00	0.05	0.09	0.04	0.09	0.02	0.02	0.11	0.02
C_6	0.06	0.02	0.01	0.01	0.01	0.00	0.06	0.04	0.02	0.03	0.05	0.03	0.03
C_7	0.06	0.02	0.06	0.05	0.06	0.05	0.00	0.05	0.05	0.08	0.06	0.07	0.04
C_8	0.06	0.05	0.03	0.02	0.06	0.05	0.10	0.00	0.04	0.11	0.10	0.10	0.09
C_9	0.03	0.07	0.03	0.09	0.06	0.05	0.09	0.06	0.00	0.05	0.03	0.11	0.03
C_{10}	0.07	0.03	0.03	0.03	0.06	0.05	0.11	0.08	0.04	0.00	0.10	0.11	0.09
C_{11}	0.08	0.02	0.04	0.03	0.06	0.05	0.11	0.11	0.05	0.07	0.00	0.11	0.11
C_{12}	0.08	0.02	0.06	0.07	0.09	0.05	0.09	0.10	0.08	0.06	0.05	0.00	0.02
C_{13}	0.03	0.01	0.02	0.01	0.02	0.05	0.07	0.10	0.02	0.09	0.07	0.07	0.00

Source: Data from Shen, K.Y., and Tzeng, G.H., Rough Sets, Fuzzy Sets, *Data Mining, and Granular Computing*, Springer, 233–244, 2015b.

Table 3.4   Total Relation Matrix T

	C_1	C_2	C_3	C_4	C_5	C_6	C_7	C_8	C_9	C_{10}	C_{11}	C_{12}	C_{13}
C_1	0.10	0.15	0.11	0.12	0.13	0.18	0.21	0.21	0.12	0.17	0.16	0.17	0.14
C_2	0.10	0.05	0.09	0.10	0.09	0.12	0.15	0.12	0.10	0.10	0.07	0.10	0.10
C_3	0.11	0.12	0.08	0.19	0.19	0.13	0.22	0.17	0.19	0.12	0.10	0.22	0.12
C_4	0.12	0.11	0.14	0.10	0.19	0.14	0.24	0.17	0.21	0.12	0.10	0.24	0.11
C_5	0.15	0.12	0.17	0.20	0.12	0.15	0.25	0.17	0.20	0.13	0.12	0.26	0.12
C_6	0.11	0.06	0.05	0.06	0.07	0.06	0.14	0.11	0.07	0.10	0.11	0.11	0.09
C_7	0.16	0.09	0.13	0.14	0.16	0.14	0.15	0.17	0.14	0.18	0.15	0.21	0.13
C_8	0.18	0.13	0.13	0.14	0.18	0.16	0.29	0.16	0.16	0.23	0.21	0.27	0.20
C_9	0.14	0.14	0.12	0.19	0.16	0.15	0.25	0.19	0.11	0.15	0.13	0.25	0.13
C_{10}	0.19	0.11	0.13	0.15	0.19	0.16	0.29	0.23	0.16	0.14	0.22	0.28	0.20
C_{11}	0.20	0.11	0.14	0.15	0.18	0.17	0.30	0.27	0.17	0.21	0.13	0.28	0.22
C_{12}	0.19	0.11	0.16	0.18	0.20	0.16	0.27	0.24	0.20	0.18	0.15	0.17	0.13
C_{13}	0.12	0.06	0.08	0.08	0.10	0.12	0.20	0.20	0.10	0.18	0.16	0.19	0.09

Source: Data from Shen, K.Y., and Tzeng, G.H., *Rough Sets, Fuzzy Sets, Data Mining, and Granular Computing*, Springer, 233–244, 2015b.

Table 3.5 Summarized Result of Cause–Effect Analysis

	C_1	C_2	C_3	C_4	C_5	C_6	C_7	C_8	C_9	C_{10}	C_{11}	C_{12}	C_{13}
r_i	1.96	1.27	1.98	2.00	2.15	1.15	1.96	2.43	2.10	2.45	2.55	2.34	1.69
c_i	1.86	1.38	1.55	1.80	1.96	1.85	2.96	2.41	1.92	2.02	1.79	2.75	1.78
$r_i - c_i$	0.10	-0.11	0.43	0.20	0.19	-0.70	-1.00	0.02	0.18	0.43	0.76	-0.41	-0.09
Cause/Effect	Cause	Effect	Cause	Cause	Cause	Effect	Effect	Cause	Cause	Cause	Cause	Effect	Cause

Source: Data from Shen, K.Y., and Tzeng, G.H., *Rough Sets, Fuzzy Sets, Data Mining, and Granular Computing*, Springer, 233–244, 2015b.

Second, by using the initial direct-relation matrix A in Section 3.3, the initial influence matrix X by the revised DEMATEL can be found by setting $\varepsilon = 10^{-5}$. Since

$$A = \begin{bmatrix} 0 & 3.5 & 1 & 0.5 \\ 4 & 0 & 0.5 & 0.5 \\ 1 & 0 & 0 & 4 \\ 0 & 1.5 & 3.5 & 0 \end{bmatrix}$$

and $\alpha^+ = 5.00001$, then X can be computed as follows:

$$X = \frac{A}{\alpha^+} = \begin{bmatrix} 0 & 0.69986 & 0.19996 & 0.09998 \\ 0.79984 & 0 & 0.09998 & 0.09998 \\ 0.19996 & 0 & 0 & 0.79984 \\ 0 & 0.29994 & 0.69986 & 0 \end{bmatrix}.$$

Based on Theorem 3.2, X^∞ will be a null matrix; then, the new total-influence matrix T^* will be $T^* = X + X^2 + \ldots + X^k + \ldots + X^\infty = X(I - X)^{-1}$. The calculations are as follows:

$$T^* = \begin{bmatrix} 0 & 0.69986 & 0.19996 & 0.09998 \\ 0.79984 & 0 & 0.09998 & 0.09998 \\ 0.19996 & 0 & 0 & 0.79984 \\ 0 & 0.29994 & 0.69986 & 0 \end{bmatrix}$$

$$\times \left(\begin{bmatrix} 1.00000 & -0.69986 & -0.19996 & -0.09998 \\ -0.79984 & 1.00000 & -0.09998 & -0.09998 \\ -0.19996 & 0.00000 & 1.00000 & -0.79984 \\ 0.00000 & -0.29994 & -0.69986 & 1.00000 \end{bmatrix} \right)^{-1}$$

$$= \begin{bmatrix} 1249.782 & 1250.147 & 1249.571 & 1249.500 \\ 1250.270 & 1249.776 & 1249.500 & 1249.454 \\ 1249.480 & 1249.467 & 1249.782 & 1250.270 \\ 1249.467 & 1249.610 & 1250.147 & 1249.776 \end{bmatrix}.$$

3.6 Conclusion

In this chapter, the infeasibility of the original DEMATEL technique is shown, and a revised DEMATEL is explained with an illustration. The revised DEMATEL can guarantee the convergence of the new total-influence matrix, which is more generalized than the original DEMATEL in certain cases.

The DEMATEL technique can decompose the attributes (factors or criteria) in complex social problems into a cause group and an effect group, which enables a systematic analysis of the addressed problem. Researchers or DMs can learn from the analytical results of DEMATEL for two major findings: (1) the INRM of the problem and (2) the influential weight of each attribute. The INRM supports the devising of a systematic improvement plan that can identify the source attributes (dimensions) for an underperforming attribute (dimension), and the influential weights can help construct an evaluation model for ranking or selection decisions.

In the following chapters, the DEMATEL technique will be further combined or integrated with the other MADM methods to form hybrid decision models. Also, several real-world applications will be provided with numerical examples in the second part of this book.

Chapter 4

DEMATEL Technique for Forming INRM and DANP Weights

As mentioned in the previous chapter, the DEMATEL and revised DEMATEL techniques can model the complex internetwork relations within a system (or a problem), which has been adopted for solving multiple attribute decision making (MADM) problems in various applications. The step-by-step calculations of the DEMATEL technique have been illustrated in Chapter 3; here, we further show how to adopt the results from DEMATEL for two analytics: (1) influential network relations maps (INRMs) and (2) DEMATEL-based analytic network process (DANP) influential weights. The advantages DEMATEL has for solving real-world problems will be discussed in Section 4.1, and the required steps for forming an INRM and DANP influential weights will be illustrated by two examples in Sections 4.2 and 4.3. Finally, a case for problem solving using hybrid dynamic multiple attribute decision making (HDMADM) is explained with a numerical example.

4.1 Methodology for Assessing Real-World Problems

A typical MADM model is a scientific analytical model for evaluating and ranking a set of alternatives based on multiple criteria (Campanella and Ribeiro 2011; Tsaur et al. 1997; Wang and Lee 2009; Loban 1997). MADM has been adopted for solving a wide variety of problems, such as supplier selection (Deng and Chan

2011), the performance evaluation of higher education (Wu et al. 2012), improving airline service quality (Kuo 2011), evaluating website quality (Chou and Cheng 2012), product design and selection (Liu 2011), evaluating hot spring hotel service quality (Tseng 2011), and prioritizing sustainable electricity production technologies (Streimikiene et al. 2012).

In addition, most MADM problems in the real world are complicated; there are often interrelations among the involved variables (i.e., dimensions or criteria). It is unrealistic to assume their independence. In the presence of those interrelated complex relations, a systematic approach with hybrid methods is often required; a single method or technique might not be enough to resolve those real-world problems. Furthermore, the traditional MCDM model is unable to capture the dynamicity and hybridity of most real-world decision problems (Campanella and Ribeiro 2011); thus, there is a need to develop a suitable HDMADM method to solve those complicated dynamic problems.

In this chapter, the DEMATEL technique is adopted as the starting point to identify the cause–effect influential relations among the attributes of a problem; next, the DEMATEL technique is further extended to calculate the influential weight of each attribute, by using the basic concept of ANP (Saaty 1996); the extended method is termed the DEMATEL-based ANP or DANP. Therefore, the emphases of this chapter aim to resolve the aforementioned in conventional MCDM applications issues (see the third and fifth points in Section 2.2).

4.2 Constructing an Influential Network Relations Map

Instead of assuming that the attributes/dimensions in a system are independent, the DEMATEL or revised DEMATEL techniques presume that all the attributes/dimensions in a system have impacts on the others, at different levels of influence; this presumption should be more reasonable for the majority of social science problems. For example, while selecting the location for a national science park, the *economic development* dimension should have influence on the *ecological environment* dimension; it would be questionable to assume that those two dimensions were independent. By using the analytical result from DEMATEL, mainly $r_i - c_i$ and $r_i + c_i$ from Equations 3.4 and 3.5, those analytics can indicate the directional influences among the dimensions of a problem; also, the directional influences among the attributes within a dimension can be shown.

Here, a case involving tourism destination competitiveness (TDC) in Taiwan (Peng and Tzeng 2012) is adopted to show how to transform the results from DEMATEL analysis into an INRM. The name of each attribute and the results

Table 4.1 DEMATEL Results of r_i and c_i

Dimensions/Criteria	r_i	c_i	$r_i + c$	$r_i - c_i$
D_1 Regulatory framework				
C_1 Policy rules and regulations	1.750	0.882	2.633	0.868
C_2 Environmental sustainability	0.865	0.933	1.798	−0.068
C_3 Safety and security	0.716	0.846	1.562	−0.131
C_4 Health and hygiene	0.764	0.886	1.651	−0.122
C_5 Prioritization of travel and tourism	1.857	1.192	3.048	0.665
D_2 Business environment and infrastructure				
C_6 Air transport infrastructure	0.726	0.935	1.661	−0.209
C_7 Ground transport	0.735	0.936	1.670	−0.201
C_8 Tourism infrastructure	0.754	1.020	1.774	−0.266
C_9 ICT infrastructure	0.734	0.884	1.618	−0.150
C_{10} Price competitiveness	0.690	1.014	1.704	−0.325
D_3 Human cultural and natural resources				
C_{11} Human resources	1.103	0.778	1.881	0.325
C_{12} Affinity for travel and tourism	0.729	0.930	1.659	−0.202
C_{13} Natural resources	0.884	0.896	1.780	−0.013
C_{14} Culture resources	0.803	0.977	1.781	−0.174

Source: Data from Peng, K.H., and Tzeng, G.H., *Intelligent Decision Technologies,* Springer, 107–115, 2012.

52 ■ *Trends of Hybrid Multiple Criteria Decision Making*

of DEMATEL (i.e., $r_i - c_i$ and $r_i + c_i$) are shown in Table 4.1; the attributes within each dimension are grouped under the corresponding dimension.

In addition, the original total relation matrix T can be rewritten as T_C^G in Equation 4.1, by the grouping of each dimension, where $\sum_{j=1}^{m} m_j = n$, $m < n$, and T_c^{ij} is a $m_i \times m_j$ matrix.

$$
T_C^G = \begin{array}{c}
 & \begin{array}{ccccc} D_1 & & D_j & & D_n \\ c_{11}\ldots c_{1m_1} & \cdots & c_{j1}\ldots c_{jm_j} & \cdots & c_{m1}\ldots c_{mm_m} \end{array} \\
\begin{array}{c} D_1 \\ \\ \\ D_i \\ \\ \\ D_n \end{array}
\begin{array}{c} c_{11} \\ c_{12} \\ \vdots \\ c_{1m_1} \\ c_{i1} \\ c_{i2} \\ \vdots \\ c_{im_j} \\ c_{m1} \\ c_{m2} \\ \vdots \\ c_{mm_m} \end{array}
\begin{bmatrix}
T_c^{11} & \cdots & T_c^{1j} & \cdots & T_c^{1m} \\
\vdots & & \vdots & & \vdots \\
T_c^{i1} & \cdots & T_c^{ij} & \cdots & T_c^{im} \\
\vdots & & \vdots & & \vdots \\
T_c^{m1} & \cdots & T_c^{mj} & \cdots & T_c^{mm}
\end{bmatrix}_{n \times n \mid m < n, \sum_{j=1}^{m} m_j = n}
\end{array}
\tag{4.1}
$$

For the case of TDC here, corresponding to Table 4.1, $m_1 = 5$, $m_2 = 5$, $m_3 = 4$, and $m = 3$ (i.e., three dimensions). T_C^G can be further normalized to become T_N^G in Equation 4.2:

$$
T_N^G = \begin{array}{c}
 & \begin{array}{ccccc} D_1 & & D_j & & D_m \\ c_{11}\ldots c_{1m_1} & \cdots & c_{j1}\ldots c_{jm_j} & \cdots & c_{m1}\ldots c_{mm_m} \end{array} \\
\begin{array}{c} D_1 \\ \\ \\ D_i \\ \\ \\ D_m \end{array}
\begin{array}{c} c_{11} \\ c_{12} \\ \vdots \\ c_{1m_1} \\ c_{i1} \\ c_{i2} \\ \vdots \\ c_{im_j} \\ c_{m1} \\ c_{m2} \\ \vdots \\ c_{mm_m} \end{array}
\begin{bmatrix}
T_c^{\alpha 11} & \cdots & T_c^{\alpha 1j} & \cdots & T_c^{\alpha 1m} \\
\vdots & & \vdots & & \vdots \\
T_c^{\alpha i1} & \cdots & T_c^{\alpha ij} & \cdots & T_c^{\alpha im} \\
\vdots & & \vdots & & \vdots \\
T_c^{\alpha m1} & \cdots & T_c^{\alpha mj} & \cdots & T_c^{\alpha mm}
\end{bmatrix}_{n \times n \mid m < n, \sum_{j=1}^{m} m_j = n}
\end{array}
\tag{4.2}
$$

Take $T_c^{\alpha 11}$, for example: it is normalized by dividing the summation of all the elements in it, as in Equations 4.3–4.4; all the other $T_c^{\alpha ij}$ (for $i, j = 1,\ldots,n$) follow the same normalization calculations.

$$
d_i^{11} = \sum_{j=1}^{m_1} t_{cij}^{11} , \quad i = 1,2,\ldots,m_1
\tag{4.3}
$$

$$T_C^{\alpha 11} = \begin{bmatrix} t_{C11}^{11}/d_1^{11} & \cdots & t_{C1j}^{11}/d_1^{11} & \cdots & t_{C1m_1}^{11}/d_1^{11} \\ \vdots & & \vdots & & \vdots \\ t_{Ci1}^{11}/d_i^{11} & \cdots & t_{Cij}^{11}/d_i^{11} & \cdots & t_{Cim_1}^{11}/d_i^{11} \\ \vdots & & \vdots & & \vdots \\ t_{Cm_11}^{11}/d_{m_1}^{11} & \cdots & t_{Cm_1j}^{11}/d_{m_1}^{11} & \cdots & t_{Cm_1m_1}^{11}/d_{m_1}^{11} \end{bmatrix}$$

$$n = \begin{bmatrix} t_{C11}^{\alpha 11} & \cdots & t_{C1j}^{\alpha 11} & \cdots & t_{C1m_1}^{\alpha 11} \\ \vdots & & \vdots & & \vdots \\ t_{Ci1}^{\alpha 11} & \cdots & t_{Cij}^{\alpha 11} & \cdots & t_{Cim_1}^{\alpha 11} \\ \vdots & & \vdots & & \vdots \\ t_{Cm_11}^{\alpha 11} & \cdots & t_{Cm_11}^{\alpha 11} & \cdots & t_{Cm_1m_1}^{\alpha 11} \end{bmatrix}. \tag{4.4}$$

The transpose of T_N^G (i.e., $(T_N^G)'$) is used to obtain the unweighted supermatrix in the DANP method, shown in Equation 4.5. Similarly, using W^{11} as an example, the submatrices in W can be obtained by referring to Equation 4.6.

$$W = (T_N^G)' = \begin{matrix} & & \begin{matrix} D_1 & & D_i & & D_m \\ c_{11}\cdots c_{1m_1} & \cdots & c_{i1}\cdots c_{im_i} & \cdots & c_{m1}\cdots c_{mm_m} \end{matrix} \\ \begin{matrix} D_1 \\ \\ \\ D_j \\ \\ \\ D_m \end{matrix} & \begin{matrix} c_{11} \\ c_{12} \\ c_{1m_1} \\ c_{j1} \\ c_{j2} \\ c_{jm_j} \\ c_{m1} \\ c_{m2} \\ c_{mm_m} \end{matrix} & \begin{bmatrix} W^{11} & \cdots & W^{i1} & \cdots & W^{m1} \\ \vdots & & \vdots & & \vdots \\ W^{1j} & \cdots & W^{ij} & \cdots & W^{mj} \\ \vdots & & \vdots & & \vdots \\ W^{1m} & \cdots & W^{im} & \cdots & W^{mm} \end{bmatrix}_{n\times n \left| m<n, \sum_{j=1}^{m} m_j = n \right.} \end{matrix}. \tag{4.5}$$

$$W^{11} = (T^{11})' = \begin{matrix} c_{11} \\ \vdots \\ c_{1j} \\ \vdots \\ c_{1m_1} \end{matrix} \begin{bmatrix} t_{c11}^{\alpha 11} & \cdots & t_{ci1}^{\alpha 11} & \cdots & t_{cm_11}^{\alpha 11} \\ \vdots & & \vdots & & \vdots \\ t_{c1j}^{\alpha 11} & \cdots & t_{cij}^{\alpha 11} & \cdots & t_{cm_1j}^{\alpha 11} \\ \vdots & & \vdots & & \vdots \\ t_{c1m_1}^{\alpha 11} & \cdots & t_{cim_1}^{\alpha 11} & \cdots & t_{Cm_1m_1}^{\alpha 11} \end{bmatrix}. \tag{4.6}$$

Next, the total effect relationship matrix of the dimensional matrix T^D can be indicated, as in Equation 4.7. The value t_D^{ij} in T^D is obtained by averaging all the elements in each submatrix W^{ij}.

54 ■ *Trends of Hybrid Multiple Criteria Decision Making*

Table 4.2 Dimensional Matrix with the Cause–Effect Analysis

Dimensions		D_1	D_2	D_3	r_i^D	c_i^D	$r_i^D + c_i^D$	$r_i^D - c_i^D$
D_1	Regulatory framework	0.305	0.825	0.782	1.912	0.916	2.828	0.996
D_2	Business environment and infrastructure	0.321	0.237	0.332	0.891	1.497	2.388	−0.606
D_3	Human cultural and natural resources	0.290	0.435	0.208	0.932	1.322	2.254	−0.389

Source: Data from Peng, K.H., and Tzeng, G.H., *Intelligent Decision Technologies,* Springer, 107–115, 2012.

$$
T^D = \begin{bmatrix}
t_D^{11} & \cdots & t_D^{1j} & \cdots & t_D^{1m} \\
\vdots & & \vdots & & \vdots \\
t_D^{i1} & \cdots & t_D^{ij} & \cdots & t_D^{im} \\
\vdots & & \vdots & & \vdots \\
t_D^{m1} & \cdots & t_D^{mj} & \cdots & t_D^{mm}
\end{bmatrix}_{m \times m}
\begin{matrix}
r_1^D \\ \vdots \\ r_i^D \\ \vdots \\ r_m^D
\end{matrix}
\quad
\begin{matrix}
c_1^D & \cdots & c_j^D \cdots & c_m^D
\end{matrix}
\tag{4.7}
$$

In Equation 4.7, $r_i^D = \sum_{j=1}^{m} t_D^{ij}$ and $c_j^D = \sum_{i=1}^{m} t_D^{ij}$, which can be used to calculate $r_i^D - c_i^D$ for dividing dimensions into a cause group and an effect group, similar to Step 3 in Section 3.5. Therefore, in this illustrated case, the dimensional matrix T^D and the corresponding $r_i^D - c_i^D$ can be shown, as in Table 4.2.

According to the total influential prominence analysis $\left(r_i^D + c_i^D \right)$, *Regulatory framework* (D_1) is the highest total influential prominence among other dimensions, which means it is the most important influencing factor; additionally, *Human cultural and natural resources* (D_3) is the factor with the weakest total influential prominence among the other factors. Next, based on the cause–effect influential relation analysis $\left(r_i^D - c_i^D \right)$, *Regulatory framework* (D_1) represents the highest degree of impact relationship and directly affects other factors; *Business environment and infrastructure* (D_2) is more likely to be influenced by the other dimensions. The total influential prominence and cause–effect analysis of the three dimensions is shown in Figure 4.1.

Here, only the directional influences among dimensions are shown in the dimensional INRM. The detailed INRM with the extended cause–effect analysis among

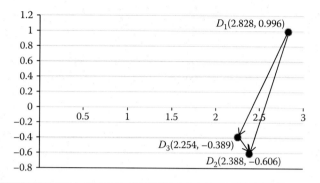

Figure 4.1 INRM of the three dimensions.

criteria will be further shown in Section 4.4; also, the target spots for improvements will be highlighted.

4.3 Determining Influential Weights Using DANP

The original ANP method assumes that each dimension has the same influence in a system, which can be adjusted by the obtained dimensional matrix T^D from the DEMATEL technique. Dimensional matrix T^D can be normalized as T_N^D by Equations 4.8–4.9:

$$d_i = \sum_{j=1}^{m} t_D^{ij}, \text{ where } j=1,2,\ldots,m. \tag{4.8}$$

$$T_N^D = \begin{bmatrix} t_D^{11}/d_1 & \cdots & t_D^{1j}/d_1 & \cdots & t_D^{1m}/d_1 \\ \vdots & & \vdots & & \vdots \\ t_D^{i1}/d_2 & \cdots & t_D^{ij}/d_2 & \cdots & t_D^{im}/d_2 \\ \vdots & & \vdots & & \vdots \\ t_D^{m1}/d_m & \cdots & t_D^{mj}/d_m & \cdots & t_D^{mm}/d_m \end{bmatrix} = \begin{bmatrix} t_D^{\alpha 11} & \cdots & t_D^{\alpha 1j} & \cdots & t_D^{\alpha 1m} \\ \vdots & & \vdots & & \vdots \\ t_D^{\alpha i1} & \cdots & t_D^{\alpha ij} & \cdots & t_D^{\alpha im} \\ \vdots & & \vdots & & \vdots \\ t_D^{\alpha m1} & \cdots & t_D^{\alpha mj} & \cdots & t_D^{\alpha mm} \end{bmatrix}. \tag{4.9}$$

The normalized dimensional matrix T_N^D is adopted to multiply with the unweighted supermatrix in Equation 4.5; thus, the weighted supermatrix (adjusted by DEMATEL) can be obtained in Equation 4.10:

$$W^\alpha = T_N^D W = \begin{bmatrix} t_D^{\alpha 11} \times W^{11} & \cdots & t_D^{\alpha i1} \times W^{i1} & \cdots & t_D^{\alpha m1} \times W^{m1} \\ \vdots & & \vdots & & \vdots \\ t_D^{\alpha 1j} \times W^{1j} & \cdots & t_D^{\alpha ij} \times W^{ij} & \cdots & t_D^{\alpha mj} \times W^{mj} \\ \vdots & & \vdots & & \vdots \\ t_D^{\alpha 1m} \times W^{1m} & \cdots & t_D^{\alpha im} \times W^{im} & \cdots & t_D^{\alpha mm} \times W^{mm} \end{bmatrix}. \tag{4.10}$$

56 ■ *Trends of Hybrid Multiple Criteria Decision Making*

The weighted supermatrix $\boldsymbol{W}^{\alpha}$ multiplies by itself several times to obtain the limit supermatrix based on the concept of the Markov chain. Then, the DANP influential weight of each criterion can be obtained by raising z for $\lim_{z \to \infty}(\boldsymbol{W}^{\alpha})^{z}$, where z is a positive integer value.

Continuing with the case of TDC, the influential weights by DANP for each dimension and criterion can be calculated as shown in Table 4.3. It can be observed that dimension D_2 has the highest influential weight among the three dimensions.

Table 4.3   Influential Weights of the Dimensions and the Criteria by DANP

Dimensions/Criteria	DANP Influential Weights	Ranking
D_1 Regulatory framework	**0.2866**	**3**
C_1 Policy rules and regulations	0.0544	3
C_2 Environmental sustainability	0.0546	2
C_3 Safety and security	0.0500	5
C_4 Health and hygiene	0.0537	4
C_5 Prioritization of travel and tourism	0.0739	1
D_2 Business environment and infrastructure	**0.3803**	**1**
C_6 Air transport infrastructure	0.0744	3
C_7 Ground transport	0.0739	4
C_8 Tourism infrastructure	0.0809	1
C_9 ICT infrastructure	0.0717	5
C_{10} Price competitiveness	0.0794	2
D_3 Human cultural and natural resources	**0.3332**	**2**
C_{11} Human resources	0.0769	4
C_{12} Affinity for travel and tourism	0.0837	3
C_{13} Natural resources	0.0841	2
C_{14} Culture resources	0.0885	1

Source: Data from Peng, K.H., and Tzeng, G.H., *Intelligent Decision Technologies,* Springer, 107–115, 2012.

Also, referring to Table 4.2 and Figure 4.1, we may learn that both D_1 and D_3 would influence D_2. That is to say, if we plan to improve D_2, the source factors (dimensions) would be D_1 and D_3. More details on improvement planning will be provided in Section 4.4.

4.4 Problem Solving for Ranking or Selection Decision by INRM and DANP

The obtained results from DEMATEL indicate the directional influences among dimensions and criteria; in addition, DANP weights show the relative importance of a dimension or criterion in a system. Once decision makers (DMs) plan to solve a ranking or selection problem, those analytics can be synthesized with the available alternatives to form the final score of each alternative. In the aforementioned TDC case, assume that the three destinations A, B, and C are evaluated by two DMs on

Table 4.4   Performance Scores of the Three Assumed Destinations

Criteria	DANP Weights	Three Destinations		
		A	B	C
C_1	0.0544	3.0	2.5	1.5
C_2	0.0546	2.5	3.5	2.5
C_3	0.0500	2.0	4.0	2.5
C_4	0.0537	1.5	4.5	2.5
C_5	0.0739	1.0	4.0	3.0
C_6	0.0744	2.5	3.5	4.5
C_7	0.0739	3.0	4.5	4.0
C_8	0.0809	3.5	3.5	2.5
C_9	0.0717	2.0	3.0	1.5
C_{10}	0.0794	2.5	2.5	1.5
C_{11}	0.0769	4.5	3.0	2.5
C_{12}	0.0837	4.0	4.0	3.0
C_{13}	0.0841	3.5	4.0	2.5
C_{14}	0.0885	3.0	3.5	1.5
Final scores		2.828	3.566	2.545

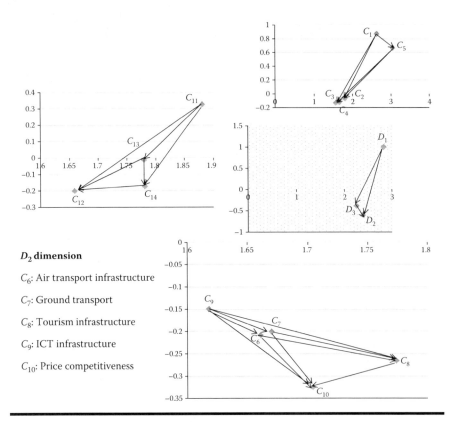

Figure 4.2 INRM for the case of TDC.

the 14 criteria using a crisp Likert scale from 1 (very bad) to 5 (very good). The averaged evaluation results from the two DMs are shown in Table 4.4.

In a ranking problem, the ranking of those three destinations should be $B \succ A \succ C$, and B would be chosen for a selection problem. Furthermore, the detailed INRM can indicate the directional influences of the subcriteria within each dimension.

For example, once DMs plan to improve the criterion *Tourism infrastructure* (C_8) for destination A, they can learn from Figure 4.2 that C_9, C_7, and C_6 all have direct influences on C_8; in other words, the source factors can be identified for destination A to plan for improvements.

4.5 Conclusion

In this chapter, the integration of two techniques—the DEMATEL and ANP methods—is introduced with an illustration of how to use the analytical results to support decisions. The concept of DEMATEL-adjusted influential weights plays a crucial role in forming HDMADM models. Why those two techniques should be integrated or combined might need additional explanation.

DEMATEL Technique for Forming INRM and DANP Weights ■ 59

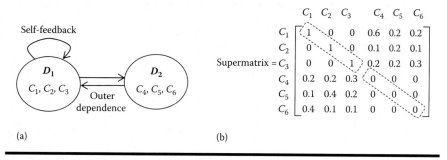

Figure 4.3 Illustration of self-feedback and outer dependence in the ANP method: (a) conceptual framework and (b) supermatrix of the ANP.

As stated in Chapter 2, real-world problems (especially complicated social problems) have to acknowledge that interrelationships among variables/factors do exist; it would be unreasonable to ignore or overlook those effects. Therefore, the key assumption of the analytic hierarchy process (AHP) (Saaty 1986b)—the independence of all the variables/criteria/dimensions/aspects—is problematic. To resolve this issue, the subsequently proposed ANP (Saaty 1996) relaxes the assumption of the outer dependence among clusters (i.e., allowing dependence and feedback among dimensions or aspects); furthermore, ANP allows for the self-feedback effect within a cluster (dimension). We use the simplified case in Figure 4.3 for illustration. Assume that there are two dimensions (D_1 and D_2) and six criteria ($C_1 \approx C_6$) for a problem; the criteria within D_1 (i.e., $C_1 \approx C_3$) have the self-feedback effect, and the criteria within D_2 do not ($C_4 \approx C_4$, shown in Figure 4.3a). The self-feedback effect in D_1 can be solved by putting 1 in the diagonal elements of this unweighted supermatrix, and 0 can be placed in the diagonal elements if no self-feedback exists (Figure 4.3b); this is how the original ANP handles the self-feedback effect on calculating weights.

The ANP method has partially resolved the issues of dependence among dimensions and the self-feedback effect within a dimension. However, two limitations remain: (1) the plausible interrelationship between any two criteria that belong to different dimensions, and (2) the weighted supermatrix is calculated by assuming the equal weight in all dimensions. As a result, the DEMATEL technique is incorporated to enhance (or adjust) the supermatrix of the original ANP method. The DANP method leverages the strength from DEMATEL by allowing the interrelationships among *all criteria*. In addition, the dimensional weights obtained by DEMATEL can relax the equal-weight assumption in ANP. The weighted supermatrix can thus be adjusted (by DEMATEL) to have the final DANP influential weights for all criteria. The calculation details of DANP have been illustrated in this chapter, and the DANP method will be further integrated or combined with the other MADM methods in Chapter 5. Certain real-world applications will be introduced in the second part of this book (e.g., Chapters 9, 10, 11, 12, and 13); interested readers can find an updated review of the DEMATEL and DANP methods in the MCDM research by Gölcük and Baykasoğlu (2016).

Chapter 5

Traditional MADM and New Hybrid MADM for Problem Solving

In this chapter, we begin with the introduction of two popular multiple attribute decision making (MADM) methods: the analytic hierarchy process (AHP) and the analytic network process (ANP), both proposed by Saaty (1977, 1988, 1996); also, the commonly observed ways to aggregate the performance scores of each alternative are discussed. The potential drawbacks and limitations of those traditional MADM and aggregation methods have led to the development of certain new hybrid MADM methods, and the ways to adopt these new hybrid MADM methods are discussed for practical problem solving.

5.1 Traditional MADM for Ranking and Selection

Traditional MADM methods focus on solving ranking and selection problems (Tzeng and Huang 2011). Usually, a weighting system that indicates the relative importance of each involved criterion needs to be constructed first; AHP or ANP might be the most prevalent methods for forming a weighting system in traditional MADM research. Next, a performance evaluation method supports decision makers (DMs) to give rating scores for all alternatives on each criterion. Finally, an aggregator is required to integrate the weighting system with the performance

62 ■ *Trends of Hybrid Mutiple Criteria Decision Making*

scores of the alternatives to form the final evaluation; the ranking or selection decision can thus be made.

5.1.1 AHP and ANP

AHP and ANP both use pairwise comparisons to form the preference structure of a DM. AHP decomposes a problem into a hierarchical system, and ANP further releases the restriction of the hierarchical structure, which is more suitable for solving real-world problems.

To construct an AHP weighting system, the following four steps are often required:

1. Define the hierarchical system of the problem. This usually comprises three levels: goal, dimensions (aspects), and criteria (attributes).
2. Compare the relative preference or importance between each pair of criteria to form a reciprocal matrix. The typical ratio scales are 1 (equal), 3 (moderate), 5 (strong), 7 (very strong), and 9 (extremely strong).
3. Collect opinions from DMs and use the averaged opinions as the averaged weight matrix $A = [a_{ij}]_{n \times n}$ for a problem with n criteria.
   In A, $a_{ij} = 1/a_{ji}$ is based on the reciprocal assumption of AHP (Saaty 1980, 1986a, 1986b, 1997, 1999, 2003). An averaged weight matrix W can be multiplied by the relative importance of its weight vector, as in Equation 5.1, and it can be rewritten as $(W - nI)w = 0$.

$$W \times w = \begin{matrix} w_1 \\ \vdots \\ w_i \\ \vdots \\ w_n \end{matrix} \begin{bmatrix} w_1/w_1 & \cdots & w_1/w_j & \cdots & w_1/w_n \\ \vdots & & \vdots & & \vdots \\ w_i/w_1 & \cdots & w_i/w_j & \cdots & w_i/w_n \\ \vdots & & \vdots & & \vdots \\ w_n/w_1 & \cdots & w_n/w_j & \cdots & w_n/w_n \end{bmatrix} \begin{bmatrix} w_1 \\ \vdots \\ w_j \\ \vdots \\ w_n \end{bmatrix} = n \begin{bmatrix} w_1 \\ \vdots \\ w_i \\ \vdots \\ w_n \end{bmatrix}. \tag{5.1}$$

4. Use the *lambda-max* method (λ_{max}) to solve $(W - nI)w = 0$. The relative importance weight can be found by finding the eigenvector that satisfies Equation 5.2:

$$(A - \lambda_{max} I)w = 0. \tag{5.2}$$

The initial steps for forming an ANP weighting system are similar to those of AHP, and the general form of a supermatrix that comprises criteria and clusters of criteria (also termed *dimensions* or *aspects*) can be indicated as follows:

Traditional MADM and New Hybrid MADM for Problem Solving ∎ 63

$$
\mathbf{W}^{ANP} = \begin{array}{c} \\ D_1 \\ \\ \\ D_j \\ \\ \\ \\ D_m \end{array}
\begin{array}{c}
\begin{array}{c} D_1 \\ c_{11}\cdots c_{1m_1} \end{array} \cdots \begin{array}{c} D_i \\ c_{i1}\cdots c_{im_i} \end{array} \cdots \begin{array}{c} D_m \\ c_{m1}\cdots c_{mm_m} \end{array}
\end{array}
\left[
\begin{array}{ccccc}
\mathbf{W}^{11} & \cdots & \mathbf{W}^{i1} & \cdots & \mathbf{W}^{m1} \\
\vdots & & \vdots & & \vdots \\
\mathbf{W}^{1j} & \cdots & \mathbf{W}^{ij} & \cdots & \mathbf{W}^{mj} \\
\vdots & & \vdots & & \vdots \\
\mathbf{W}^{1m} & \cdots & \mathbf{W}^{im} & \cdots & \mathbf{W}^{mm}
\end{array}
\right]_{n \times n \mid m < n,\ \sum_{j=1}^{m} m_j = n}, \quad (5.3)
$$

where:

    m equals the number of dimensions

    n equals the number of criteria in $\mathbf{W}^{ANP}$

In addition, $\mathbf{W}^{ij}$ denotes the principle eigenvector of the influences of the elements in the jth dimension compared with the ith dimension.

The supermatrix is further used to calculate the weighted supermatrix by transforming the sums all of columns to unity; this idea is very similar to the concept of a Markov chain. Then, as z approaches infinity in Equation 5.4, a stable weighted supermatrix $\mathbf{W}_S^{ANP}$ can be obtained.

$$
\mathbf{W}_S^{ANP} = \lim_{z \to \infty} \left(\mathbf{W}^{ANP} \right)^z. \quad (5.4)
$$

In cases where the effect of cyclicity exists (i.e., more than one limiting supermatrix), the Cesàro sum can be used to get the final weight for each criterion, as follows:

$$
\lim_{z \to \infty} \left(\frac{1}{N} \right) \sum_{k=1}^{N} \left(\mathbf{W}_k^{ANP} \right)^z, \quad (5.5)
$$

where $\mathbf{W}_k^{ANP}$ denotes the kth limiting supermatrix, and there are total N limiting supermatrices. The extensions of AHP and ANP consider the impreciseness of judgments from DMs; therefore, fuzzy set theory has been incorporated to form fuzzy AHP (FAHP) and fuzzy ANP (FANP) for a variety of applications. The details of FAHP and FANP can be found in Tzeng and Huang (2011).

5.2 New Hybrid Modified MADM

Although certain traditional MADM models have addressed the issue of independence among the criteria of a system (e.g., DEMATEL and ANP), the ranking or selection decision might not be sufficient for DMs to resolve practical problems. For example, if a company hopes to select among four marketing plans to promote

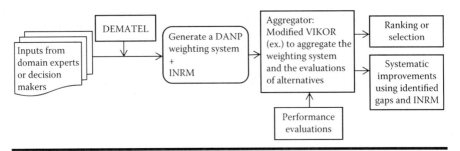

Figure 5.1 New framework of hybrid MCDM for problem solving.

its new e-commerce service, the selected plan (with the highest performance evaluation) might not be fully satisfied by this company. In other words, a ranking or selection problem in traditional MADM might force DMs to accept a relatively outperformed alternative from a group of inferior candidates. Traditional MADM research are lacking adequate tools to support DMs on improving existing alternatives to reach the desired or aspired level, in a systematic approach based on INRM.

To bridge this gap, a new hybrid MADM method with a modified performance evaluation approach is required. The insufficiency of improvement planning in traditional MADM should be acknowledged, and a new framework (Figure 5.1) is proposed to resolve this issue.

5.2.1 DEMATEL-Based ANP Instead of AHP and ANP

In this new framework, DEMATEL-based ANP (DANP) is adopted to replace AHP or ANP in traditional MADM weighting systems. It is obvious that the hierarchical assumption of AHP makes it unsuitable for modeling interdependent criteria in practice.

Besides, although ANP can measure the feedbacks and influences among dimensions (also denoted as clusters) and criteria (inner dimensions/clusters), in diagonal matrices under the assumption that they are independent (zero matrix, W^{ii} or exhibit self-relation (Identity matrix, I). The weighted supermatrix is obtained to assume the equal weights in calculating dimensional influences, it is not reasonable in the original ANP. Saatys' ANP eliminates the limitations of the Analytic Hierarchy Process (AHP) that assumes that inner criteria and outer dimensions/clusters are all independent. The difference between these methods is that the ANP is applied to decision-making problems for interrelationships only in outer dimensions. Therefore, the normalized dimensional influence relation matrix T_N^D in DEMATEL can be used to adjust dimensional weights of DANP to be suggested in the new hybrid MCDM approach. The detailed steps and formulae of DANP have been introduced in Chapter 4; a conceptual illustration is provided in Figure 5.2 to show the procedure of adjusting dimensional weights in DANP.

Traditional MADM and New Hybrid MADM for Problem Solving ■ 65

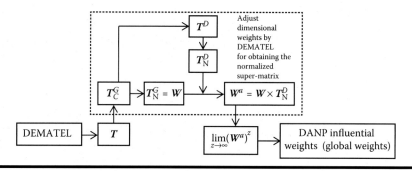

Figure 5.2 Adjusting dimensional weights to form a weighted supermatrix by DANP. (Note: Refer to Chapter 4 for the symbols used in this figure.)

5.2.2 Modified VIKOR for Measuring Performance Gaps

Opricovic proposed the VIKOR compromise-ranking method in MADM (Opricovic 1986; Opricovic and Tzeng 2004, 2007), involving both performance evaluation and aggregation. Assume that the available alternatives for a problem are $A_1,\ldots,A_k,\ldots,A_K$. The performance scores of alternative A_k on the jth criterion are denoted as f_{kj} ($k = 1,2,\ldots,K;\ j = 1,2,\ldots,n$); w_j is the relative importance of the jth criterion in a weighting system, where $j = 1,2,\ldots,n$ and n is the number of total criteria of a decision problem. The original VIKOR used the min-max $\left(f_j^*\ \text{and}\ f_j^-\right)$ values from the available alternatives on each criterion to calculate the total performance gap for each alternative. The original L_p-metric in Equation 5.6 is a way to reach compromised solutions by setting different values in p, where $f_j^* = \max_k\{f_{kj}\ |\ k = 1,2,\ldots,K\}$ and $f_j^- = \min_k\{f_{kj}\ |\ k = 1,2,\ldots,K\}$ in the traditional min-max approach.

$$L_k^p = \left\{\sum_{j=1}^{n}\left[w_j\left(f_j^* - f_{kj}\right)/\left(f_j^* - f_j^-\right)\right]^p\right\}^{\frac{1}{p}}. \tag{5.6}$$

While $p = 1$ and $p = \infty$, $L_k^{p=1}$ and $L_k^{p=\infty}$ are calculated to find the E_k and Q_k indices for the kth alternative. The definitions of E_k and Q_k are shown in Equations 5.7 and 5.8:

$$E_k = L_k^{p=1} = \sum_{j=1}^{n}[w_j(|f_j^* - f_{kj}|)/(|f_j^* - f_j^-|)]. \tag{5.7}$$

66 ■ *Trends of Hybrid Mutiple Criteria Decision Making*

$$Q_k = L_k^{p=\infty} = \max_j \{(|\ f_j^* - f_{kj}\ |)\ /\ (|\ f_j^* - f_j^-\ |)\ |\ j = 1, 2, \ldots, n\} \qquad (5.8)$$

The min-max approach in traditional VIKOR can form the synthesized indices E_k, Q_k, and U_k for the kth alternative. The definition of U_k is as follows:

$$U_k = \delta \left[(E_k - E^*)\ /\ (E^- - E^*) \right] + (1 - \delta) \left[(Q_k - Q^*)\ /\ (Q^- - Q^*) \right]. \qquad (5.9)$$

where:

$0 \le \delta \le 1$

$E^- = \max\{E_k\ |\ k = 1, \ldots, K\}$

$E^* = \min\{E_k\ |\ k = 1, \ldots, K\}$

$Q^- = \max\{Q_k\ |\ k = 1, \ldots, K\}$

$Q^* = \min\{Q_k\ |\ k = 1, \ldots, K\}$

The index Q_k indicates the synthesized gap for the kth alternative; in addition, the group utility (or *average gap*) is emphasized when the value of δ is high (e.g., 0.9); on the contrary, if δ is low (e.g., 0.05), the individual maximum regrets and gaps gain prominence.

Although the min–max approach for performance evaluation can be adopted for the traditional VIKOR to synthesize the indices for ranking or selection, it also implies that DMs might have to choose a relatively outperformed alternative from a group of inferior candidates; this limitation of the original VIKOR was not present in traditional MADM. To address the aforementioned limitation, a modified VIKOR is devised by using the aspired value (i.e., f_j^{aspired}) and the worst value (i.e., f_j^{worst}) on the jth criterion to replace the original min–max values; in addition, the relative importance w_j on the jth criterion is replaced by the DANP influential weight w_j^{DANP} to form a new L_p-metric, as follows:

$$L_k^p = \{ \sum_{j=1}^n [w_j^{\text{DANP}} (|\ f_j^{\text{aspired}} - f_{kj}\ |)\ /\ (|\ f_j^{\text{aspired}} - f_j^{\text{worst}}\ |)]^p \}^{1/p}. \qquad (5.10)$$

The ideal or aspired value is also called the *aspiration level* (Simon 1959, 1972); the performance gap (weighted by DANP), measured against the aspired level, can show the improvement priority for the jth alternative. Once DMs hope to improve an alternative, this improvement priority can be used with an INRM from the DEMATEL analysis to identify the influential cause–effect relations among the

Traditional MADM and New Hybrid MADM for Problem Solving ■ 67

dimensions. The required steps of the modified VIKOR method are summarized as follows.

Step 1: Obtain or define the aspiration level and the worst value.

It is commonly observed that the 5-point Likert scale or 10-point evaluation scale is used for performance evaluation. Here, only crisp set–based evaluation is discussed. Taking the 10-point scale evaluation as an example, DMs can set the aspiration level as 10 and the worst value as 0. As a result, the original min–max performance-rating matrix can be converted into a normalized gap-rating score r_{kj} by using the aspiration level f_j^{aspired} and the worst value f_j^{worst}, as follows:

$$r_{kj} = (|f_j^{\text{aspired}} - f_{kj}|)/(|f_j^{\text{aspired}} - f_j^{\text{worst}}|) = (|10 - f_{kj}|)/10, \text{ for } k = 1,\dots,K. \quad (5.11)$$

Step 2: Calculate the DANP weighted average group utility S_k and maximal regret R_k.

Referring to Equation 5.10, the indices for the DANP weighted average group utility S_k and maximal regret R_k, for the kth alternative, can be calculated as follows:

$$S_k = \sum_{j=1}^{n} w_j^{\text{DANP}} r_{kj}, \text{ for } k = 1,\dots,K. \quad (5.12)$$

$$R_k = \max_j \{r_{kj} \mid j = 1, 2,\dots,n\}, \text{ for } k = 1,\dots,K. \quad (5.13)$$

Step 3: Set δ to define the relative emphases on S_k and R_k.

DMs can determine the relative emphases on the average group utility and the maximal individual regret by forming a synthesized index V_k. Referring to Equation 5.9, U_k can be changed to V_k by using the aspired/worst approach; therefore, V_k can be formed as follows:

$$V_k = \delta(S_k - S^{\text{aspired}})/(S^{\text{worst}} - S^{\text{aspired}}) + (1-\delta)(R_k - R^{\text{aspired}})/(R^{\text{worst}} - R^{\text{aspired}}). \quad (5.14)$$

In addition, since $S^{\text{worst}} = R^{\text{worst}} = 1$ and $S^{\text{aspired}} = R^{\text{aspired}} = 0$, by using the 10-point evaluation, V_k can be rewritten as follows:

$$V_k = \delta(S_k - S^{\text{aspired}}) + (1-\delta)(R_k - R^{\text{aspired}}), \text{ for } k = 1,\dots,K. \quad (5.15)$$

Step 4: Obtain the improvement priority based on weighted performance gaps.

Once DMs plan to improve the overall performance of the kth alternative, the normalized gap-rating score r_{kj} (for $j = 1,\dots,n$) can be multiplied with the DANP influential weight on the jth criterion (i.e.,

68 ■ *Trends of Hybrid Mutiple Criteria Decision Making*

$g_{kj} = w_j^{\mathrm{DANP}} \times n_{kj}$, for $j = 1,\ldots,n$) to form an improvement priority. DMs can rank g_{kj} (for $j = 1,\ldots,n$), from high to low, as their improvement priority for the kth alternative.

Step 5: Make ranking or selection decisions based on the synthesized index V_k. Finally, the synthesized index V_k is applied to rank (or select) the alternatives, from low to high, by the synthesized performance gaps.

The modified VIKOR method is applied to determine the compromise solution for two main purposes: (1) to make ranking or selection decisions and (2) to form a priority for a specific alternative to plan for systematic improvement. In brief, the modified VIKOR with the aspired/worst performance evaluation approach has paved a new way for traditional MADM to shift the research focus from ranking/selection to improvement planning, and can be used for only one (single) alternative performance improvement planning.

5.3 Additive and Nonadditive Types of Aggregators

In Figure 5.1, after forming a weighing system and obtaining performance evaluations for alternatives, an aggregator is needed to synthesize those two parts for the final evaluation for each alternative. Aggregators can be roughly categorized as two types: additive and nonadditive; furthermore, DMs can use hard (i.e., crisp set) or soft computing (e.g., fuzzy set or gray set) techniques in these aggregators.

5.3.1 Additive-Type Aggregators

Additive-type aggregators assume that there is no synergy effect among dimensions or criteria, which require fewer calculation efforts. The simple additive weighting (SAW) method might be the most prevalent aggregator in MADM studies. In the modified VIKOR method, the synthesized index V_k also belongs to the additive-type aggregator. In Section II of this book, both crisp and fuzzy additive-type aggregators will be introduced in certain empirical cases, and we focus on discussing the non-additive-type aggregator in this chapter.

5.3.2 Nonadditive-Type Aggregators (Fuzzy Integrals)

MADM is based on the utility theory in economics (multiple attribute utility theory, MAUT) to model the preference structure of a DM; however, conventional economic studies mainly adopt statistical models, such as regressions, to form utility functions, which is constrained by the linear presumption in statistics. As a result, MAUT has to bear the questionable preferential independence assumption (Grabisch 1995). To resolve this issue, decision science researchers sought help from nonadditive-type aggregators to model the synergy effects among dimensions or criteria, and we introduce a representative one here: *fuzzy integrals*.

The fuzzy integral technique was proposed by Gustave Choquet (1953), and is hence also called the *Choquet integral*. Later, Michio Sugeno (1974) introduced the idea of the *fuzzy measure*, to be integrated with fuzzy integrals for modeling the imprecise synergy effect among variables. How to adopt fuzzy integrals for MADM research is introduced here, and the details of fuzzy measures and fuzzy integrals can be found in Dubois and Parade (1980) and Grabisch (1995). The required steps for adopting fuzzy integrals in MADM problems can be summarized as follows.

Step 1: Construct the λ-measure and fuzzy density functions for a MADM problem.

Let g be a λ-measure function defined on a power set $P(C)$ for a finite set C comprised of n variables. Then, g should satisfy the properties in Equations 5.16 through 5.18:

$$g : P(C) \rightarrow [0,1], \, g(\varnothing) = 0, \, g(C) = 1. \tag{5.16}$$

$$\forall c_A, c_B \in P(C), c_A \cap c_B = \varnothing. \tag{5.17}$$

$$g_\lambda(c_A \cup c_B) = g_\lambda(c_A) + g_\lambda(c_B) + \lambda g_\lambda(c_A) g_\lambda(c_B), \, \text{for} \, -1 \leq \lambda < \infty. \tag{5.18}$$

In Equation 5.18, λ indicates the aforementioned non-additive (synergy)-type effect; when $\lambda < 0$, $\lambda = 0$, and $\lambda > 0$, it denotes the substitutive (negative synergy), additive (zero synergy), and multiplicative (positive synergy) effects, respectively. Next, considering the λ-measure on multiple criteria, the g_i fuzzy density function can be defined as follows:

$$g_\lambda(\{c_1, c_2, \ldots, c_n\}) = \sum_{i=1}^{n} g_i + \lambda \sum_{i_1=1}^{n-1} \sum_{i_2=i_1+1}^{n} g_{i_1} g_{i_2} + \ldots + \lambda^{n-1} g_1 g_2 \ldots g_n$$

$$\tag{5.19}$$

$$= \frac{1}{\lambda} \left(\prod_{i=1}^{n} (1 + \lambda g_i) - 1 \right) = 1, \, \text{for} \, -1 \leq \lambda < \infty.$$

Step 2: Calculate the fuzzy measure based on the concept of MAUT (Keeney and Raiffa 1993).

The aggregated utility function regarding C can be indicated as follows, based on the concept of MAUT.

70 ■ *Trends of Hybrid Mutiple Criteria Decision Making*

$$u(c_1, c_2, \ldots, c_n) = \sum_{i=1}^{n} w_i u(c_i) + \lambda \sum_{i=1, j>i}^{n} w_i w_j u(c_i) u(c_j) + \ldots$$

(5.20)

$$+ \lambda^{n-1} w_1 w_2 \ldots w_n u(c_1) u(c_2) \ldots u(c_n).$$

In Equation 5.20, $u(c_1^0, c_2^0, \ldots, c_n^0) = 0$ and $u(c_1^*, c_2^*, \ldots, c_n^*) = 1$; $u(c_i)$ is a conditional utility function of c_i, and $u(c_i^0) = 0$ and $u(c_i^*) = 1$, for $i = 1, 2, \ldots, n$; $w_i = u(c_i^*, c_{-i}^0)$; λ is a solution of $1 + \lambda = \prod_{i=1}^{n}(1 + \lambda w_i)$.

Thus, Equation 5.19 can be redefined as Equation 5.21, based on the concept of MAUT, whe re $g_\lambda^{(i)}$ denotes that this fuzzy density function measures i criteria simultaneously.

$$g_\lambda^{(n)}(\{c_1, c_2, \ldots, c_n\}) = \sum_{i=1}^{n} g_\lambda^{(1)}(\{c_i\}) + \lambda \sum_{i=1, j>i}^{n} g_\lambda^{(2)}(\{c_i\}) g_\lambda(\{c_j\})$$

(5.21)

$$+ \ldots + \lambda^{n-1} g_\lambda^{(n)}(\{c_1\}) g_\lambda(\{c_2\}) \ldots g_\lambda(\{c_n\}),$$

where:
$g_\lambda^{(n)}(\{c_1^*, c_2^*, \ldots, c_n^*\}) = g_\lambda^{(n)}(\{c_1, c_2, \ldots, c_n\}) = 1$ $g_\lambda^{(1)}(\{c_i^*\}) = 1$ and $g_\lambda^{(1)}(\{c_i^0\}) = 0$
(for $i = 1, 2, \ldots, n$)
$w_i = u(c_i^*, c_{-i}^0) = g_\lambda^{(1)}(\{c_i\})$

$$1 + \lambda = \prod_{i=1}^{n}(1 + \lambda g_\lambda^{(1)}(\{c_i\}))$$

Step 3: Incorporate DANP influential weights as the initial weights for fuzzy measures.

To provide guidance for the initial weight of each criterion in a MADM model, DMs can adopt DANP influential weights (introduced in Chapter 4).

$$g_\lambda(\{c_1\}), \ldots, g_\lambda(\{c_i\}), \ldots, g_\lambda(\{c_n\}) = \gamma(w_1^{DANP}, \ldots, w_i^{DANP}, \ldots, w_n^{DANP})$$

(5.22)

$$= (\gamma w_1^{DANP}, \ldots, \gamma w_i^{DANP}, \ldots, \gamma w_n^{DANP}).$$

where:
γ is the adjustment coefficient
w_i^{DANP} is the DANP influential weight of the ith criterion

Step 4: Aggregate the performance score from the criteria in each dimension by fuzzy integral.

Traditional MADM and New Hybrid MADM for Problem Solving ■ 71

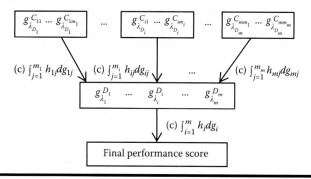

Figure 5.3 Fuzzy integral for the performance aggregation of a MADM problem.

For this step, we assume the structure of dimensions and criteria can be denoted as a fuzzy integral DANP matrix $W_{\text{FuzzInt}}^{\text{DANP}}$ with m dimensions and n criteria, as follows:

$$W_{\text{FuzzInt}}^{\text{DANP}} = \begin{array}{c} \\ D_1 \\ \vdots \\ D_j \\ \vdots \\ D_m \end{array} \begin{array}{c} \overset{D_1}{c_{11}\ldots c_{1m_1}} \quad \cdots \quad \overset{D_i}{c_{i1}\ldots c_{im_i}} \quad \cdots \quad \overset{D_m}{c_{m1}\ldots c_{mm_m}} \\ \left[\begin{array}{ccccc} W^{11} & \cdots & W^{i1} & \cdots & W^{m1} \\ \vdots & & \vdots & & \vdots \\ W^{1j} & \cdots & W^{ij} & \cdots & W^{mj} \\ \vdots & & \vdots & & \vdots \\ W^{1m} & \cdots & W^{im} & \cdots & W^{mm} \end{array} \right]_{n \times n \Big| m < n, \; \sum_{j=1}^{m} m_j = n} \end{array}. \quad (5.23)$$

Let h be a measurable set function for performance, defined on the fuzzy measurable space. Assume that $h(p_{i1}) \geq h(p_{i2}) \geq \ldots \geq h(p_{im_i})$ in the ith dimension; then, the fuzzy integral (e.g., $(c)\int hdg$) of fuzzy measure $g_\lambda(\cdot)$ with respect to $h(\cdot)$ can be defined, as in Equation 5.23 for the ith dimension in Equation 5.24:

$$(c)\int h_i dg_i = h_i(p_{im_i})g_i(H_{im_i}) + [h_i(p_{im_{i-1}}) - h_i(p_{im_i})]g_i(H_{im_{i-1}}) + \cdots \quad (5.24)$$
$$+ [h_i(p_{i1}) - h(p_{i2})]g_i(H_{i1}),$$

where:
$H_{i1} = \{c_{p_{i1}}\}$
$H_{i2} = \{c_{p_{i1}}, c_{p_{i2}}\}$
...
$H_{im_i} = \{c_{p_{i1}}, c_{p_{i2}}, \ldots, c_{p_{im_i}}\}$ (i.e., the total subcriteria in the ith dimension)

Step 5: Aggregate the final performance score from all dimensions by fuzzy integral.

The final performance for an alternative has to be aggregated from all the dimensions (assume that there are a total of m dimensions in a MADM problem), which is illustrated in Figure 5.3.

5.4 Conclusion

In this chapter, a new hybrid modified MADM framework is proposed that comprises three parts: (1) a weighting system, (2) a performance evaluation method, and (3) an additive or nonadditive-type aggregator. DMs can choose a suitable combination from these three parts based on the confronted problem. Two important new concepts are highlighted here: (1) the traditional min–max approach should be replaced by the modified aspired/worst performance evaluation for improvement planning; and (2) interdependent relations and certain synergy effects often exist in practical decision problems. Therefore, a hybrid MADM model is more flexible than a single traditional MADM method for modeling the complexity in this kind of problem.

Chapter 6

MODM with De Novo and Changeable Spaces

In this chapter, the focus shifts from multiple attribute decision making (MADM) to multiple objective decision making (MODM). In general, MODM aims to search for the optimal allocation of limited resources in the presence of multiple objectives or goals. MODM originates from multiobjective programming (MOP), a branch of mathematical programming that has been used to deal with various optimization problems. In conventional MODM studies, researchers or decision makers (DMs) attempt to achieve the efficient frontier by using various mathematical modeling or AI techniques, such as genetic algorithms (GAs) (Hsieh and Liu 2004).

The conventional MODM approach is constrained by fixed or given solution spaces (including decision and objective spaces); however, in the new era of knowledge economy, numerous innovations and creative business models have enabled the possibility of expanding or enhancing those previously fixed solution spaces. Hence, conventional MOP models (or MODM problems) should be extended from the concept of fixed parameters to changeable parameters, also called *changeable spaces* (Tzeng and Huang 2014). In this chapter, the essential ideas of MODM and three types of MOP models are introduced with changeable parameters to help DMs achieve the desired point (aspiration level), which opens a new window for DMs to pursue continuous improvements.

73

6.1 Basic Concepts and Trends of MODM

The early development of MODM can be traced back to Vilfredo Pareto (1848–1923), who initiated the study of aggregating conflicting criteria into a single composite index. A fundamental idea in MODM was also named after him: *Pareto optimality*, which indicates the allocation of resources to the optimal state when it is not possible to improve any criterion without deteriorating at least one other. Later on, as proposed by Dantzig in 1947, linear programming was formally applied as a mathematical tool for handling optimization problems (Dantzig 1951; von Neumann 1947). The original motivation of mathematical programming was the need to solve complex planning problems in wartime operations. After the Second World War, this technique was extended to solve various industrial optimization problems and resulted in great successes. Nevertheless, traditional mathematical programming models, whether single-objective programming (SOP) or MOP, involve optimizing a fixed objective function(s) under given constraints or resources. Therefore, it is constrained by optimizing an existing system; how to refine or redesign the existing system is unexplored. More specifically, most of the progress in traditional optimization methods focuses on algorithms or convergence properties to derive the optimal solution within a system (Zeleny 1998). In the early developments of MODM, roughly 1950–1975 (see Figure 1.3), two traditional optimization approaches attracted the highest attention: (1) goal-programming (GP) and (2) compromise solution. Those two approaches are briefly explained here.

Initially, GP was conceived as an application of single-objective linear programming by Charnes and Cooper (1955, 1961), and then gained popularity in the 1960s and 1970s through the works of Ijiri (1965), Lee (1972), and Ignizio (1976). GP is an analytical approach devised to address decision-making problems where targets have been assigned to all the attributes/criteria and where the DM is interested in minimizing the nonachievement of the corresponding goals (Romero 2004). There are four main characteristics that distinguish GP from the other linear programming techniques:

1. The conceptualization of objectives as goals
2. The assignment of priorities or weights to the achievement of the goals
3. The presence of deviational variables and to measure overachievement and underachievement from target or (threshold) levels
4. The minimization of weighted sums of deviational variables to find solutions that best satisfy the goals

Among the various GP models, three types might be the most widely adopted: (1) weighted GP (WGP), (2) lexicographic GP (LGP), and (3) min–max GP (MGP) models, the essential ideas of which are as follows.

MODM with De Novo and Changeable Spaces ■ 75

1. Weighted GP model
Most MOP problems can be mathematically denoted as follows:

$$\max\left[f_1(\boldsymbol{x}), f_2(\boldsymbol{x}), \ldots, f_k(\boldsymbol{x})\right], \tag{6.1}$$

$$s.t. \quad \boldsymbol{A}\boldsymbol{x} \le \boldsymbol{b},$$

$$\boldsymbol{x} \ge \boldsymbol{0}.$$

As for the mathematical programming of a WGP model, it can be expressed as follows (Ignizio 1976):

$$\min \sum_i (a_i d_i^- + \beta_i d_i^+), \tag{6.2}$$

$$s.t. \quad f_i(\boldsymbol{x}) + d_i^- - d_i^+ = g_i, \quad i = 1, 2, \ldots, k,$$

$$\boldsymbol{A}\boldsymbol{x} \le \boldsymbol{b}$$

$$d_i^- \cdot d_i^+ = 0, \quad d_i^- \ge 0, \quad d_i^+ \ge 0,$$

where:
$a_i = w_i^- / k_i$ if d_i^- is unwanted; otherwise, $\alpha_i = 0$
$\beta_i = w_i^+ / k_i$ if d_i^+ is unwanted; otherwise, $\beta_i = 0$

The parameters w_i^-, w_i^+ and $k_i = f_i^* - f_i^-$ are the weights reflecting the preferential and normalizing purposes attached to the achievement of the ith goal; f_i^* can be set as the positive ideal point or as the aspiration level of the ith goal, and f_i^- can be set as the negative ideal point or as the worst value of the ith goal, respectively.

$$\min \sum_{i=1}^{q} a_i (d_i^- + d_i^+), \tag{6.3}$$

$$s.t. \quad f_i(\boldsymbol{x}) + d_i^- - d_i^+ = g_i, i = 1, 2, \ldots, q,$$

$$\boldsymbol{A}\boldsymbol{x} \le \boldsymbol{b},$$

$$d_i^- \cdot d_i^+ = 0,$$

$$d_i^- \ge 0, \quad d_i^+ \ge 0.$$

76 ■ *Trends of Hybrid Multiple Criteria Decision Making*

In Equation 6.3,

$$a_i = \frac{w_i}{f_i^* - f_i^-}$$

and w_i can be obtained by the AHP, ANP, or DEMATEL-based ANP (DANP); w_i denotes the relative importance or influential weight of the ith goal for the overall evaluation system.

2. Lexicographic GP model

The achievement function of the LGP model is based on the Q number of priority levels established in the model (as a vector), which reflects the preferential sequence of each goal. Each component in this vector represents the unwanted deviation variables of the goals placed in the corresponding priority level. The simplified mathematical model (Ignizio 1976) can be indicated as follows:

$$Lex \ \min a = \left[\sum_{i \in h_1} \left(a_i d_i^- + \beta_i d_i^+ \right), ..., \sum_{i \in h_r} \left(a_i d_i^- + \beta_i d_i^+ \right), ..., \sum_{i \in h_Q} \left(a_i d_i^- + \beta_i d_i^+ \right) \right], \quad (6.4)$$

$$s.t. \quad f_i(\boldsymbol{x}) + d_i^- - d_i^+ = g_i \quad i \in \{1,...,q\} \quad i \in h_r \quad r \in \{1,...,Q\},$$

$$\boldsymbol{x} \in F, \quad d_i^- \geq 0, \quad d_i^+ \geq 0.$$

where:

h_r represents the index set of goals placed in the rth priority state

$\boldsymbol{x} \in F$ denotes the feasible solution (decision space)

3. Min–max GP model

The achievement function of an MGP model seeks for the minimization of the maximum deviation from any single goal. If D represents this maximum deviation, the mathematical programming of an LGP model is as follows (Flavell 1976):

$$\min_x D, \quad (6.5)$$

$$s.t. \quad a_i d_i^- + \beta_i d_i^+ \leq D,$$

$$f_i(\boldsymbol{x}) + d_i^- - d_i^+ = g_i \quad i \in \{1,...,q\},$$

$$\boldsymbol{x} \in F, \quad d_i^- \geq 0, \quad d_i^+ \geq 0.$$

In Equation 6.5, $\boldsymbol{x} \in F$ also denotes the feasible solution (decision space). The second mainstream in traditional MOP is the *compromise solution*, proposed by Yu and Zeleny in 1972. The idea of the compromise solution is straightforward; once there are trade-offs between objects, an ideal point will not be attainable in an MOP problem. Yu (1973) proposed an approach that calculates the L_p-norm distance to find the optimal solution that is closest to the ideal point. A generalized MOP problem can be denoted as in Equation 6.1; it can be solved to find the Pareto-optimal solutions by using traditional methods such as the WGP in Equation 6.6:

$$\max \sum_{i=1}^{k} w_i f_i(\boldsymbol{x}),$$

(6.6)

$$s.t. \ \boldsymbol{Ax} \leq \boldsymbol{b},$$

$$\boldsymbol{x} \geq \boldsymbol{0}, \ \sum_{i=1}^{k} w_i = 1.$$

By using the L_p-norm distance approach to define the distance between a point and an ideal point, it can be shown as follows:

$$d_p = \left\| \boldsymbol{f}^* - \boldsymbol{f} \right\|_p,$$

(6.7)

$$p = 1,2,\ldots,\infty.$$

If $f_i(\boldsymbol{x})$ indicates the ith objective, we may solve the MOP for each objective, as in Equation 6.8, to get the ideal point of each objective as $f^* = \left(f_1^*, f_x^*, \ldots, f_n^* \right)$.

$$\max f_i(\boldsymbol{x}),$$

(6.8)

$$s.t. \quad \boldsymbol{Ax} \leq \boldsymbol{b},$$

$$\boldsymbol{x} \geq \boldsymbol{0}.$$

Then, the concept of the L_p-norm distance can be used to measure the distance between the objective values and the ideal point, as in Equation 6.9; thereby, a normalized compromise solution can be reached.

$$\min \ d^p = \left\{ \sum_{i=1}^{k} \left(w_i \left(\frac{| f_i^* - f_i(\boldsymbol{x}) |}{| f_i^* - f_i^- |} \right) \right)^p \right\}^{1/p} \text{ for } p = 1,2,\ldots,\infty,$$

(6.9)

$$s.t. \quad \boldsymbol{Ax} \leq \boldsymbol{b},$$

$$\boldsymbol{x} \geq \boldsymbol{0},$$

where:

f_i^* and f_i^- denote the ideal and the worst values of the ith goal, respectively

w_i denotes the relative importance of the ith goal

Traditional mathematical programming techniques might work well in a fixed and less dynamic environment; however, the growing complexity and fast pace of the business environment have led researchers to seek out an advanced way of thinking, in order to overcome the existing constraints. For example, the existence of dispatched workers may provide the possibility of extending the current bottleneck of human resources in a company. Or, in the global supply chain, a manufacturer may outsource or subcontract certain productions to its suppliers to overcome the limitations of its own production capacity. Therefore, MOP or MODM nowadays should involve the possibility of redesigning or reshaping the problem confronted in a system (Zeleny 1998).

As one of the early pioneers, Zeleny (1982, 1986) proposed De Novo programming, which might be the first method to enhance traditional mathematical programming by relaxing the assumption of fixed resources in a system. By using information on the unit prices of resources in a system, De Novo programming further eliminates the trade-offs among objects in order to reach optimal resource allocation (Zeleny 1990; Huang et al. 2005a; Huang et al. 2006b, 2006c; Chen and Tzeng 2009). The extended ideas of De Novo programming, proposed by Yu and colleagues (2006), introduced the concept of *changeable space* for a typical production resource allocation problem. There are multiple ways to change the existing parameters in a production resource allocation problem, such as further investment, outsourcing, and the improvement of the technical efficiency of the production lines, and these new ideas in MOP can migrate an existing system by changing previously fixed constraints. The following sections in this chapter will give further explanations using several numerical examples.

6.2 De Novo Programming

De Novo programming can be regarded as an approach to the optimization of resource allocation problems. A typical resource allocation problem can be expressed as follows:

$$\max \; z = Cx, \tag{6.10}$$

$$s.t. \quad Ax \leq b,$$

$$x \geq 0,$$

where:

C	comprises the objective coefficient matrix
$\boldsymbol{x}$	denotes the variable vector
A	denotes the technological coefficient matrix
Vector $\boldsymbol{b}$	reflects the maximum limitations for resources

Nevertheless, the traditional resource allocation model is fully fixed by the existing technical coefficients and the limitations of resources, which ignores one important fact: the unit price of each resource can be considered when redesigning the resource allocation model. Considering the unit prices of resources and the total budget, De Novo programming breaks down the existing framework of fixed resource limitations.

Zeleny (1982, 1986) proposed De Novo programming to enhance traditional resource allocation problems by reforming a given system, considering the unit prices of resources. He claimed that a system can be enhanced to reach a superior ideal point by using this new way of thinking. The following MOP problem from Zeleny (1982) shows how to enhance the traditional optimization outcome by using De Novo programming.

Let's assume that two products, suits (in quantities x_1) and dresses (in quantities x_2), are manufactured by a company, and these two products need to consume five kinds of resources, from nylon to golden thread. The required technical coefficients for each resource are shown in Table 6.1.

In this case, two objectives, the profit (f_1) and quality indices (f_2), are the only considered objectives by this company; from Table 6.1, those two objectives and constraints can be modeled as the following linear system:

$$\max \quad f_1 = 400x_1 + 300x_2,$$

$$\max \quad f_2 = 6x_1 + 8x_2,$$

$$s.t. \quad 4x_1 \leq 20,$$

$$2x_1 + 6x_2 \leq 24,$$

$$12x_1 + 4x_2 \leq 60,$$

$$3x_2 \leq 10.5,$$

$$4x_1 + 4x_2 \leq 26,$$

$$x_1, x_2 \geq 0,$$

where f_1 and f_2 denote the two objective indices: profit and quality. Assume that the two objectives are equally important. Next, setting $p = 2$ using the compromise

80 ■ *Trends of Hybrid Multiple Criteria Decision Making*

Table 6.1   A Case from Zeleny's Example

Unit Price	Resource	Technological Coefficients		No. of Units
		$x_1 = 1$	$x_2 = 1$	
30	Nylon	4	0	20
40	Velvet	2	6	24
9.5	Silver thread	12	4	60
20	Silk	0	3	10.5
10	Golden thread	4	4	26

Source: Data from Huang, J.J., and Tzeng, G.H., *Technological and Economic Development of Economy*, 20(2), 242–261, 2014.

solution method, the optimal solution can be obtained: $x_1 = 3.9837$, $x_2 = 2.5163$, $f_1 = 2348.37$, and $f_2 = 44.03$; this is also the outcome of the so-called traditional MOP approach.

By introducing the idea of De Novo programming, DMs have a chance to further improve the optimized result from the traditional approach. Zeleny (1998) proposed the concept of an optimal portfolio of resources, in which resources should be regarded as a portfolio; by considering the required unit price of each resource, the whole system can be reformulated. In this research line, certain extended studies, such as fuzzy technique–supported coefficients (Li and Lee 1990, 1993), optimum-path ratios (Shi 1995) and the 0–1 programming problem (Kim et al. 1993), have been proposed to cover various scenarios with more complicated problem settings.

To conduct De Novo programming, the required steps can be summarized as follows:

1. Identify the ideal level vector ($\boldsymbol{z}^o$) for each objective function separately, considering the budget constraint.

$$\max \ z_k^o = \boldsymbol{c}_k \boldsymbol{x} \qquad k = 1,...,m, \tag{6.11}$$

$$s.t. \ \boldsymbol{Vx} \le \boldsymbol{B},$$

$$\boldsymbol{x} \ge \boldsymbol{0}.$$

The ideal level vector $z^o = \begin{bmatrix} z_1^o & z_2^o & \cdots & z_m^o \end{bmatrix}$ is comprised of m objectives. $V = p \times A$ represents the unit cost vector; p is the unit price vector of each resource, and B stands for the total budget.

2. The ideal level vector z^o can support to obtain the minimum budget B^* and the associated resource allocation (i.e., x^* and b^*). This step can be expressed as follows:

$$\min \ B^* = Vx^*, \tag{6.12}$$

$$s.t. \ Cx = z^*,$$

$$x \geq 0.$$

The outcome B^* indicates the required minimal budget to reach the ideal states for each objective, which is often larger than the original total budget.

3. Calculate the optimum-path ratio (r) by dividing the original budget B with B^* (i.e., $r = B/B^*$). The corresponding solution for z, x, and b can thus be obtained as follows:

If De Novo programming is adopted for the previous example, the final result for f_1 and f_2 will be further improved from 2348.37 and 44.03 to 2375 and 44.5 under the total budget $B = 2387$; the associated x and b will be (4.03, 2.54) and (16.12, 23.3, 58.52, 7.62, 26.28), respectively. The improvements on the two objective functions are due to the refinement of the appropriate resource allocation.

The original formulae of De Novo programming (Zeleny 1982) derived just one possible Pareto solution; it can be formulated as in Equation 6.13 to generalize De Novo programming.

$$\max \ Cx, \tag{6.13}$$

$$s.t. \ p'Ax \leq B,$$

$$Ax \leq b,$$

$$x \geq 0.$$

The preceding problem can thus be solved by traditional MOP methods; even if the included objectives are not equally weighted, this problem may apply the WGP or compromise solution to get the optimal solution of De Novo programming.

The aforementioned De Novo programming assumes that the technological and objective coefficients (parameters) of models are fixed; however,

82 ■ *Trends of Hybrid Multiple Criteria Decision Making*

if certain exogenous factors are considered, those fixed parameters can be changed as well. For example, Chiang Lin et al. (2007) proposed a SOP model in which the parameters for the objective functions and available resources can all be changed by taking new actions.

$$\max \ \sum_{j=1}^{m} (c_{j1} + c_{j2} y) x_j, \qquad (6.14)$$

$$s.t. \ \sum_{j=1}^{m} a_{kj} x_j \le (d_{k1} + d_{k2} z), \ k = 1, \ldots, r,$$

$$0 \le y \le y', \ 0 \le z \le z', \ x_j \ge 0, \ j = 1, 2, \ldots, m,$$

where:

c_{j1} and c_{j2}	denote the original and the new jth objective coefficient caused by y
d_{k1} and d_{k2}	are the original resources and the extra kth available resources caused by z

For example, if y and z are assumed to be the additional investment on advertising and seeking external strategic alliances/partners, then the associated c_{j2} and d_{k2} might denote the increased brand awareness and extra distribution channels. Nevertheless, the upper constraints for y and z still have to be considered.

The previous example from Zeleny (1982) shows how De Novo programming can be adopted to redesign or reshape a problem with given constraints. Nevertheless, the original idea of De Novo programming is to reallocate production resources so that a system's trade-offs can be eliminated, and the new ideal point under the given total budget can thus be reached. However, if the new ideal point still cannot meet the requirements of DMs, what can be done based on their expectations? Taking the case of two objective functions as an example, Figure 6.1 illustrates the concept of three levels of optimization. The first optimized level can be reached through the traditional compromise solution method by setting $p = 2$, and the initial optimized point will lay on the upper-right curve of the objective space. Second-level optimization by De Novo programming might reach the point y^* under a given total budget, which implies that the optimized point might be between the point y^* and the upper-right curve of the objective space. Third-level optimization is rooted in the essential notion of MODM; once DMs can point out their desired point (such as the point y^{**} in Figure 6.1), clear guidance should be provided to indicate how to work in the most efficient direction.

To meet the requirements or expectations of DMs in Figure 6.1, the old ideas about fixed parameters need to be changed. De Novo programming has taken one step in this direction, but much more can be done to move toward

DMs' desired points (or *aspiration levels*). In the context of business operations, it is common for companies to consider various approaches, such as making new investments or outsourcing certain portions of productions, to seek out performance improvements. Nevertheless, how to choose the most efficient way, considering the current constraints and the plausible alternatives/approaches, is not answered by the previous MOP methods. To provide clear guidance for DMs to move closer to their aspiration level, a new approach called *changeable spaces* is introduced in the next section.

6.3 MOP with Changeable Parameters

The previous two sections discuss the developments of MOP models in MODM: from traditional approaches (such as GP and compromise solution) to De Novo programming. Nevertheless, if DMs are still unsatisfied with the results obtained by De Novo programming, they will need new approaches to help them achieve their desired outcomes. In this section, three commonly adopted approaches for businesses to consider are discussed, to further enhance the traditional MOP models so that they are closer to DMs' expectations. Those three approaches are (1) increasing budgets, (2) improving objective coefficients, and (3) enhancing production efficiency. Three short numerical examples will be illustrated, and the generalized model of changeable spaces will be provided at the end of this section.

Assume that n objectives need to be achieved and m products are produced by company A; the desired point or the aspiration level of the ith objective function is denoted as $f_i^{**}(x)$, for $i = 1,\ldots,n$. The first approach can be implemented by increasing financial budgets, and the original MOP problem can be reformulated as Equation 6.15.

Model 1: MOP with changeable budgets

$$\min \ \hat{B}, \tag{6.15}$$

$$s.t. \ \sum_{j=1}^{m} c_{ij} x_{ij} \geq f_i^{**}(x), \ i = 1,\ldots,n,$$

$$p'Ax \leq B + \hat{B}, a \ \text{<extra conditions for } \hat{B} >,$$

$$x \geq 0,$$

where:

c_{ij}	denotes the jth coefficient of the ith objective function
p	denotes the unit price vector of resources
B and $\hat{B}$	are the original extra budgets obtained from additional financing

Example 1: Reconsider the case of producing suits and dresses in Section 6.2 (also refer to Table 6.1). If the objective functions and constraints are unchanged by De Novo programming, the two optimized objectives are $f_1 = 2375$ and $f_2 = 44.5$, respectively. If company *A* hopes to increase f_1 (profit) from 2375 to 2600 and f_2 (quality) from 44.5 to 60 (i.e., its aspiration levels), it can conduct extra financing to meet this end. The extra budget is denoted as $\hat{B}$, and the MOP problem of Example 1 can be expressed as follows by minimizing the extra budget under the existing constraints.

$$\min \hat{B},$$

$$s.t.\ 400x_1 + 300x_2 \geq 2600,$$

$$6x_1 + 8x_2 \geq 60,$$

$$30 \times 4x_1 + 40 \times (2x_1 + 6x_2) + 9.5 \times (12x_1 + 4x_2) + 20 \times 3x_2$$

$$+ 10 \times (4x_1 + 4x_2) \leq 2600 + \hat{B},$$

$$x_1, x_2 \geq 0.$$

The result of the preceding linear system shows $\hat{B} = 376$, while producing $x_1 = 2$ and $x_2 = 6$; here, the corresponding profit and quality indices reach 2600 and 60, respectively. More precisely, this extra financing would require additional interest expenses ($\hat{B} \times i\%$) and certain transaction costs; those accompanied costs can be calculated separately. Therefore, by increasing the total budget, the solution space is expanded from De Novo programming optimized point y^* to desired point y^{**} (Figure 6.1) in Model 1.

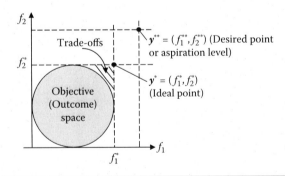

Figure 6.1 Concept of the three levels of optimization. (Reprinted with permission from Huang, J.J., and Tzeng, G.H., *Technological and Economic Development of Economy*, 20(2), 242–261, 2014.)

MODM with De Novo and Changeable Spaces ■ 85

Aside from borrowing money from capital markets, the second approach to achieving a company's desired goal is to improve the objective coefficients of a system. In practice, there are various ways to improve objective coefficients by taking new actions while considering the associated improvement costs. For example, to improve the quality objective (i.e., f_2) of the original problem, company A may add one more inspection station for production or adopt a new quality control mechanism; as a result, it will yield additional improvement costs. The management team of company A should evaluate the plausible actions or strategies associated with their addressed objectives; how to select the best action plan can be regarded as another MADM problem. In this regard, the changes in the objective coefficients are expressed in Equation 6.16.

Model 2: MOP with changeable objective coefficients

$$\min \hat{B}, \tag{6.16}$$

$$s.t. \sum_{j=1}^{m} (c_{ij} + \hat{c}_{ij})x_{ij} \geq f_i^{**}(\boldsymbol{x}), \, i = 1,\dots,n,$$

$$\boldsymbol{p}'\boldsymbol{A}\boldsymbol{x} + \sum_{i=1}^{n} \sum_{j=1}^{m} p_{ij}^c \hat{c}_{ij} \leq B + \hat{B},$$

$$\boldsymbol{x} \geq \boldsymbol{0},$$

where:

p_{ij}^c denotes the unit cost with respect to the jth product coefficient of the ith objective function

$\hat{c}_{ij}$ is the jth upgrading product coefficient of the ith objective function

Example 2: In this case, assume that the company does not intend to conduct additional financing but still hopes to increase f_1 (profit) from 2375 to 2600 and f_2 (quality) from 44.5 to 60; the second approach of changing the objective coefficients can be considered. For example, company A plans to increase its average sale prices (ASPs) of the two products to increase its profitability (f_1) by enhancing its quality control (f_2) mechanism. The assumed improvement unit costs are listed in Table 6.2.

The following MOP system denotes how to achieve the desired point (f_1 = 2600 and f_2 = 60) by considering the accompanying improvement costs, and $\hat{c}_{ij}$ denotes the increased contribution brought by each x_j for the ith objective function.

86 ■ *Trends of Hybrid Multiple Criteria Decision Making*

Table 6.2 Associated Unit Improvement Costs for the Two Objectives

Two Objectives	$x_1 = 1$	$x_2 = 1$
f_1 coefficients (Unit improvement costs)	400 ($0.200)	300 ($0.289)
f_2 coefficients (Unit improvement costs)	6 ($2.225)	8 ($2.487)

Source: Data from Huang, J.J., and Tzeng, G.H., *Technological and Economic Development of Economy*, 20(2), 242–261, 2014.

$$\min \hat{B},$$

$$s.t. \ (400 + \hat{c}_{11})x_1 + (300 + \hat{c}_{12})x_2 \geq 2600,$$

$$(6 + \hat{c}_{21})x_1 + (8 + \hat{c}_{22})x_2 \geq 60,$$

$$30 \times 4x_1 + 40 \times (2x_1 + 6x_2) + 9.5 \times (12x_1 + 4x_2) + 20 \times 3x_2$$

$$+10 \times (4x_1 + 4x_2) + (0.200\hat{c}_{11} + 0.289\hat{c}_{12} + 2.225\hat{c}_{21} + 2.487\hat{c}_{22}) \leq 2600 + \hat{B},$$

$$x_1, x_2 \geq 0.$$

The result of the preceding system is $\hat{B} = 0$; no extra budget is required. Then, the new result of the changes in the objective coefficients are $x_1 = 4.43$, $x_2 = 2.70$, $\hat{c}_{11} = 3.51$, $\hat{c}_{12} = 0.18$, $\hat{c}_{21} = 1.44$, and $\hat{c}_{22} = 2.00$. However, if $\hat{B} \neq 0$, the desired point cannot be reached by improving the objective coefficients; company A might need to consider the other approach to reach its aspiration level without increasing the total budget.

The third model discussed here is to enhance the outcome space by improving the technology coefficients of a system. In the so-called knowledge economy era, innovations can often increase the existing technology coefficients without changing the existing budget. For example, a company might increase its profit index by upgrading the settings of certain production machines to use fewer resources, or it may redesign the production procedures or flows to enhance its profitability and quality indices without increasing the budget. This approach can be explained by Model 3.

Model 3: MOP with Changeable technological coefficients

$$\min \hat{B}, \tag{6.17}$$

MODM with De Novo and Changeable Spaces ■ 87

$$s.t. \sum_{j=1}^{m} c_{ij} x_{ij} \geq f_i^{**}(\boldsymbol{x}), i = 1, \ldots, n,$$

$$\boldsymbol{p}'(\boldsymbol{A} - \hat{\boldsymbol{A}})\boldsymbol{x} + \sum_{k=1}^{r} \sum_{j=1}^{m} p_{kj}^a \hat{a}_{kj} \leq B + \hat{B},$$

$$\boldsymbol{x} \geq \boldsymbol{0},$$

where:

$\hat{A} = \left[\hat{a}\right]_{kj}$ is the upgraded technological coefficient matrix

p_{kj}^a is the unit-upgrading cost with respect to the jth technology coefficient of the kth constraint

The case can be explained by continuing with the previous example.

Example 3: The example of company A producing suits and dresses is used for the following problem. If company A still cannot achieve its desired point by adding extra budget or improving objective coefficients, the third approach could be to upgrade the technological coefficients of the system. The unit-upgrading costs for the associated technological coefficients are defined and shown in Table 6.3.

Incorporating information on the unit-upgrading costs of the technological coefficients, the following MOP system can be formulated for company A to reach its desired point $y^{**} = (2600, 60)$.

$$\min \hat{B},$$

$$s.t. \ 400x_1 + 300x_2 \geq 2600,$$

$$6x_1 + 8x_2 \geq 60,$$

$$30 \times (4 - \hat{a}_{11})x_1 + 40 \times ((2 - \hat{a}_{21})x_1 + (6 - \hat{a}_{22})x_2) + 9.5 \times ((12 - \hat{a}_{31})x_1$$

$$+ (4 - \hat{a}_{32})x_2) + 20 \times (3 - \hat{a}_{42})x_2 + 10 \times ((4 - \hat{a}_{51})x_1 + (4 - \hat{a}_{52})x_2)$$

$$+ 0.5\hat{a}_{11} + 0.5\hat{a}_{21} + 0.27\hat{a}_{22} + 12\hat{a}_{31} + 4\hat{a}_{32} + 3\hat{a}_{42} + 4\hat{a}_{51} + 4\hat{a}_{52} \leq 2600 + \hat{B},$$

$$x_1, x_2 \geq 0.$$

The preceding MOP system also does not require extra budget (i.e., $\hat{B} = 0$). The outcomes are derived as follows: $x_1 = 2.42$, $x_2 = 5.69$, $\hat{a}_{11} = 2.03$, $\hat{a}_{21} = 1.27$, $\hat{a}_{22} = 0.30$, $\hat{a}_{31} = 0.27$, $\hat{a}_{32} = 0.25$, $\hat{a}_{42} = 0.26$, $\hat{a}_{51} = 0.25$, and $\hat{a}_{52} = 0.26$.

88 ■ *Trends of Hybrid Multiple Criteria Decision Making*

Table 6.3 Information Table for Example 3

Objective Coefficients		Unit Price	Resource	Technological Coefficients	
				$x_1 = 1$	$x_2 = 1$
Unchanged		Unchanged			Upgraded
400	300	30	Nylon	4 ($0.5)[a]	0
6	8	40	Velvet	2 ($0.5)	6 ($0.27)
		9.5	Silver thread	12 ($0.27)	4 ($0.26)
		20	Silk	0	3 ($0.25)
		10	Golden thread	4 ($0.25)	4 ($0.25)

Source: Data from Huang, J.J., and Tzeng, G.H., *Technological and Economic Development of Economy*, 20(2), 242–261, 2014.

[a] The figure within a bracket denotes the unit upgrading cost for each technological coefficient.

In the previous cases, using the three types of changeable spaces, company A can evaluate the plausible ways in which it may achieve its desired point. In practice, those three scenarios might coexist. Therefore, a generalized model of changeable parameters can be illustrated as follows to show the three scenarios in one MOP system.

$$\min \hat{B}, \tag{6.18}$$

$$s.t. \sum_{j=1}^{m} (c_{ij} + \hat{c}_{ij})x_{ij} \geq f_i^{**}(\boldsymbol{x}), \ i = 1,\dots,n,$$

$$\boldsymbol{p}'(A - \hat{A})\boldsymbol{x} + \sum_{i=1}^{n}\sum_{j=1}^{m} p_{ij}^c \hat{c}_{ij} + \sum_{i=1}^{n}\sum_{j=1}^{m} p_{ij}^a \hat{a}_{ij} \leq B + \hat{B},$$

$$\boldsymbol{x} \geq \boldsymbol{0}.$$

In Equation 6.18, $f_i^{**}(\boldsymbol{x})$ denotes the desired values of the ith objective function; if $\hat{B} = 0$, it means that the desired point can be achieved without increasing the budget. The DM can improve the objective coefficients (when $\hat{c}_{ij} \neq 0$), update the technological coefficients (when $\hat{a}_{ij} \neq 0$), or both (when $\hat{c}_{ij} \neq 0$ and $\hat{a}_{ij} \neq 0$) to achieve the desired point. However, once $\hat{B} \neq 0$ to reach the desired point, a company still needs to borrow money from capital markets.

6.4 Discussion

In the traditional MOP approach, a given system is considered to solve optimal problems, in which the budget, objective functions, and technological coefficients are all fixed. However, in the context of MODM, the desired points of DMs often lay beyond the optimized result from a fixed system; more flexible approaches should be considered to help DMs attain their aspiration levels.

The main concern of problem solving in practice is how to reach the desired outcome, not just to identify the limitations or shortages of current problems. On the one hand, in MADM problems, we propose a hybrid approach—integrating the DEMATEL, DANP, and modified VIKOR methods (Chapter 5)—to identify performance gaps in all criteria; DMs can be guided regarding how to improve their current alternatives or status in different scenarios. On the other hand, MODM problems can also be redesigned to meet the expectations of DMs using the concept of changeable spaces, as described in this chapter.

Next, the differences between GP (or compromise solution) and the proposed approach were briefly discussed. GP was proposed by Charnes and Cooper (1961) to deal with linear MOP problems. GP searches for a solution that is nearest to the ideal point. Thus, the target or goal for each objective needs to be assigned first, and then the *distance* (in the form of L_p-norm distance) from the targets to the objectives is minimized to find the optimized solution; the obtained result is located in Pareto solutions, closest to the ideal point. However, GP only searches within the existing spaces; only trade-offs among objective functions are considered.

De Novo programming incorporates information on the unit prices of resources into traditional MOP to design a better system, which moves toward the DM's desired point within the limitations of the total budget. In Section 6.2, we regarded the optimized result by De Novo programming as the second level of optimization. The essential idea of De Novo programming inspired the subsequent new concept/approach of changeable spaces for enhancing a given system (Tzeng and Huang 2011, 2014). A typical MOP problem comprises of three spaces—namely, the budget, objective, and technological coefficients; by relaxing these three previously fixed spaces, DMs can learn how to move toward their desired points. This level of optimization (third-level optimization) not only indicates an optimized outcome for DMs, but also "inspires" them to overcome the existing limitations of a system, which has a profound impact on problem solving in practice. The idea of adopting changeable spaces programming for three-level optimization is illustrated

90 ■ *Trends of Hybrid Multiple Criteria Decision Making*

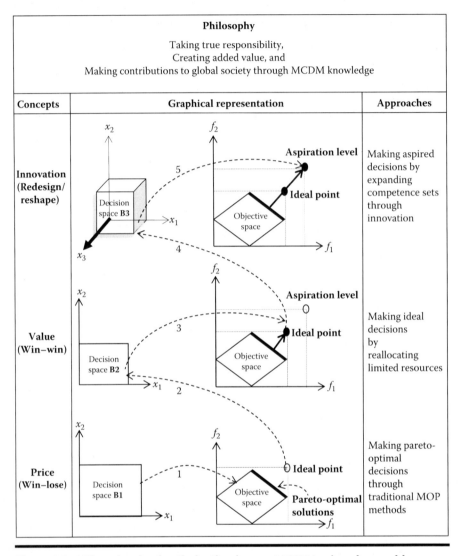

Figure 6.2 Three levels of optimization in new MODM using changeable spaces programming.

in Figure 6.2, and a case of automated factory planning using this new concept is discussed in Chapter 16.

6.5 Conclusion

While facing a given MOP problem, such as the example of producing clothes, DMs should not be satisfied by the optimized outcome from current limitations. A broader range of possibilities should be considered, and those possibilities often come from domain knowledge, creativities, or innovations; in other words, a new MODM theory has been formed, moving from optimizing the existing system to redesigning it, to reach a set of desired goals. Although only the case of optimizing production is adopted in this chapter, other problems can be solved by using the concept of changeable spaces to search for aspiration levels. Certain empirical examples will be provided in Section II of this book.

in Figure 6.2, and a case of automated factory planning using this new concept is discussed in Chapter 16.

6.5 Conclusion

While facing a given MOP problem, such as the example of producing clothes, DMs should not be satisfied by the optimized outcome from current limitations. A broader range of possibilities should be considered, and those possibilities often come from domain knowledge, creativities, or innovations; in other words, a new MODM theory has been formed, moving from optimizing the existing system to redesigning it, to reach a set of desired goals. Although only the case of optimizing production is adopted in this chapter, other problems can be solved by using the concept of changeable spaces to search for aspiration levels. Certain empirical examples will be provided in Section II of this book.

Chapter 7

Multiple Rules-Based Decision Making for Solving Data-Centric Problems

In recent years, experts in informatics have proclaimed the current era to be the so-called Big Data era. Firms have assigned higher priority to the identification (or exploration) of the complex relations among valuable data sets (such as financial or marketing data for business decisions). In the current text, we term the class of problems that involve high numbers of attributes with numerous historical records (data) *data-centric problems*. Although decision makers (DMs) or domain experts can give direct opinions to construct a complex multiple attribute decision making (MADM) model depicting the interrelations among its attributes (refer to Chapters 4 and 5), a human brain will encounter obstacles to evaluating problems that involve too many attributes; certain imprecise and vague patterns in data-centric problems can only be identified by advanced machine learning techniques. A field that is gaining prominence in multiple criteria decision making (MCDM), multiple rules-based decision making (MRDM), holds great promise for resolving these kinds of data-centric decision problems.

The original concept of MRDM was proposed in a series of studies by Greco et al. (1997, 1999, 2001, 2002b, 2005). It was developed by adopting the dominance-based rough set approach (DRSA) to induce rough dominance relations for supporting decision making. The essential concepts of classical rough set theory and DRSA are introduced in Chapter 1; in this chapter, the extended variable-consistency

93

94 ■ *Trends of Hybrid Multiple Criteria Decision Making*

DRSA (VC-DRSA) (Błaszczyński et al. 2011, 2013) is described. In addition, three approaches in MRDM—namely, the reference point-based, core attribute-based, and hybrid bipolar approaches—are briefly discussed.

7.1 Variable-Consistency Dominance-Based Rough Set Approach

VC-DRSA is an extension of DRSA. Similar to DRSA, VC-DRSA begins with a 4-tuple information system $IS = (U^{vc}, A, V, f)$ that involves the extended consideration of dominance relationships and a controlled level of consistency. The set U^{vc} is a finite universe of discourse. The set A is a finite set of m attributes, $A = \{a_1, \ldots, a_m\}$. The set A is composed of a multiple-condition-attributes set C and a decision set D. In addition, V_a is the value domain of attribute a, where $A = \bigcup_{a \in A} V_a$ and $f: U^{vc} \times A \to V$ is a total function, in which $f(x, a) \in Va$ for each $a \in A$ and $x \in U^{vc}$. In a typical data-centric problem, condition attributes can be specified in the form of numbers or discretized by DMs to denote the *granules of knowledge* relevant to the problem.

Similar to DRSA, $\succeq_a$ is defined as a weak preference relationship on U^{vc} relative to an attribute $a \in A$, in which $x \succeq_a y$ denotes that "x is at least as good as y with respect to attribute a for alternatives x and y." The decision set D can be defined as $Cl = \{Cl_t, t = 1, \ldots, n\}$, which is a set of decision classes (DCs) of U. For example, given the preferred order of DCs, if $k \succ m$, then the DC Cl_k is preferred to Cl_m. The upward union and downward union of DCs can be defined as $Cl_t^\geq = \bigcup_{s \geq t} Cl_s$ and $Cl_t^\leq = \bigcup_{s \leq t} Cl_s$, respectively. The upward and downward unions of DCs support the definition of dominance relations for $K \subseteq C$ as D_k. If an alternative x K-dominates y relative to K, then $x \succeq_a y$ for all $a \in K$, written as $xD_K y$. For a partial set K in C (i.e., $K \subseteq C$) and $x, y \in U^{vc}$, the K-dominating and K-dominated sets can be denoted as follows: $D_K^+(x) = \{y \in U^{vc} : yD_K x\}$ and $D_K^-(x) = \{y \in U^{vc} : xD_K y\}$, respectively. Then, the K-lower and K-upper approximations of $Cl_t^\geq$ and $Cl_t^\leq$ can be denoted as $\left[\underline{K}(Cl_t^\geq), \overline{K}(Cl_t^\geq)\right]$ and $\left[\underline{K}(Cl_t^\leq), \overline{K}(Cl_t^\leq)\right]$, respectively, which can be defined as follows in Equations 7.1 through 7.4 for $t = 2, \ldots, n$:

$$\underline{K}\left(Cl_t^\geq\right) = \{x \in U^{vc} : D_K^+(x) \subseteq Cl_t^\geq\}, \tag{7.1}$$

$$\overline{K}\left(Cl_t^\geq\right) = \{x \in U^{vc} : D_K^-(x) \cap Cl_t^\geq \neq \varnothing\}, \tag{7.2}$$

$$\underline{K}\left(Cl_t^\leq\right) = \{x \in U^{vc} : D_K^+(x) \subseteq Cl_t^\leq\}, \tag{7.3}$$

$$\overline{K}\left(Cl_t^\leq\right) = \{x \in U^{vc} : D_K^-(x) \cap Cl_t^\leq \neq \varnothing\}. \tag{7.4}$$

In addition, the K-boundary regions of $Cl_t^{\geq}$ and $Cl_t^{\leq}$ can be denoted as $Bn_K\left(Cl_t^{\geq}\right)$ and $Bn_K\left(Cl_t^{\leq}\right)$; K-boundary regions are calculated by deducting the lower approximations from the upper approximations (i.e., $Bn_K\left(Cl_t^{\geq}\right)=\overline{K}\left(Cl_t^{\geq}\right)-\underline{K}\left(Cl_t^{\geq}\right)$ and $Bn_K\left(Cl_t^{\leq}\right)=\overline{K}\left(Cl_t^{\leq}\right)-\underline{K}\left(Cl_t^{\leq}\right)$, for $t=2,\ldots,n$), which denote the doubtful regions.

In DRSA, $\underline{K}\left(Cl_t^{\geq}\right)$ or $\underline{K}\left(Cl_t^{\leq}\right)$ contain only consistent objects that comply with the dominance relations. However, in VC-DRSA, acceptably consistent objects (i.e., within controlled levels of consistency) are included in the K-lower approximation. There are two types of consistency measure: the gain type (i.e., the higher the value is, the higher the consistency is) and the cost type (i.e., the lower the value is, the higher the consistency is). Through the use of upward approximations to illustrate these measures, for $Cl_t^{\geq}\subseteq U^{vc}$ and $z\in U^{vc}$, the object consistency measure and a fixed gain threshold can be denoted as Θ_W and θ_W, respectively, and the $\underline{K}\left(Cl_t^{\geq}\right)$ with the gain threshold θw can be defined as follows:

$$\underline{K}^{\theta_W}\left(Cl_t^{\geq}\right)=\{z\in Cl_t^{\geq}:\Theta_W(z)\geq\theta_W\}. \tag{7.5}$$

Use W to denote $Cl_t^{\geq}$, and $\neg W\subseteq U$ (where $\neg W=U-W$); the K-upper approximation of set W (i.e., $Cl_t^{\geq}$) is defined in Equation 7.6, and the K-boundary of set W is defined in Equation 7.7.

$$\overline{K}^{\theta_W}\left(Cl_t^{\geq}\right)=\overline{K}^{\theta_W}(W)=U-\underline{K}^{\theta_W}(\neg W). \tag{7.6}$$

$$Bn_K^{\theta_W}=\overline{K}^{\theta_W}(W)-\underline{K}^{\theta_W}(W). \tag{7.7}$$

The accuracy of the approximation for W can be defined as $a_K^{\theta_W}(W)$, as shown in Equation 7.8:

$$a_K^{\theta_W}(W)=\frac{\left|\underline{K}^{\theta_W}(W)\right|}{\left|\overline{K}^{\theta_W}(W)\right|}. \tag{7.8}$$

The ratio $a_K^{\theta_W}(W)$ denotes all of the correctly classified objects for attributes $P\subseteq C$ with the consistency threshold θ_W. Each minimal subset $P\subseteq C$ that can cause $\gamma_P^{\theta_W}(W)=\gamma_C^{\theta_W}(W)$ is termed a *REDUCT* of W. In addition, the intersection of all REDUCTs is called the *CORE* ($CORE_W$). The dominance-based relationship in W can generate a set of decision rules in the form of "If *antecedents* (premises), then *consequence* (conclusion)." According to Błaszczyński et al. (2013), a decision rule in VC-DRSA suggests that the categorization into set W is denoted

96 ■ *Trends of Hybrid Multiple Criteria Decision Making*

as $rule_W$. The condition part and decision part of $rule_W$ are indicated by $\Phi(rule_W)$ and $\Psi(rule_W)$, respectively. In addition, $\|\Phi(rule_W)\|$ denotes the set of objects that satisfy the condition part of $rule_W$. The consistency measure (μ-consistency) of $rule_W$ can then be defined as follows:

$$\hat{\mu}_W(rule_W) = \frac{\|\Phi(rule_W)\| \cap \underline{W}|}{\|\Phi(rule_W)\|}. \tag{7.9}$$

7.2 Basic Notions of the Reference Point-Based MRDM Approach

The early idea of adopting DRSA for forming decision supports might have originated from Greco et al. (1997, 1999); this renowned rough set research group attempted to resolve ranking and selection problems by introducing two critical tools: the pairwise comparison table (PCT) and net flow score (NFS).

Forming a PCT is the first step in reorganizing the original information system of DRSA into a new decision table; this table is formed by requesting a DM to rank a partial set of alternatives from those available. In a PCT, rows denote pairs of objects (alternatives) with the associated evaluations of all attributes, and a PCT is supposed to capture the pairwise comparisons between each pair of objects on each attribute in order to form multigraded dominance relations. After multigraded dominance relations are obtained from the PCT, multiple decision rules will be available to calculate the preference relations for each object, and an NFS can be calculated to support the ranking or selection decision. A brief introduction to this reference point-based approach is provided in later passages of this chapter.

Referring to the information system of DRSA introduced in Chapter 1 and the associated essential concepts, we assume that the available objects (alternatives) form a set A°, and R° is a partial set of A° ($R^\circ \subseteq A^\circ$) that denotes the reference objectives; a DM has the necessary confidence to rank this partial set R°. Accordingly, each pair of objects (r_x, r_y) in R° can be expressed by a preferential function P as $P(r_x, r_y)$. The statements $P(r_x, r_y) > 0$ and $P(r_x, r_y) < 0$ denote that r_x is preferred to r_y and r_y is preferred to r_x, respectively, to a certain degree. For all $q_i \in C$ (where C denotes the set of condition attributes of a DRSA *IS*, and $i = 1,2,\ldots,n$), P can be used to express the preferential relation between (r_x, r_y) on the jth criterion as $P_j[q_j(r_x), q_j(r_y)]$. Similarly, the function $P_j(\cdot)$ can be used to express the preferential degree (or grade) of (r_x, r_y) on the jth criterion. For example, if we define the preferential degrees *dislike* $= -1$, *neutral* $= 0$, and *like* $= +1$, and if we assume that $q_j(r_x) = 1$ and $q_j(r_y) = -1$, then $P_j[q_j(r_x), q_j(r_y)] = 1 - (-1) = 2$, which means that r_x is preferred to r_y for 2 degrees on the jth criterion. Once a DM gives his or her ranking for R°, the corresponding preference degrees of each pair of objects on each criterion can

Multiple Rules-Based Decision Making ■ 97

be shown in a PCT. To illustrate the concept of a PCT, a simple example is shown in Table 7.1. Assume that the preferential degrees of each attribute are identical to the three previously mentioned levels (*dislike* = −1, *neutral* = 0, and *like* = +1). In a decision to purchase one of four alternative cars (*A*, *B*, *C*, and *D*) by considering five attributes (q_1: *brand*; q_2: *price*; q_3: *size*; q_4: *acceleration*; q_5: *quality*), the PCT might look like Table 7.1.

With the decision table from a PCT and a partial condition attribute set $P \subseteq C$, the pair dominance relation can be defined as *P*-dominating and *P*-dominated pair sets for $r_x, r_y, r_a, r_b \in R^\circ$, as shown in Equations 7.10 and 7.11, respectively:

$$D_P^+\left(r_x, r_y\right) = \left\{\left(r_a, r_b\right) \in R^\circ : \left(r_a, r_b\right) D_P\left(r_x, r_y\right)\right\}. \tag{7.10}$$

$$D_P^-\left(r_x, r_y\right) = \left\{\left(r_a, r_b\right) \in R^\circ : \left(r_x, r_y\right) D_P\left(r_a, r_b\right)\right\}. \tag{7.11}$$

With those considered pairs of reference points, *P*-upper and *P*-lower approximations for the upward unions can be respectively defined in Equations 7.12 and 7.13 for preference degrees of at least *m*; similarly, *P*-upper and *P*-lower approximations can be respectively defined for the downward unions in Equations 7.14 and 7.15 for preference degrees of at most *m*.

$$\underline{P}\left(Pair_m^\geq\right) = \left\{\left(r_x, r_y\right) \in R^\circ : D_P^+\left(r_x, r_y\right) \subseteq Pair_m^\geq\right\}. \tag{7.12}$$

$$\overline{P}\left(Pair_m^\geq\right) = \bigcup_{\left(r_x, r_y\right) \in R^\circ} D_P^+\left(r_x, r_y\right) = \left\{\left(r_x, r_y\right) \in R^\circ : D_P^-\left(r_x, r_y\right) \cap Pair_m^\geq \neq \varnothing\right\}. \tag{7.13}$$

$$\underline{P}\left(Pair_m^\leq\right) = \left\{\left(r_x, r_y\right) \in R^\circ : D_P^-\left(r_x, r_y\right) \subseteq Pair_m^\geq\right\}. \tag{7.14}$$

Table 7.1 PCT for the Car Purchase Example

Pairs	Δq_1	Δq_2	Δq_3	Δq_4	Δq_5	Overall
(*A*, *B*)	+1	+1	0	+2	+1	+1
(*A*, *C*)	−1	0	−2	+1	0	−1
⋮	⋮	⋮	⋮	⋮	⋮	⋮
(*D*, *A*)	−1	0	−1	0	+1	−1

Note: Δ denotes the *delta* between two alternatives on the associated criterion.

$$\overline{P}\left(Pair_m^{\leq}\right)= \bigcup_{\left(r_x,r_y\right)\in R^{\circ}} D_{\overline{P}}^{-}\left(r_x,r_y\right)=\left\{\left(r_x,r_y\right)\in R^{\circ}:D_P^{+}\left(r_x,r_y\right)\cap Pair_m^{\geq}\neq\varnothing\right\}. \qquad (7.15)$$

Therefore, the rough boundary approximations for $Pair_m^{\geq}$ and $Pair_m^{\leq}$ can be defined, respectively, as follows:

$$Bn_P\left(Pair_m^{\geq}\right)=\overline{P}\left(Pair_m^{\geq}\right)-\underline{P}\left(Pair_m^{\geq}\right). \qquad (7.16)$$

$$Bn_P\left(Pair_m^{\leq}\right)=\overline{P}\left(Pair_m^{\leq}\right)-\underline{P}\left(Pair_m^{\leq}\right). \qquad (7.17)$$

Similar to the classical DRSA, the quality of the approximations, REDUCTs, and CORE can be inductively reasoned for $Pair_m^{\geq}$ and $Pair_m^{\leq}$. Furthermore, suppose that there are ω pairs of objects ($|W| = \omega$, where W denotes the set comprises all the pairs from R°) in a PCT; the quality of approximation of comprehensive preference $P(r_x,r_y)$ can then be defined by criteria from $P \subseteq C$, as shown in Equation 7.18:

$$\eta_P = \frac{\left|W-\bigcup_m Bn_P\left(Pair_m^{\geq}\right)\right|}{|W|}=\frac{\left|W-\bigcup_m Bn_P\left(Pair_m^{\leq}\right)\right|}{|W|}. \qquad (7.18)$$

This quality of approximation ratio η_P indicates all pairs $(r_x,r_y) \in W$ whose preference degree of x over y is correctly evaluated by P in the whole set W. Those rough approximations of $Pair_m^{\geq}$ and $Pair_m^{\leq}$ can yield three types of rough decision rules: $D_{\geq}$, $D_{\leq}$, and $D_{\geq\leq}$. For example, a $D_{\leq}$-decision rule might be similar to "If Car A is preferred to Car C relative to $q_1 \wedge$ Car A is strongly preferred to Car C relative to q_3, then Car A is comprehensively worse than Car C." Each type of rule is supported by the associated pairs of objects. This reference point-based MRDM approach assumes that these three types of decision rules calculated from the rough approximations of $Pair_m^{\geq}$ and $Pair_m^{\leq}$ would form a preference model for a DM (who can give a precise preference ranking for R°, $R^{\circ} \subseteq A^{\circ}$). These calculated decision rules can thus be applied for the whole available objects in set A°.

With the assumed car purchase as an example, $P(r_x,r_y) = -1$ denotes an outranking relation for only two values (-1 and $+1$); if $P(r_x,r_y) = +1$, then r_x is at least as preferable as r_y; if $P(r_x,r_y) = -1$, then r_x is not at least as preferable as r_y. Subsequently, if $P(a_x,a_y) = +1$, it can be denoted as $a_x \, S \, a_y$; if $P(a_x,a_y) = -1$, it can be denoted as $a_x \, S^c \, a_y$. In this setting, all pairs of objects in A° can match those decision rules in four ways:

1. At least one $D_{\geq}$-decision rule and neither $D_{\leq}$- nor $D_{\geq\leq}$-decision rules
2. At least one $D_{\leq}$-decision rule and neither $D_{\geq}$- nor $D_{\geq\leq}$-decision rules
3. At least one $D_{\geq}$-decision rule and at least one $D_{\leq}$-decision rule, or at least one $D_{\geq\leq}$-decision rule
4. No decision rule

These four ways correspond to four types of outranking: (1) true, (2) false, (3) contradictory, and (4) unknown. These four situations compose what was termed *four-value outranking* by Greco et al. (1998), in which the presence of positive, negative, contradictory, and unknown supports for outranking were considered. A final ranking score is defined as an NFS for each $a_x \in A^\circ$ as follows:

$$S^{NFS}\left(a_x\right) = S^{++}\left(a_x\right) - S^{+-}\left(a_x\right) + S^{-+}\left(a_x\right) - S^{--}\left(a_x\right), \qquad (7.19)$$

where:

$S^{++}(a_x)$ = cardinality $(\{a_y \in A^\circ : \text{at least one decision rule affirms } a_x Sa_y\})$
$S^{+-}(a_x)$ = cardinality $(\{a_y \in A^\circ : \text{at least one decision rule affirms } a_y Sa_x\})$
$S^{-+}(a_x)$ = cardinality $(\{a_y \in A^\circ : \text{at least one decision rule affirms } a_y S^C a_x\})$
$S^{--}(a_x)$ = cardinality $(\{a_y \in A^\circ : \text{at least one decision rule affirms } a_x S^C a_y\})$

Although there are other measures for calculating the final ranking score for each alternative, NFS might be the most prevalent index in this approach. This approach has solid theoretic background regarding outranking relations and decision science; however, it still has limitations with certain practical data-centric decision problems, such as the following:

1. When the involved criteria in a system are numerous (e.g., >15 criteria), it is often difficult for a DM to rank a partial set of objects precisely;
2. If multiple DMs' opinions must be included for a complex data set, the obtained *IS* and decision rules from different DMs (or domain experts) are often inconsistent; how to synthesize the preference models from multiple DMs can be problematic.

In addition, if the set W is relatively large (e.g., >10 elements) in a complex problem (with a high number of condition attributes), a high number of contradictory preference relations is often yielded; consequently, it would be problematic for DMs to interpret the exact meaning of the obtained result with high confidence. To address the aforementioned issue for resolving data-centric problems by MRDM, two other approaches are briefly discussed in the following sections.

7.3 Core Attribute-Based MRDM Approach

The core attribute-based MRDM approach adopts a CORE set, calculated by DRSA or VC-DRSA, from a data-centric decision problem as the critical criteria (with the associated dimensions). In addition, the strong decision rules obtained from DRSA/VC-DRSA can be integrated with DEMATEL (Chapter 3), DANP (Chapter 4), modified VIKOR, or fuzzy integrals (Chapter 5) for improvement planning. Shen and Tzeng have proposed this approach to diagnose financial performance (Shen and Tzeng 2014a, 2014b) and to solve investment decision problems (Shen and Tzeng 2015c,d), and the potential of this approach has been noted recently in certain financial studies based on MCDM (Zopounidis et al. 2015).

The prominent behavioral theory of bounded rationality proposed by the Nobel Prize Laureate H.A. Simon (1972) implies that complex or imprecise patterns in data-centric decision problems often impede DMs who seek to identify the minimal but critical criteria precisely. To overcome this difficult challenge, the concept of the core-attribute-based MRDM approach was originated from reducing a problem's dimensional complexity by understandable logical rules to aid decisions.

Furthermore, strong decision rules play a crucial role in depicting the logical relations among a partial set of criteria; this fact has been overlooked in traditional MADM research. A decision rule often includes multiple antecedents under the associated criteria and a consequence; in other words, some crucial criteria might appear only in a specific context (with the other antecedents). The meaning of this type of context-specific criteria is similar to *latent variables* in social sciences research. Traditional MADM research only model the relations among a group of selected and fixed criteria or attributes; therefore, this type of contextual relationship has been overlooked in previous research.

To solve practical problems, the obtained decision rules can be integrated with the DEMATEL analysis from a CORE set, which is called a *directional flow graph* (DFG) (Shen and Teng 2015e,f). A DFG further indicates the directional influences among strong decision rules; once DMs plan to improve the performance gap on a specific criterion in a rule, a DFG can indicate the source factors (criteria) that can systematically guide improvement planning.

7.4 Hybrid Bipolar MRDM Approach

The proposed bipolar decision model is based on a combination of DRSA and performance evaluation, and it can be illustrated in four main parts, as shown in Figure 7.1. This bipolar model is devised to retrieve rough knowledge from a complex data set to aid decisions; therefore, it begins by analyzing a finite, nonempty set U in the form of a 4-tuple information system (i.e., IS $= \langle U, A, V, f \rangle$).

In this *IS*, U is a finite set of alternatives or objects; furthermore, the proposed bipolar approach assumes that DMs can provide concrete opinions to categorize

Multiple Rules-Based Decision Making ▪ 101

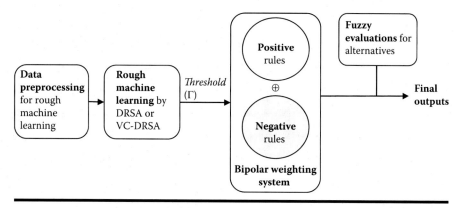

Figure 7.1 Framework of a bipolar decision model.

every confirmed alternative or object O_i in U ($O_i \in U$, $i = 1, 2, \ldots, n^c$) into the negative (*NEG*) or the positive (*POS*) region; and the remaining unconfirmed objects O_j ($O_j \in U$, $j = 1, 2, \ldots, n^u$) can be categorized into the neutral (*NEU*) region (i.e., there are a total of $n^c + n^u$ alternatives in U). Those three disjoint regions are presumed to have a preferential order relation: $NEG \prec NEU \prec POS$, defined by DMs.

Next, A is a finite set of attributes or criteria, which comprises condition attributes C and a decision attribute D. In a typical MCDM problem, there are multiple conditional attributes and a single decision attribute $\{d\}$. The term V_a is the value domain of attribute a; f is a total function, and $f: U \times A \to V$ holds for $f(o,a) \in V_a$. The DCs of D a can be denoted as $Cl = \{Cl_{NEG}, Cl_{NEU}, Cl_{POS}\}$ in this bipolar model. The following step requires the adoption of a DRSA algorithm to form two groups (bipolar) of rough knowledge: positive and negative decision rules.

7.4.1 Dominance-Based Rough Set Approach

To proceed, $\succeq_{a_i}$, is defined as a weak preference relation on C considering attribute a_i, for $a_i \in A$ and $i = 1, 2, \ldots, n$. For objects $x, y \in U$, if $x \succeq_{a_1} y$ holds, then x is at least as preferable as y considering attribute a_1. In addition, decision attributes are defined as three classes (*Cl*): $Cl_{NEG} \prec Cl_{NEU} \prec Cl_{POS}$. Generally, if $p \succ q$ ($p, q \in U$), then a set comprising the upward and downward unions of the *Cl*s can be defined as follows:

$$Cl_p^{\succeq} = \bigcup_{p \succeq q} Cl_q, \tag{7.20}$$

$$Cl_p^{\preceq} = \bigcup_{p \preceq q} Cl_q. \tag{7.21}$$

102 ■ *Trends of Hybrid Multiple Criteria Decision Making*

The upward and downward unions of the *Cl*s on *C* are used to define the dominance relation for any partial criteria set *P* in *C* ($P \subseteq C$). The term Dom_p denotes the dominance relation with respect to *P*, and $xDom_py$ means that *x* *P*-dominates *y* on any partial conditional attribute set in *P*. Accordingly, the *P*-dominating set and *P*-dominated set (for $P \subseteq C$) of *x* ($x \in U$) can be defined as $Dom_P^{\uparrow}(x)$ and $Dom_P^{\downarrow}(x)$, as shown in Equations 7.22 and 7.23, respectively:

$$Dom_P^{\uparrow}(x) = \{ y \in U : yDom_Px \}, \tag{7.22}$$

$$Dom_P^{\downarrow}(x) = \{ y \in U : xDom_P y \}. \tag{7.23}$$

Let $Cl = \{ Cl_j, j \in T \}$, where $T = \{ NEG, NEU, POS \}$ for the decision attribute. The *P*-dominating and *P*-dominated sets are then used to define the subsequent *P*-lower, *P*-upper, and boundary approximations. For example, regarding the upward unions of the *Cl*s, the *P*-lower and *P*-upper approximations are shown in Equations 7.24 and 7.25, respectively:

$$\underline{P}\left(Cl_j^{\succeq}\right) = \left\{ x \in U : Dom_P^{\uparrow}(x) \subseteq Cl_j^{\succeq} \right\}, \tag{7.24}$$

$$\overline{P}\left(Cl_j^{\succeq}\right) = \left\{ x \in U : \bigcup_{x \in Cl_j^{\succeq}} Dom_P^{\uparrow}(x) \cap Cl_j^{\succeq} \neq \varnothing \right\} \tag{7.25}$$

The boundary approximation can be defined by excluding the *P*-lower approximations (the certain part) from the *P*-upper approximations (the uncertain part) as follows:

$$Bn_P\left(Cl_j^{\succeq}\right) = \overline{P}\left(Cl_j^{\succeq}\right) - \underline{P}\left(Cl_j^{\succeq}\right). \tag{7.26}$$

Therefore, the boundary region denotes the doubtful region, also termed the *P*-boundary region. DRSA measures the quality of approximations for ordinal DCs using γ_P, as in Equation 7.27; both the upward and downward unions of the *Cl*s are included.

$$\gamma_P(Cl) = \frac{\left| U - \left(\left(\bigcup_{j \in \{NEU, POS\}} Bn_P\left(Cl_j^{\succeq}\right) \right) \cup \left(\bigcup_{j \in \{NEG, NEU\}} Bn_P\left(Cl_j^{\preceq}\right) \right) \right) \right|}{|U|}, \tag{7.27}$$

where $|\bullet|$ denotes the cardinality of a set. Each minimal subset P ($P \subseteq C$) that enables $\gamma_P(Cl) = \gamma_C(Cl)$ is a REDUCT of C, and the intersection of all REDUCTs is a CORE set. The DRSA CORE set-based hybrid decision-making approach was recently referred to as MRDM (Shen and Tzeng 2015g).

Using the rule-induction mechanism of DRSA, yielded from the approximations of the dominance relations in U, a group of decision rules can be calculated. Although three types of decision rules can be calculated (i.e., $D_\geq$, $D_\leq$, and $D_{\geq\leq}$), only $D_\geq^{POS}$ and $D_\leq^{NEG}$ rules are used in the subsequent bipolar decision model.

The general form of $D_\geq^{POS}$ and $D_\leq^{NEG}$ are as decision rules: "If *antecedents*, then *consequent*." In the following bipolar model, assume that there are n_+ and n_- numbers of certain $D_\geq^{POS}$- and $D_\leq^{NEG}$-decision rules obtained from a set of data A_n in U, respectively. The general form of the upward unions of a DRSA decision rule can be denoted as follows:

$$\text{Rule}_i^+ \equiv \text{"If } f_{c_1}(x) \geq v_{c_1} \wedge f_{c_2}(x) \geq v_{c_2} \wedge ... \wedge f_{c_g}(x) \geq v_{c_g} \text{, then } x \in Cl_r^{\geq} \text{"}$$

where $f_{c_1}(x)$ denotes the value of alternative x on criterion C_1, and $v_{c_1} \subseteq V_{c_1}$, $v_{c_2} \subseteq V_{c_2}, ..., v_{c_g} \subseteq V_{c_g}$; $\{C_1, C_2, ..., C_g\} \subseteq C$, and $r \in \{NEG, NEU, POS\}$.

The two groups of decision rules denote the bipolar (i.e., positive and negative) rough knowledge calculated from a set of data A_n, which DMs have confirmed to be the positive and negative references. The alternatives or objectives that are consistent with each decision rule will be calculated as the support of a rule (i.e., support weight), which denotes the degree of confidence suggested by that rule.

To select the covered references from those two groups of rough decision rules, a threshold-based mechanism is described. DMs must assign a threshold value θ^B that denotes the covered percentage in positive and negative decision rules, which must satisfy Equations 7.28 and 7.29:

$$\theta^B \leq \frac{\sum_{i=1}^{n_+^z} s_i^+}{\left| A_n^{POS} \right|}, \tag{7.28}$$

$$\theta^B \leq \frac{\sum_{i=1}^{n_-^z} s_i^-}{\left| A_n^{NEG} \right|}, \tag{7.29}$$

104 ■ *Trends of Hybrid Multiple Criteria Decision Making*

where n_+^z and n_-^z denote the minimal numbers that satisfy Equations 7.28 and 7.29, respectively. Two points are crucial here. First, $R_1^+ \cdots R_{n_+}^+$ and $R_1^- \cdots R_{n_-}^-$ are ordered from high supports to low supports, which implies that $s_1^+ \geq s_2^+ \geq \cdots \geq s_{n_+}^+$ and $s_1^- \geq s_2^- \geq \cdots \geq s_{n_-}^-$. Second, certain alternatives or instances tend to appear in multiple decision rules. The two selected group of rules thus form the weighting system of a bipolar decision model, in which $R_1^+ \cdots R_{n_+}^+$ and $R_1^- \cdots R_{n_-}^-$ are regarded as the new criteria (in the form of rules), and the support weights of these two new criteria can be defined as shown in Equations 7.30 and 7.31, respectively:

$$w_{\text{sup}^+}^i = \frac{s_i^+}{n_+^z + n_-^z}, \text{ for } i = 1,...,n_+^z \tag{7.30}$$

$$w_{\text{sup}^-}^j = \frac{s_j^-}{n_+^z + n_-^z}, \text{ for } j = 1,...,n_-^z, \tag{7.31}$$

where $w_{\text{sup}^+}^i$ and $w_{\text{sup}^-}^i$ denote the support weights for the ith positive and jth negative rules (new criteria), respectively.

7.4.2 Evaluations for an Aggregated Bipolar Decision Model

A bipolar decision model comprises three parts according to the proposed aggregation approach: (a) a bipolar weighting system, (b) an evaluation system, and (c) a method for aggregating the performance evaluations of alternatives by the weighting system. The weighting system can be derived from Equations 7.30 and 7.31. An evaluation system can be measured by either crisp set evaluations or fuzzy set evaluations (e.g., hesitant fuzzy sets). The third aggregation method can be either linear or nonlinear, and numerous prevailing methods can be adopted (e.g., SAW, VIKOR, and TOPSIS) in practice. An additive-type aggregator that uses fuzzy sets for the aggregation approach is indicated in Equation 7.32:

$$F_i = d_{fuzzy}\left({}^i w_{\text{sup}^+}^1 \times {}^i \tilde{p}_+^1 \oplus \cdots \oplus {}^i w_{\text{sup}^+}^{n_+^z} \times {}^i \tilde{p}_+^{n_+^z}\right)$$
$$- d_{fuzzy}\left({}^i w_{\text{sup}^-}^1 \times {}^i \tilde{p}_-^1 \oplus \cdots \oplus {}^i w_{\text{sup}^-}^{n_-^z} \times {}^i \tilde{p}_-^{n_-^z}\right), \tag{7.32}$$

where:

F_i is the aggregated performance evaluation for the ith alternative

$d_{fuzzy}(\cdot)$ denotes a defuzzification function

${}^i\tilde{p}_+^{n_+^z}$ and ${}^i\tilde{p}_-^{n_-^z}$ respectively indicate the fuzzy performance evaluations for the n_+^zth positive and n_-^zth criteria from the ith alternative

An application of this bipolar model for solving a financial modeling problem is illustrated in Chapter 20.

7.5 Conclusion

The rising trend of using multiple rules and reserving imprecise logical relations as decision aids has been noted in MCDM research. The importance of machine learning and soft computing should be recognized when analyzing complex social problems with observable data. Thus, how to fuse two heterogeneous inputs (i.e., rules or knowledge learned from data and subjective judgments or preferences from DMs) to enhance the quality of decisions is a challenging but valuable research direction.

One renowned MRDM method originated from DRSA (Greco et al. 1999, 2000, 2005), which has inspired the two approaches introduced in this chapter: the core attribute–based and hybrid bipolar approaches. Interested readers can find several financial or investment applications using MRDM models in Section II of this book. Despite the advantages of MRDM, there are certain limitations that can be enhanced in the existing models, such as the discretization of attributes. How to preprocess data adequately to obtain the desired or ideal inputs (for rule induction) is a critical issue in MRDM; another fast-growing field, *granular computing* (Pedrycz et al. 2008; Pedrycz 2013), might provide plausible tools and techniques for answering this question.

After introducing the essential ideas and the recent theoretical developments in hybrid MCDM research in Section I, the following list of studies would be a useful reference list for those who want to explore the related research of hybrid MCDM (mainly from the Prof. Tzeng's research's group and some eminent researchers) in a more comprehensive way:

Andy et al. 2014; Belton and Stewart 2002; Chan and Tzeng 2008, 2009, 2011; Chang et al. 2009, 2015a, 2015b; Chang and Tzeng 2010; Chen and Tzeng 1999, 2000, 2003, 2004, 2011, 2015; Chen et al. 2000, 2001a, 2001b, 2002, 2003, 2005, 2005a, 2005b, 2008, 2011, 2011a, 2011b, 2015, ; Chen and Dhillon 2003; Chen Chen Tzeng 2001; Cheng et al. 2004, 2013;; Chiang et al. 2000; Chiang and Tzeng 2000a, 2000b; Chin et al. 2010; Chiou and Tzeng 2002, 2003; Chiou et al. 2005; Chiu et al. 2006, 2014; Chiu, Y.C., Shyu and Tzeng 2004; Chiu and Tzeng 1999; Chou et al. 2006, 2016; Chu et al. 2004, 2007a, 2007b; Fan and Tzeng 2006; Fan et al. 2007; Fang et al. 2012; Hsieh et al. 2004, 2010; Hsu et al. 2003, 2004, 2005, 2007, 2012, 2016; Hu et al. 2002, 2003a, 2003b, 2003c, 2003d, 2003e, 2003f, 2003g, 2004a, 2004b, 2012, 2015, 2016; ; Hu and Tzeng 2002, 2003; Huang et al. 2005a, 2005b, 2005c, 2005d, 2006a, 2006b, 2006c, 2006d, 2006e, 2007, 2008, 2010, 2011, 2012, 2012a, 2012b, 2012c, 2016; Huang and Tzeng 2007, 2008; Ishii et al. 2012; Jeng and Tzeng 2012; Kahneman et al. 1991; Kahneman and Tversky 1979, 1984, 2000; Kahraman et al. 2006; Ko et al. 2013a, 2013b, 2014; Koopmans 1951; Kuan et al. 2012; Kuo et al. 2007; Lai et al. 1994; Larbani et al. 2011; Lee et al. 2001, 2005a, 2005b, 2011 Li and Tzeng 2009a, 2009b; Li and Lee 1993; Lin et al. 2006, 2010a, 2010b, ,2011, 2016; Lin and Tzeng 2010; Liou et al. 2011, 2016; Liou and Tzeng 2007, 2010;

Liou et al.2008, 2010, 2011, 2012; Liu et al.1992, 2001, 2003, 2004a, 2004b, 2012, 2013; Lu et al. 2004, 2013, 2015a, 2015b, 2016; Markowitz 1952, 1959, 1987; Michalewicz and Schoenauer 1996; Mon et al. 1995; Mulpuru et al. 2012; Opricovic 1998; Opricovic and Tzeng 2002, 2003a, 2003b, 2003c, 2003d; Ou Yang et al. 2009; Ozaki et al. 2010, 2011, 2013; Pareto 1906; Pawlak 1984, 2002a, 2002b; Pawlak and Slowinski 1994, Piotroski and So 2012; Pourahmad et al. 2015; Roy 1971, 1976, 1981; Roy and Bertier 1973; Rust et al. 2004; Shee et al. 2000, 2003, 2012; Shee and Tzeng 2002; Shyng et al. 2011; Shyng, J.Y., Tzeng and Wang 2007; Simon 1977; Słowiński et al. 2005; Sousa and Kaymak 2002; Su et al. 2015; Sugeno et al. 1995; Tang et al. 1999, 2002; Teng and Tzeng 1996, 1998; Ting et al. 2013; Ting and Tzeng 2003, 2004, 2016; Tsai et al. 2006; Tsaur and Tzeng 1996; Tsaur and 1997; Tseng and Tzeng 1999, 2002; Tseng et al. 1999a, 1999b, 2001a, 2001b, 2001c; Tversky and Kahneman 1986; Tzeng et al. 1992, 1994, 1996, 1997, 1998, 2000, 2001, 2002, 2005, 2005a, 2005b, 2006, 2007a, 2007b, 2010; Tzeng and Chen 1993, 1997, 1998, 1999; Tzeng and Hu 1996; Tzeng and Huang 1997, 2012; Tzeng and Kuo 1996; Tzeng and Lee 2001; Tzeng and Lin 2000; Tzeng and Teng 1998; Tzeng and Tsaur, 1993, 1994, 1997; Tseng et al. 2002; Wang et al. 2007, 2010, 2014; Wang and Tzeng 2012; Wu et al. 2014; Yang et al. 2005, 2007, 2008; Yang and Tzeng 2011; Yu and Tzeng 2009; Yu et al. 1999, 2001, 2004, 2005, 2007; Yu and Chianglin 2006; Yu and Zeleny 1975; Yu and Tzeng 2006; Yuan et al. 2005; Zeleny 1972, 2012.

APPLICATIONS OF MCDM

Chapter 8

The Case of DEMATEL for Assessing Information Risk

Information risk has become one of the main concerns for governmental and business operations; the threat of information security incidents not only jeopardizes the valuable information held by organizations but might even cause the failure of the daily operations of governments/businesses, which cannot be overlooked in this information era. To show how to analyze this problem, a case involving information risk assessment using the decision-making trial and evaluation laboratory (DEMATEL) and DEMATEL-based analytic network process (DANP) methods is illustrated in this chapter.

Although information security risk assessments and ways of controlling risk have been examined in previous research, few studies calculate the integrated risks or assess the performance of the implemented controls while considering the dependence among criteria. In practice, information risk factors usually influence each other; the traditional risk assessment methods are constrained by the assumption of independence among the criteria (factors). To resolve this issue, an information security risk control assessment model (Rene 2005) that combines DEMATEL and ANP is adopted, which falls into the category of multiple attribute decision making (MADM). MADM methods are often used when problems are characterized by several noncommensurable and conflicting (competing) criteria, and there may be no solution that satisfies all criteria simultaneously. Here, the DEMATEL and DANP methods are expected to solve the problem of conflicting criteria with dependence and feedback on modeling.

109

8.1 Background of the Case and the Research Framework

In January 2001, the Taiwanese government launched the National Information and Communication Security Taskforce (NICST); its primary intention was to set up an integrated information and communication security defense system for the related governmental organizations to control potential information risks. In 2002, it further expedited an information security project throughout the government bureaucracy. This project enforced four levels of information security; all governmental units/departments have to examine their operations in various aspects, including size, type of task, and amount of investment, to meet the corresponding standards. Each unit/department has to check its own information security controls regularly to ensure the safety of operations. Once a unit has found that it is underperforming on a certain criterion or criteria, it has to identify the source factors (criteria or dimensions) that are causing the problem.

The aforementioned information control checks and management processes can be structured into four steps, as proposed by Biery (2006): (1) risk assessment, (2) risk remediation, (3) risk monitoring and review, and (4) risk management enhancement, as shown in the upper part of Figure 8.1. Risk assessment involves identifying and analyzing the vulnerability to exploitation by a threat, and the second step involves using controls to address the risk (also called *risk treatment*). The third step involves monitoring and measuring the risk controls for effectiveness. The fourth step is a continuous or consecutive improvement process based on observations from each of the previous steps, which serves as a feedback for the risk management cycle.

The monitoring and measuring of risk controls plays a critical role in the whole risk management process; therefore, this research framework is developed to enhance the third step by proposing a *risk control assessment system*. The process of risk control and this research framework are illustrated in Figure 8.1.

Referring to Figure 8.1, the inputs from *risk remediation* will be analyzed by this risk control assessment system, and the outputs of this system will be provided for regular checks (*risk monitoring and review*). This research framework adopts the audit items from Taiwan's Research Development and Evaluation Commission (RDEC), and the audit items from ISO/IEC 17799 (BS 7799-1) are adopted as the 11 critical criteria for control in two dimensions: *Organizational/management* (D_1) and *Operational/technical* (D_2). The 11 control factors (criteria) for information security management are *Security policy* (C_1), the *Organization of information security* (C_2), *Asset management* (C_3), *Human resources security* (C_4), *Physical and environmental security* (C_5), *Communications and operations management* (C_6), *Access control* (C_7), *Information systems acquisition, development, and maintenance* (C_8), *Information security incident management* (C_9), *Business continuity management* (C_{10}), and *Compliance* (C_{11}). The structure of this MADM model for risk control assessment is illustrated in Figure 8.2.

The Case of DEMATEL for Assessing Information Risk ■ 111

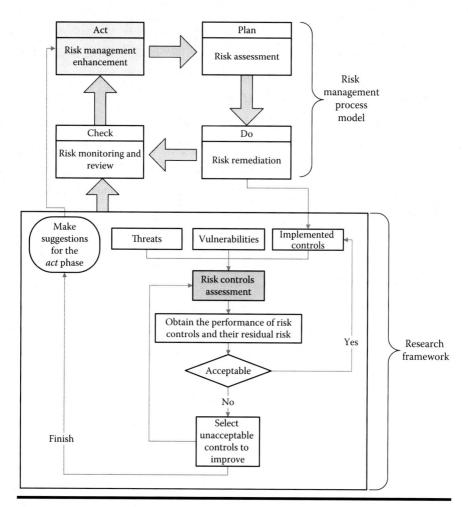

Figure 8.1 Overview of the information risk assessment process. (Reprinted with permission from OuYang, Y.P., Shieh, H.M., and Tzeng, G.H., *Information Sciences*, 232, 482–500, 2013.)

Based on the structure of this model, a questionnaire was devised to collect the opinions of domain experts for the DEMATEL analysis and DANP influential weights. In the questionnaire, a scale of 0–5 represents the range from "no influence" to "very high influence," with respondents proposing the degree of direct influence that each criterion exerts on another, and the dimensional influence between D_1 and D_2 is aggregated by averaging the associated criteria in each dimension (refer to Equation 4.7 in Chapter 4). The averaged opinions of the domain experts (from the questionnaires) are used in the DEMATEL analysis. One thing needs to be noted here: there are six criteria (C_1, C_2, C_3, C_4, C_{10}, and C_{11}) in D_1 and five criteria (C_5, C_6, C_7, C_8, and C_9) in D_2, respectively. Next, the grades of importance of

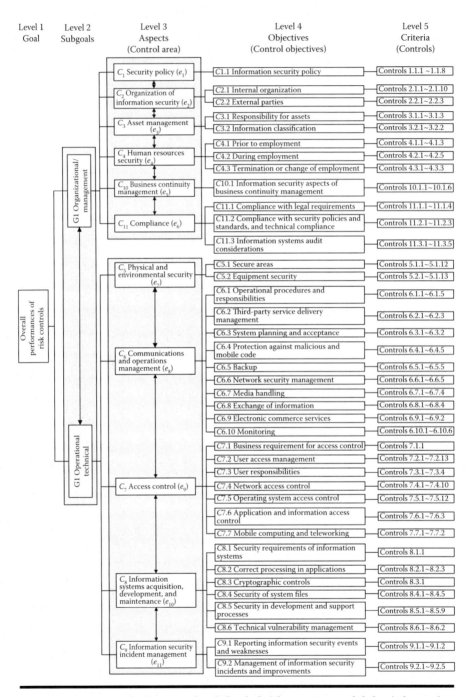

Figure 8.2 Research framework of the hybrid MADM model for information risk assessment. (Reprinted with permission from OuYang, Y.P., Shieh, H.M., and Tzeng, G.H., *Information Sciences*, 232, 482–500, 2013.)

the dimensions and criteria were calculated by adopting the inputs from the initial average matrix A of DEMATEL. Similarly, a scale of 0 to 5 represents the range from "absolutely unimportant" to "absolutely important." The corresponding data were used to obtain the DANP influential weights in Section 8.3.

8.2 DEMATEL Analysis with INRM

Referring to Figure 8.2, $[C_1, C_2, C_3, C_4, C_{10}, C_{11}]$ and $[C_5, C_6, C_7, C_8, C_9]$ are renamed $[e_1, e_2, e_3, e_4, e_5, e_6]$ and $[e_7, e_8, e_9, e_{10}, e_{11}]$ for the subsequent analyses. The DEMATEL analysis of the experts' questionnaires generates the dimensional matrix T^D for the two dimensions D_1 and D_2, as follows:

$$T^D = \begin{bmatrix} 3.52 & 4.52 \\ 3.52 & 3.52 \end{bmatrix} \text{ and the normalized } T_N^D = \begin{bmatrix} 0.43 & 0.56 \\ 0.50 & 0.50 \end{bmatrix}.$$

Therefore, the dimensional DEMATEL cause–effect analysis can be shown, as in Table 8.1.

Furthermore, the total-influence matrices T_{D_1} and T_{D_2} for D_1 and D_2 are

$$T_{D_1} = \begin{bmatrix} 1.31 & 1.42 & 1.34 & 1.41 & 1.49 & 1.43 \\ 1.41 & 1.18 & 1.26 & 1.32 & 1.42 & 1.34 \\ 1.21 & 1.15 & 0.97 & 1.15 & 1.22 & 1.17 \\ 1.21 & 1.17 & 1.11 & 1.02 & 1.23 & 1.19 \\ 1.36 & 1.29 & 1.22 & 1.28 & 1.20 & 1.30 \\ 1.36 & 1.28 & 1.20 & 1.27 & 1.35 & 1.14 \end{bmatrix} \text{ and}$$

$$T_{D_2} = \begin{bmatrix} 1.76 & 2.15 & 2.23 & 2.00 & 1.96 \\ 2.05 & 2.09 & 2.37 & 2.14 & 2.10 \\ 2.07 & 2.31 & 2.16 & 2.15 & 2.10 \\ 1.83 & 2.06 & 2.12 & 1.74 & 1.87 \\ 1.83 & 2.06 & 2.11 & 1.91 & 1.71 \end{bmatrix},$$

respectively. The corresponding cause–effect analyses for D_1 and D_2 are combined as a criteria analysis, shown in Table 8.2.

From Table 8.2, the dimension analysis by DEMATEL shows that D_1 is the source dimension that has direct influence on D_2. Combining the cause–effect analyses of the dimensions and criteria (e_1 through e_{11}), the influential network

114 ■ *Trends of Hybrid Multiple Criteria Decision Making*

Table 8.1 DEMATEL Cause–Effect Analysis among Dimensions

Dimensions		D_1	D_2	r_i^D	c_i^D	$r_i^D + c_i^D$	$r_i^D - c_i^D$
D_1	Organizational/ management	3.52	4.52	8.04	7.04	15.08	1.00
D_2	Operational/ technical	3.52	3.52	7.04	8.04	15.08	−1.00

Source: Data from OuYang, Y.P., Shieh, H.M., and Tzeng, G.H. *Information Sciences*, 232, 482–500, 2013.

relations map (INRM) can be obtained, and the directional influences between the two subgoals are shown in Figure 8.3.

8.3 DANP Influential Weights for Criteria

Referring to the required steps for calculating DANP influential weights in Chapter 4, the dimensional matrix T^D can be used to adjust the unweighted super-matrix W; the weighted supermatrix for DANP can be obtained by $W^\alpha = T_N^D W$. The dimensional matrix T^D and the unweighted supermatrix W are as follows:

$$T_N^D = \begin{bmatrix} 0.44 & 0.56 \\ 0.50 & 0.50 \end{bmatrix} \text{ and}$$

$$W = \begin{array}{c} e_1 \\ e_2 \\ e_3 \\ e_4 \\ e_5 \\ e_6 \\ e_7 \\ e_8 \\ e_9 \\ e_{10} \\ e_{11} \end{array}
\begin{array}{c}
\begin{array}{ccccccccccc}
e_1 & e_2 & e_3 & e_4 & e_5 & e_6 & e_7 & e_8 & e_9 & e_{10} & e_{11}
\end{array} \\
\begin{bmatrix}
0.186 & 0.179 & 0.157 & 0.152 & 0.175 & 0.187 & 0.176 & 0.180 & 0.174 & 0.168 & 0.174 \\
0.168 & 0.198 & 0.143 & 0.157 & 0.164 & 0.152 & 0.138 & 0.152 & 0.148 & 0.140 & 0.162 \\
0.155 & 0.146 & 0.223 & 0.146 & 0.141 & 0.143 & 0.188 & 0.156 & 0.164 & 0.162 & 0.134 \\
0.155 & 0.150 & 0.150 & 0.220 & 0.157 & 0.148 & 0.148 & 0.152 & 0.174 & 0.178 & 0.164 \\
0.166 & 0.177 & 0.170 & 0.159 & 0.202 & 0.168 & 0.184 & 0.188 & 0.168 & 0.184 & 0.198 \\
0.170 & 0.150 & 0.157 & 0.166 & 0.161 & 0.202 & 0.166 & 0.172 & 0.172 & 0.168 & 0.168 \\
0.191 & 0.191 & 0.221 & 0.182 & 0.182 & 0.180 & 0.252 & 0.170 & 0.186 & 0.162 & 0.168 \\
0.202 & 0.202 & 0.198 & 0.193 & 0.209 & 0.211 & 0.196 & 0.242 & 0.204 & 0.194 & 0.202 \\
0.214 & 0.202 & 0.220 & 0.234 & 0.198 & 0.214 & 0.212 & 0.216 & 0.240 & 0.208 & 0.192 \\
0.191 & 0.184 & 0.195 & 0.191 & 0.193 & 0.196 & 0.170 & 0.186 & 0.190 & 0.262 & 0.176 \\
0.202 & 0.221 & 0.166 & 0.200 & 0.218 & 0.198 & 0.170 & 0.186 & 0.180 & 0.174 & 0.262
\end{bmatrix}
\end{array}.$$

The Case of DEMATEL for Assessing Information Risk ■ **115**

Table 8.2 DEMATEL Cause–Effect Analyses for the Criteria

Dimensions	Symbols	Criteria	$r_i^C + c_i^C$	$r_i^C - c_i^C$
	C_1 (e_1)	Security policy	16.27	0.54
	C_2 (e_2)	Organization of information security	15.41	0.44
D_1	C_3 (e_3)	Asset management	13.96	−0.23
	C_4 (e_4)	Human resources security	14.37	−0.52
	C_{10} (e_5)	Business continuity management	15.55	−0.25
	C_{11} (e_6)	Compliance	15.18	0.03
	C_5 (e_7)	Physical and environmental security	19.63	0.55
	C_6 (e_8)	Communications and operations management	21.42	0.08
D_2	C_7 (e_9)	Access control	21.79	−0.20
	C_8 (e_{10})	Information systems acquisition, development, and maintenance	19.56	−0.31
	C_9 (e_{11})	Information security incident management	19.35	−0.12

Source: Data from OuYang, Y.P., Shieh, H.M., and Tzeng, G.H. *Information Sciences*, 232, 482–500, 2013.

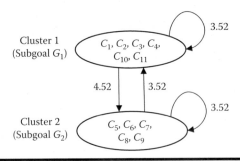

Figure 8.3 Structure of the subgoals of this information risk assessment problem. (Reprinted with permission from OuYang, Y.P., Shieh, H.M., and Tzeng, G.H., *Information Sciences*, 232, 482–500, 2013.)

To proceed with the multiplication of T_N^D and W to adjust the dimensional weighting in DANP, the criteria e_1 through e_6 in dimension D_1 are used to explain how to operate (the upper-left region in W with bold and italic figures, named T_C^{11}, referring to Equation 4.2 in Chapter 4). Thus, W^α is obtained by multiplying $0.44 \left(\dfrac{3.52}{3.52 + 4.52} \cong 0.44, \text{ at the upper-left corner of } T_N^D \right)$ by T_C^{11}.

$$T_C^{\alpha 11} = 0.44 \times \begin{bmatrix} 0.186 & 0.179 & 0.157 & 0.152 & 0.175 & 0.187 \\ 0.168 & 0.198 & 0.143 & 0.157 & 0.164 & 0.152 \\ 0.155 & 0.146 & 0.223 & 0.146 & 0.141 & 0.143 \\ 0.155 & 0.150 & 0.150 & 0.220 & 0.157 & 0.148 \\ 0.166 & 0.177 & 0.170 & 0.159 & 0.202 & 0.168 \\ 0.170 & 0.150 & 0.157 & 0.166 & 0.161 & 0.202 \end{bmatrix}$$

$$= \begin{bmatrix} 0.081 & 0.078 & 0.069 & 0.067 & 0.077 & 0.082 \\ 0.074 & 0.087 & 0.063 & 0069 & 0.072 & 0.067 \\ 0.068 & 0.064 & 0.098 & 0.064 & 0.062 & 0.063 \\ 0.068 & 0.066 & 0.066 & 0.096 & 0.069 & 0.065 \\ 0.073 & 0.077 & 0.074 & 0.070 & 0.088 & 0.074 \\ 0.074 & 0.066 & 0.069 & 0.073 & 0.070 & 0.088 \end{bmatrix}.$$

The stable DANP influential weight for each criterion can be obtained by raising z, multiplying W^{α} with itself several times (i.e., $W^* = \lim_{z \to \infty} \left(W^{\alpha} \right)^z$), as follows:

$$
W^* = \begin{array}{c} \\ e_1 \\ e_2 \\ e_3 \\ e_4 \\ e_5 \\ e_6 \\ e_7 \\ e_8 \\ e_9 \\ e_{10} \\ e_{11} \end{array}
\begin{array}{ccccccccccc}
e_1 & e_2 & e_3 & e_4 & e_5 & e_6 & e_7 & e_8 & e_9 & e_{10} & e_{11} \\
\left[\begin{array}{c} 0.082 \end{array}\right. & 0.082 & 0.082 & 0.082 & 0.082 & 0.082 & 0.082 & 0.082 & 0.082 & 0.082 & \left.\begin{array}{c} 0.082 \end{array}\right] \\
0.073 & 0.073 & 0.073 & 0.073 & 0.073 & 0.073 & 0.073 & 0.073 & 0.073 & 0.073 & 0.073 \\
0.075 & 0.075 & 0.075 & 0.075 & 0.075 & 0.075 & 0.075 & 0.075 & 0.075 & 0.075 & 0.075 \\
0.077 & 0.077 & 0.077 & 0.077 & 0.077 & 0.077 & 0.077 & 0.077 & 0.077 & 0.077 & 0.077 \\
0.085 & 0.085 & 0.085 & 0.085 & 0.085 & 0.085 & 0.085 & 0.085 & 0.085 & 0.085 & 0.085 \\
0.080 & 0.080 & 0.080 & 0.080 & 0.080 & 0.080 & 0.080 & 0.080 & 0.080 & 0.080 & 0.080 \\
0.100 & 0.100 & 0.100 & 0.100 & 0.100 & 0.100 & 0.100 & 0.100 & 0.100 & 0.100 & 0.100 \\
0.108 & 0.108 & 0.108 & 0.108 & 0.108 & 0.108 & 0.108 & 0.108 & 0.108 & 0.108 & 0.108 \\
0.113 & 0.113 & 0.113 & 0.113 & 0.113 & 0.113 & 0.113 & 0.113 & 0.113 & 0.113 & 0.113 \\
0.103 & 0.103 & 0.103 & 0.103 & 0.103 & 0.103 & 0.103 & 0.103 & 0.103 & 0.103 & 0.103 \\
0.104 & 0.104 & 0.104 & 0.104 & 0.104 & 0.104 & 0.104 & 0.104 & 0.104 & 0.104 & 0.104
\end{array}.
$$

8.4 Discussion and Conclusion

In this case of information risk assessment, the two dimensions are highly interrelated; the effect of D_1 (*Organizational/management*) on D_2 (*Operational/technical*) is 4.52, whereas the effect of D_2 on D_1 is 3.52.

Two facts are clear from Table 8.1 and Figure 8.3: (1) each cluster has feedback and dependence; (2) the effect of Cluster 1 on Cluster 2 is 4.52, whereas the effect of Cluster 2 on Cluster 1 is 3.52. In other words, the degree to which Cluster 2 is affected is higher than that of Cluster 1. Therefore, Cluster 2 should be paid more attention than the other clusters in the real world; that is, it should be given additional weight, whereas Cluster 1 should have its weight reduced. Since e_1 through e_6 belong to Cluster 1 and e_7 through e_{11} belong to Cluster 2, the criteria e_7 through e_{11} should be paid more attention than e_1 through e_6. Using the traditional normalization method implies that each cluster has the same weight (each criterion in a column is divided by the number of clusters to normalize the unweighted supermatrix). However, there are different degrees of influence among the clusters of factors/criteria in this empirical case. Thus, as stated in Section 4.5 in Chapter 4, by using DEMATEL to adjust the normalization of ANP in the unweighted supermatrix, these results should be closer to the real situations.

The risk management process comprises four stages: (1) plan, (2) do, (3) check, and (4) act (PDCA), as shown in Figure 8.1. Since each department is requested to regularly examine and check its vulnerability to external threats, the plausible underperforming subgoals/aspects/criteria should be highlighted after each evaluation. Furthermore, the source factors can be identified by DEMATEL analysis (Table 8.2), and the management team may plan accordingly for the required actions. Although the concept of PDCA in management is prevalent, conventional methods have difficulty depicting the directional influences among variables with a systematic approach; using a combination of the DEMATEL and ANP methods, managers can find an effective way to enhance internal information control.

Chapter 9

E-Store Business Evaluation and Improvement Using a Hybrid MADM Model

In the Internet era, the purchase behavior of consumers has shifted from shopping at conventional channels (such as supermarkets and department stores) to buying on virtual e-stores.

The online retail sales of European e-stores increased by 18% from 2009 to 2010 (Carini et al. 2011). Forrester research forecasts that European online sales will have a compound annual growth rate of 12%, from €96.7 billion (£82.0 billion) in 2011 to €171.9 billion (£145.8 billion) in 2016 (Chloe 2012; Gill et al. 2012); also, according to the estimation from the Center of Retail Research, the online e-commerce sales in Western Europe and Poland might increase to €265.68 billion (£230.62 billion) in 2017 (http://www.retailresearch.org/). Also, the UK Office for National Statistics reports that e-store sales increased by 13.1% in March 2011. Forrester research forecasts that UK online sales will have compound annual growth rate of 11%, from £30.1 billion in 2011 to £51.0 billion in 2016, and the United Kingdom's proportion of online shoppers will increase from 75% of the population in 2011 to 85% in 2016. In the United States, online retail sales are expected to grow, according to Forrester research, from $176.2 billion in 2010 to $278.9 billion in 2015, which is an increase of more than 10% (Mulpuru et al. 2011). Forrester research forecasts note that US online shoppers will have a compound annual growth rate of 15%, from 167 million people in 2012 to 192 million people in 2016, and it is predicted that each consumer's spending will grow by 44%,

119

120 ■ *Trends of Hybrid Multiple Criteria Decision Making*

from $1207 in 2012 to $1738 in 2016. All these reports show that sales on e-stores are growing fast in all areas. The convenience and flexibility of online shopping have caused the growth of this high-potential industry; however, competition among the e-stores is also increasing rapidly. Therefore, how to analyze the strengths and weaknesses of an e-store to enhance its market position is a critical and important issue.

In this empirical case, we focus on analyzing the needs and requirements of consumers shopping at e-stores. The important dimensions and criteria for evaluating e-stores were identified by a literature review and a preliminary survey, from which the framework of a multiple attribute decision making (MADM) weighting system can be formed. Referring to the framework for problem solving in Figure 5.1 (Chapter 5), this case adopts the DEMATEL-based ANP (DANP) method to find the influential weight of each criterion, and the modified VIKOR is incorporated to identify the weighted performance gaps where the assessed e-stores might plan for improvements. Furthermore, the DEMATEL technique supports the analysis of the cause–effect relations among the key dimensions and criteria, which can be used to construct an influential network relations map (INRM) (Chapter 4) so that decision makers (DMs) may plan for systematic improvements. Three famous online e-stores are analyzed and compared in the following sections.

9.1 Background of the Case and the Research Framework

Chang and Chen (2009) surveyed the majority of Taiwan-based e-commerce websites, and Yahoo.com, Yahoo auction, PChome.com, and Books.com were found to be the top-four online shopping e-stores for domestic consumers. Another study (To et al. 2007) found that the most popular website was Yahoo (45.1%), followed by PChome (7.8%), Books (6.8%), and Ezfly (5.3%), which accounted for 65% of all responses. Therefore, using an empirical study of three e-stores—Yahoo, PChome, and Books—this chapter aims to demonstrate the proposed new hybrid MADM model for problem solving.

Two sides of these e-stores' operation problems are considered. On the one hand, e-store consumer behavior has been examined in numerous e-commerce studies; most research suggests that knowing the needs of customers and how to meet those needs is important to running a successful e-store (Ažman and Gomišček 2012). In other words, e-stores should meet customers' needs (Porter 2001) to create customer value in consumer marketing (Chen and Dubinsky 2003). On the other hand, customer satisfaction is important mostly because of its indirect influence on the profitability of companies (Ažman and Gomišček 2012). Satisfied customers tend to make not only more purchases but also repeat purchases (Bearden and Teel 1983; Siddiqi 2011; Szymanski and Hise 2000), and so customer satisfaction becomes a key element of many companies' business strategies (Ellinger et al. 2012). Therefore, understanding how to influence customer satisfaction should also be considered. In addition, the purchase behavior of e-store shoppers can be divided into three stages: the preliminary stage, the evaluation stage, and the post-purchase stage (Chon 1990; Gan et al. 2006;

E-Store Business Evaluation Using a Hybrid MADM Model ■ 121

Wong and Hsu 2008). At the preliminary stage, e-store operators need to understand the needs of potential consumers based on the e-store's market positioning; also, information on those online products should be provided to the "right" customers. At the evaluation stage, how those potential consumers make the purchase decision should be examined to strengthen the e-store's brand awareness. At the final stage, the consumers will provide feedback based on the purchase experience, such as blog writing, Facebook reviews, and the tendency to repurchase. Based on the aforementioned three stages, this chapter adopts the dimensions and associated criteria to form the research framework. Brief definitions are listed in Table 9.1.

The emergence of the e-store has lowered the entrance barrier for many new vendors to compete in the retail market, and it has also had a huge impact on existing business players. Through e-store channels, companies have the chance to reach new markets, new customers, and new information, improve customer services, distribute

Table 9.1 Research Framework and the Associated Criteria

Need Recognition (D_1)	
Product availability (C_1)	Products that consumers want or need to be available
Information Search (D_2)	
Ease of finding (C_2)	How easily consumers were able to find information about the products
Overall look/design (C_3)	How consumers feel about the overall look and design of the website
Clarity of product information (C_4)	How clear and understandable the information about the products is
Evaluation of Alternatives (D_3)	
Number of reviews (C_5)	How many reviews there are of other people's overall rating and experience of this purchase
Brand (C_6)	How much consumers trust the website's brand
Relative price (C_7)	How the prices compare with other websites
Choice/Purchase (D_4)	
Selection (C_8)	Types of products available
Variety of shipping options (C_9)	The desired shipping options were available
Shipping charges (C_{10})	Cost of shipping/delivery
Charges statement (C_{11})	Total purchase amount (including shipping/handling charges, etc.) displayed before order submission
Postpurchase Behavior (D_5)	
Order tracking (C_{12})	Ability to track orders until delivered
On-time delivery (C_{13})	Products were in stock at time of expected delivery
Product expectations met (C_{14})	Correct products were delivered and everything worked as described/depicted

Source: Data from Chiu, W.Y., Tzeng, G.H., and Li, H.L., *Knowledge-Based Systems*, 37, 48–61, 2013.

122 ■ *Trends of Hybrid Multiple Criteria Decision Making*

products more quickly, communicate more effectively, and eventually increase their profitability (Asllani and Lari 2007; Ellinger et al. 2003). Analyses of this problem can provide suggestions for e-stores to improve their businesses; furthermore, vendors can apply this decision model to evaluate and select their partners.

9.2 DANP for Finding Influential Weights

The modeling of the e-store business evaluation problem comprises three parts (refer to Section 5.4 in Chapter 5): (1) a weighting system, (2) a performance evaluation method, and (3) an aggregator. In this case, the DANP method was adopted to find the influential weight for each criterion from eight domain experts (the first questionnaire). Next, the evaluation of the three target e-stores was done by collecting opinions (the second questionnaire) from experienced online shoppers (1018 consumers who often purchase products from the three target e-stores; the reliability is 93.9% and the validity is 92.9% by KMO [Kaiser-Meyer-Olkin] test).

The first questionnaire was conducted on the basis of pairwise comparisons to evaluate the effect and influence of the criteria, using a five-point scale ranging from 4 (extremely influential) to 0 (completely noninfluential). The second questionnaire was the e-stores' performance questionnaire, which used an 11-point scale/score ranging from 0 (very poor performance) to 10 (excellent performance)—that is, very dissatisfied or very bad to very satisfied or very good.

The DEMATEL technique is used to model influential relationships among dimensions and criteria and to establish an INRM for those dimensions and criteria, using pairwise comparisons. The eight domain experts were asked to determine the influence of the relationships among the criteria. The averaged initial direct-relation matrix A, obtained using pairwise comparisons in terms of influences and directions between criteria, is shown in Table 9.2.

In matrix A (refer to Equation 3.2 in Chapter 3), $\alpha = 29.75$, as shown in the total influence of C_1; therefore, the normalized direct-relation matrix D (termed T_C^G in Chapter 4) can be calculated and shown in Table 9.3.

Next, referring to Equations 4.3 and 4.4 from Chapter 4, the dimensional influence matrix T^D is as follows:

$$
T^D = \begin{bmatrix}
0.389 & 0.443 & 0.448 & 0.400 & 0.395 \\
0.406 & 0.385 & 0.365 & 0.343 & 0.334 \\
0.385 & 0.360 & 0.349 & 0.323 & 0.330 \\
0.361 & 0.345 & 0.344 & 0.304 & 0.311 \\
0.386 & 0.365 & 0.377 & 0.339 & 0.330
\end{bmatrix}.
$$

The normalized total-influence matrix T_N^G can be obtained by referring to Equations 4.1 through 4.4 in Chapter 4, as shown in Table 9.4. Then, the associated $r_i + c_i$ and $r_i - c_i$ from T_N^G and the associated $r_j^D + c_j^D$ and $r_j^D - c_j^D$ from T^D can be

Table 9.2 Initial-Influential Matrix A

Dimensions/criteria			D_1	D_2			D_3			D_4				D_5			
			C_1	C_2	C_3	C_4	C_5	C_6	C_7	C_8	C_9	C_{10}	C_{11}	C_{12}	C_{13}	C_{14}	Total
D_1	C_1	Product availability	0.000	2.125	1.875	3.000	3.375	1.250	3.250	2.125	2.250	2.375	2.000	1.375	1.875	2.875	29.750
D_2	C_2	Ease of finding	1.875	0.000	2.625	2.625	1.875	1.375	1.375	2.500	1.375	1.125	1.500	1.375	1.250	2.000	22.875
	C_3	Overall look/ design	2.000	2.750	0.000	3.375	1.875	1.375	1.500	2.875	1.500	1.125	1.500	1.500	1.250	1.625	24.250
	C_4	Clarity	3.125	3.625	3.125	0.000	1.875	1.375	1.750	3.250	2.000	1.500	1.625	1.375	1.250	2.625	28.500
D_3	C_5	Number of reviews	2.125	1.875	1.625	1.875	0.000	2.750	2.250	1.500	1.125	1.500	1.250	1.250	1.500	2.750	23.375
	C_6	Brand	1.250	1.750	2.125	2.125	2.875	0.000	2.000	1.750	1.750	1.750	2.000	1.750	1.875	2.125	25.125
	C_7	Relative price	3.125	1.250	1.250	2.000	2.500	1.375	0.000	1.500	1.375	2.500	1.250	1.250	1.500	2.625	23.500
D_4	C_8	Selection	2.000	3.000	2.750	3.000	2.000	1.625	1.750	0.000	2.125	1.625	1.750	1.125	1.375	1.625	25.750
	C_9	Variety of shipping	1.750	1.250	1.500	1.625	1.750	1.625	2.000	1.625	0.000	2.750	1.875	1.750	2.250	1.625	23.375
	C_{10}	Shipping charges	2.000	1.125	1.250	1.625	1.500	2.250	3.000	1.250	2.875	0.000	1.375	1.875	2.125	1.625	23.875
	C_{11}	Charge statement	1.750	1.375	1.125	1.625	1.250	1.625	1.500	1.500	1.625	1.625	0.000	1.375	1.125	1.375	18.875
D_5	C_{12}	Order tracking	1.250	1.125	1.375	1.500	1.625	1.625	1.250	1.250	3.125	2.250	1.250	0.000	3.500	2.125	23.250
	C_{13}	On-time delivery	1.500	1.250	1.125	1.500	2.250	2.375	1.750	1.375	3.125	2.750	1.000	2.875	0.000	1.875	24.750
	C_{14}	Met expectations	3.000	2.000	2.000	3.125	3.000	2.500	2.500	1.500	1.375	1.625	0.875	1.500	1.750	0.000	26.750
		Total	26.750	24.500	23.750	29.000	27.750	23.125	25.875	24.000	25.625	24.500	19.250	20.375	22.625	26.875	—

Source: Data from Chiu, W.Y., Tzeng, G.H., and Li, H.L., *Knowledge-Based Systems*, 37, 48–61, 2013.

Table 9.3   Normalized Direct-Relation Matrix D

Dimensions/Criteria		D_1	D_2			D_3			D_4				D_5		
		C_1	C_2	C_3	C_4	C_5	C_6	C_7	C_8	C_9	C_{10}	C_{11}	C_{12}	C_{13}	C_{14}
D_1	C_1 Product availability	0.000	0.071	0.063	0.101	0.113	0.042	0.109	0.071	0.076	0.080	0.067	0.046	0.063	0.097
D_2	C_2 Ease of finding	0.063	0.000	0.088	0.088	0.063	0.046	0.046	0.084	0.046	0.038	0.050	0.046	0.042	0.067
	C_3 Overall look/ design	0.067	0.092	0.000	0.113	0.063	0.046	0.050	0.097	0.050	0.038	0.050	0.050	0.042	0.055
	C_4 Clarity	0.105	0.122	0.105	0.000	0.063	0.046	0.059	0.109	0.067	0.050	0.055	0.046	0.042	0.088
D_3	C_5 Number of reviews	0.071	0.063	0.055	0.063	0.000	0.092	0.076	0.050	0.038	0.050	0.042	0.042	0.050	0.092
	C_6 Brand	0.042	0.059	0.071	0.071	0.097	0.000	0.067	0.059	0.059	0.059	0.067	0.059	0.063	0.071
	C_7 Relative price	0.105	0.042	0.042	0.067	0.084	0.046	0.000	0.050	0.046	0.084	0.042	0.042	0.050	0.088
D_4	C_8 Selection	0.067	0.101	0.092	0.101	0.067	0.055	0.059	0.000	0.071	0.055	0.059	0.038	0.046	0.055
	C_9 Variety of shipping	0.059	0.042	0.050	0.055	0.059	0.055	0.067	0.055	0.000	0.092	0.063	0.059	0.076	0.055
	C_{10} Shipping charges	0.067	0.038	0.042	0.055	0.050	0.076	0.101	0.042	0.097	0.000	0.046	0.063	0.071	0.055
	C_{11} Charge statement	0.059	0.046	0.038	0.055	0.042	0.055	0.050	0.050	0.055	0.055	0.000	0.046	0.038	0.046
D_5	C_{12} Order tracking	0.042	0.038	0.046	0.050	0.055	0.055	0.042	0.042	0.105	0.076	0.042	0.000	0.118	0.071
	C_{13} On-time delivery	0.050	0.042	0.038	0.050	0.076	0.080	0.059	0.046	0.105	0.092	0.034	0.097	0.000	0.063
	C_{14} Met expectations	0.101	0.067	0.067	0.105	0.101	0.084	0.084	0.050	0.046	0.055	0.029	0.050	0.059	0.000

Source: Data from Chiu, W.Y., Tzeng, G.H., and Li, H.L., *Knowledge-Based Systems, 37*, 48–61, 2013.

Table 9.4 Normalized Total-Influence Matrix T_N^G

	C_1	C_2	C_3	C_4	C_5	C_6	C_7	C_8	C_9	C_{10}	C_{11}	C_{12}	C_{13}	C_{14}	r_i
Product availability (C_1)	0.389	0.422	0.403	0.505	0.500	0.372	0.472	0.412	0.427	0.421	0.339	0.333	0.377	0.476	**0.389**
Ease of finding (C_2)	0.363	0.283	0.354	0.408	0.370	0.303	0.335	0.353	0.324	0.307	0.266	0.270	0.289	0.365	**1.046**
Overall look/design (C_3)	0.385	0.385	0.290	0.448	0.388	0.317	0.355	0.380	0.344	0.323	0.279	0.286	0.303	0.373	**1.123**
Clarity (C_4)	0.470	0.459	0.432	0.404	0.443	0.361	0.414	0.437	0.406	0.380	0.321	0.322	0.346	0.453	**1.294**
Number of reviews (C_5)	0.377	0.345	0.328	0.391	0.321	0.350	0.370	0.325	0.323	0.325	0.263	0.272	0.303	0.395	**1.041**
Brand (C_6)	0.367	0.356	0.357	0.415	0.424	0.281	0.376	0.348	0.358	0.349	0.297	0.301	0.329	0.392	**1.082**
Relative price (C_7)	0.411	0.328	0.318	0.398	0.401	0.313	0.305	0.327	0.335	0.360	0.265	0.275	0.307	0.395	**1.019**
Selection (C_8)	0.401	0.407	0.388	0.455	0.409	0.340	0.379	0.306	0.378	0.353	0.299	0.289	0.321	0.389	**1.336**
Variety of shipping (C_9)	0.361	0.320	0.318	0.377	0.370	0.316	0.359	0.324	0.287	0.362	0.279	0.287	0.326	0.357	**1.252**
Shipping charges (C_{10})	0.375	0.322	0.316	0.384	0.370	0.339	0.395	0.319	0.382	0.285	0.269	0.296	0.328	0.365	**1.255**
Charge statement (C_{11})	0.307	0.275	0.260	0.320	0.299	0.267	0.290	0.273	0.286	0.279	0.181	0.233	0.245	0.295	**1.018**
Order tracking (C_{12})	0.343	0.313	0.312	0.370	0.364	0.316	0.334	0.310	0.383	0.349	0.258	0.233	0.363	0.369	**0.965**
On-time delivery (C_{13})	0.368	0.333	0.321	0.389	0.401	0.353	0.367	0.330	0.399	0.378	0.265	0.334	0.272	0.381	**0.987**
Met expectations (C_{14})	0.448	0.391	0.380	0.476	0.459	0.381	0.420	0.367	0.371	0.370	0.283	0.313	0.348	0.357	**1.017**
c_i	**0.389**	**1.127**	**1.076**	**1.260**	**1.146**	**0.945**	**1.051**	**1.222**	**1.332**	**1.280**	**1.028**	**0.879**	**0.983**	**1.107**	

Source: Data from Chiu, W.Y., Tzeng, G.H., and Li, H.L., *Knowledge-Based Systems*, 37, 48–61, 2013.

reorganized/summarized as in Table 9.5, in which the directional influence relations among dimensions or criteria can be identified using $r_j^D + c_j^D$ or $r_i - c_i$, respectively.

By transposing the normalized total-influence matrix T_N^G, the unweighted supermatrix W for the DANP method can be obtained (i.e., $W = (T_N^G)'$). Furthermore, T^D can be used to adjust the dimensional weights in the unweighted supermatrix: $W^\alpha = T_N^D W$ (refer to Equation 4.10 in Chapter 4). After raising z for $\lim_{z \to \infty} (W^\alpha)^z$, a set of stable influential weights can be obtained using the DANP method, shown in Table 9.6.

9.3 Performance Measures and Modified VIKOR for Evaluations

After forming the weighting system of this e-store evaluation problem using the DANP method, the next step requires a performance evaluation of the three target e-stores—Yahoo (A_1), PChome (A_2), and Book (A_3)—using the 14 criteria. Questionnaires were collected from 1018 experienced consumers who have bought products from e-stores; those consumers gave a performance rating (i.e., their level of satisfaction) for the three e-stores on the 14 criteria, ranging from 0 (the worst) to 10 (the best or ideal). The average performance scores for each alternative are summarized with the DANP influential weights in Table 9.6.

Using the modified VIKOR method, the ideal or aspired level of satisfaction (i.e., performance) was set as 10 to calculate the performance gaps of the three alternatives. For example, since the performance score of A_1 (Yahoo) on C_1 (*Product availability*) is 7.212, the performance gap is $(10 - 7.212)/(10 - 0) = 0.279$. Referring to Equation 5.12 in Chapter 5, the averaged and weighted gap (i.e., index S) for each alternative can be obtained, as shown at the bottom of Table 9.6. The weighted gap denotes the DANP weighted average group utility, and the weighted gaps for A_1, A_2, and A_3 are 0.246, 0.253, and 0.240, respectively, which indicates $A_3 > A_1 > A_2$. This result is consistent with the ranking result by calculating the weighted performance score (i.e., simple additive weighting [SAW]), which suggests the consistency of this analysis.

9.4 Discussion

This case study provides rich insights for the management teams of e-stores to refer to. First, using the cause–effect analysis of DEMATEL, shown in Table 9.5, the directional influences among the five dimensions can be depicted as in Figure 9.1 (INRM); the complicated interrelations of dimensions can be used as a guide for e-store managers to plan for marketing strategies.

Table 9.5   Cause–Effect Analyses of Dimensions and Criteria by DEMATEL Technique

Dimensions		r_j^D	c_j^D	$r_j^D + c_j^D$	$r_j^D - c_j^D$	Criteria		r_i	c_i	$r_i + c_i$	$r_i - c_i$
D_1	Need recognition	2.075	1.927	4.002	0.148	C_1	Product availability	0.389	0.389	0.778	0.000
D_2	Information search	1.833	1.898	3.731	−0.065	C_2	Ease of finding	1.046	1.127	2.173	-0.081
						C_3	Overall look/ design	1.123	1.076	2.199	0.047
						C_4	Clarity	1.294	1.26	2.554	0.034
D_3	Evaluation of alternatives	1.746	1.884	3.630	−0.138	C_5	Number of reviews	1.041	1.146	2.187	-0.105
						C_6	Brand	1.082	0.945	2.027	0.137
						C_7	Relative price	1.019	1.051	2.070	-0.032
D_4	Choice/ purchase	1.665	1.708	3.373	−0.043	C_8	Selection	1.336	1.222	2.558	0.114
						C_9	Variety of shipping	1.252	1.332	2.584	-0.080
						C_{10}	Shipping charges	1.255	1.28	2.535	-0.025
						C_{11}	Charge statement	1.018	1.028	2.046	-0.010
D_5	Postpurchase behavior	1.797	1.700	3.497	0.097	C_{12}	Order tracking	0.965	0.879	1.844	0.086
						C_{13}	On-time delivery	0.987	0.983	1.970	0.004
						C_{14}	Met expectations	1.017	1.107	2.124	-0.090

Source: Data from Chiu, W.Y., Tzeng, G.H., and Li, H.L., *Knowledge-Based Systems*, 37, 48–61, 2013.

Table 9.6   Performance Scores and Gaps for the Three e-Stores with DANP Influential Weights

Alternatives / Dimensions/Criteria	Local Weights (DANP)	Global Weights (DANP)	Yahoo (A_1)		PChome (A_2)		Books (A_3)	
			Scores (A_1)	Gaps (VIKOR)	Scores (A_2)	Gaps (VIKOR)	Scores (A_3)	Gaps (VIKOR)
Need recognition (D_1)	**0.079**		**7.212**	**(0.279)**	**7.261**	**(0.274)**	**7.040**	**(0.296)**
C_1 Product availability	1.000	0.079	7.212	(0.279)	7.261	(0.274)	7.040	(0.296)
Information search (D_2)	**0.226**		**7.336**	**(0.266)**	**7.341**	**(0.266)**	**7.605**	**(0.239)**
C_2 Ease of finding	0.318	0.072	7.586	(0.241)	7.481	(0.252)	7.484	(0.252)
C_3 Overall look/ design	0.310	0.070	7.133	(0.287)	7.108	(0.289)	7.668	(0.233)
C_4 Clarity	0.372	0.084	7.292	(0.271)	7.417	(0.258)	7.656	(0.234)
Evaluation of alternatives (D_3)	**0.223**		**7.584**	**(0.242)**	**7.510**	**(0.249)**	**7.491**	**(0.251)**
C_5 Number of reviews	0.359	0.080	7.735	(0.226)	7.511	(0.249)	7.508	(0.249)
C_6 Brand	0.300	0.067	7.644	(0.236)	7.628	(0.237)	8.114	(0.189)
C_7 Relative price	0.341	0.076	7.373	(0.263)	7.405	(0.259)	6.927	(0.307)

(Continued)

Table 9.6 (Continued) Performance Scores and Gaps for the Three e-Stores with DANP Influential Weights

Alternatives / Dimensions/Criteria	Local Weights (DANP)	Global Weights (DANP)	Yahoo (A_1)		PChome (A_2)		Books (A_3)	
			Scores (A_1)	Gaps (VIKOR)	Scores (A_2)	Gaps (VIKOR)	Scores (A_3)	Gaps (VIKOR)
Choice/purchase (D_4)	**0.270**		**7.759**	**(0.224)**	**7.550**	**(0.245)**	**7.710**	**(0.229)**
C_8 Selection	0.259	0.070	7.983	(0.202)	7.687	(0.231)	7.892	(0.211)
C_9 Variety of shipping	0.270	0.073	7.776	(0.222)	7.511	(0.249)	7.666	(0.233)
C_{10} Shipping charges	0.263	0.071	7.104	(0.290)	7.188	(0.281)	7.255	(0.274)
C_{11} Charge statement	0.208	0.056	8.285	(0.171)	7.889	(0.211)	8.115	(0.188)
Postpurchase behavior (D_5)	**0.202**		**7.544**	**(0.246)**	**7.546**	**(0.245)**	**7.791**	**(0.221)**
C_{12} Order tracking	0.292	0.059	7.605	(0.240)	7.484	(0.252)	7.626	(0.237)
C_{13} On-time delivery	0.322	0.065	7.610	(0.239)	7.582	(0.242)	7.746	(0.225)
C_{14} Met expectations	0.386	0.078	7.442	(0.256)	7.564	(0.244)	7.953	(0.205)
Total performance scores (ranking)	**1.000**	**1.000**	7.538 (2)	0.246 (2)	7.470 (3)	0.253 (3)	7.601 (1)	0.240 (1)

Source: Data from Chiu, W.Y., Tzeng, G.H., and Li, H.L., *Knowledge-Based Systems*, 37, 48–61, 2013.

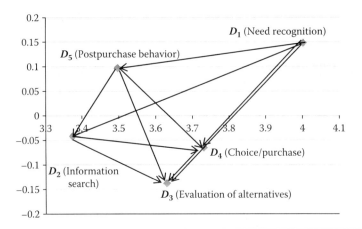

Figure 9.1 Internetwork relationship map (INRM) of the five dimensions.

In Figure 9.1, it is obvious that D_1 (*Need recognition*) has a direct influence on the other four dimensions, and also plays an essential role in consumers recognizing an e-store. Next, D_5 (*Postpurchase behavior*) has an influence on D_2 (*Information search*), D_4 (*Choice/purchase*), and D_3 (*Evaluations of alternatives*). In other words, once customers feel satisfied with their purchase experience, this yields positive postpurchase behavior, influencing the subsequent potential customers' perceptions on D_2, D_4 and D_3. From the other perspective, let's look at the bottom of Figure 9.1: D_3 (*Evaluations of alternatives*) is influenced by the other four dimensions. We may further delve into D_3 to explore the directional influences among the three criteria (i.e., C_5, C_6, and C_7) as an example. Referring to Table 9.5, the sub-INRM within dimension D_3 is illustrated in Figure 9.2.

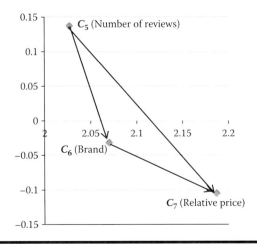

Figure 9.2 Internetwork relationship map (INRM) of the three criteria in D_3.

To facilitate the *Evaluations of alternatives* (D_3), e-stores can utilize former customers' experiences—namely, word of mouth (chat rooms and blogs) and the services offered—to improve the image of the *Brand* (C_6); they can display references to other e-stores or relative prices to improve the *Relative price* (C_7); and they can correct any unfavorable perceptions to improve the *Number of reviews* (C_5) (Figure 9.2).

The second results are the DANP influential weights, as seen in Table 9.6. *Clarity* (C_4) and *Number of reviews* (C_5) were the two highest-weighted criteria, assessed at 0.084 and 0.080, respectively. This result indicates that the *Clarity* (C_4) of the product information is the highest influential factor for an e-store to attract online customers, and a high *Number of reviews* (C_5) will cause a positive feedback loop to strengthen the e-store's competitiveness. Thus, it is extremely important to deliver clear and correct information to customers on an e-store website because this parameter can contribute to positive reviews. The lowest weighted criterion was *Charge statement* (C_{11}). This result might owe to the fact that most e-stores have clearly displayed prices, quantities, shipping fees, and total fee information; therefore, the *Charge statement* (C_{11}) is regarded as a minor factor for the evaluation of e-stores.

Third, referring to Table 9.6, the synthesized final performance of each e-store can be used for the ranking or selection of alternatives. Both the SAW and modified VIKOR methods suggest the same performance ranking: $A_3 > A_1 > A_2$. The underperforming e-stores can plan for improvements using the weighted performance gaps from the modified VIKOR method. Take Yahoo (A_1), for example: its weighted performance gaps on the five dimensions are D_1 (0.279) > D_2 (0.266) > D_5 (0.246) > D_3 (0.242) > D_4 (0.224). This ranking of performance gap on each dimension can be used as the improvement priority for the management team of A_1 to refer to. If A_1 plans to improve the top-two underperforming dimensions, it should expand its *Product availability* (C_1) to satisfy its potential customers; also, it should enhance its *Information search* (D_2) to improve the web-browsing or -searching experience of its customers.

Referring to the DEMATEL analysis in Table 9.5 and the INRM in Figure 9.1, the source dimensions that have direct influences on D_2 are D_1 (*Product availability*) and D_5 (*Postpurchase behavior*). As a result, a marketing strategy that will generate positive postpurchase behavior in customers and improve the *Information search* (D_2) will be required. Furthermore, if A_1 examines the weighted performance gaps of criteria in D_2, the top-priority gap emerges: C_3 (*Overall look/design*) = 0.287. If A_1 only has a limited budget for improving the *Information search* (D_2), then the overall look or design should be its top priority in this dimension. The example of Yahoo (A_1) shows how e-stores can adopt the hybrid MADM model, the DANP weighting system, and the modified VIKOR aggregator to conduct systematic evaluations and improvement planning by using the opinions of domain experts and consumers.

9.5 Conclusion

An in-depth performance evaluation for e-stores is provided in this chapter, and the findings can be used by e-store managers to strengthen their weaknesses in meeting customers' expectations. The competition among e-stores is dynamic and fast changing; since customers can switch between online retailers easily, e-stores should observe and evaluate their relative competitiveness in this industry. The proposed hybrid MADM model not only identifies the relative importance of each dimension/criterion to evaluating performance but also offers a combined tool—DANP and modified VIKOR—for e-stores to assess their relatively under-performing dimensions/criteria in a dynamic business environment.

The analytical results in Table 9.6 show that the improvement priorities in the respective dimensions are not the same for the three e-stores.

- Yahoo (A_1): Need recognition (D_1) > Information search (D_2) > Postpurchase behavior (D_5) > Evaluation of alternatives (D_3) > Choice/purchase (D_4)
- PChome (A_2): Need recognition (D_1) > Information search (D_2) > Evaluation of alternatives (D_3) > Choice/purchase (D_4) > Postpurchase behavior (D_5)
- Books (A_3): Need recognition (D_1) > Evaluation of alternatives (D_3) > Information search (D_2) > Choice/purchase (D_4) > Postpurchase behavior (D_5)

The aforementioned results imply that the managers of e-stores can use this approach to understand their customers' perceptions and to devise marketing/improvement plans accordingly. After executing those plans for a period of time, e-stores should go back to the performance evaluation again to examine the effects of those improvement. This process is in line with the prevailing management mechanism: plan, do, check, and act (PDCA). In other words, this hybrid approach acts as a dynamic evaluation system to support the *plan* and *check* steps for e-stores, and is capable of solving their practical performance evaluation problems.

Chapter 10

Improving the Performance of Green Suppliers in the TFT-LCD Industry

Numerous 3C (computers/communications/consumer-electronics) appliances, ranging from PC monitors to tablet devices, have been developed and manufactured to meet consumer needs and demand for information in the Internet era. One of the key components of these 3C appliances, the display panel, is expected to grow with strong momentum in the coming years (IEK 2014), and likewise the optoelectronics industry. In Taiwan, the thin film transistor liquid crystal display (TFT-LCD) industry had the highest output value in the optoelectronics industry in 2012, which reached approximately \$31.48 billion in 2014, and it was expected to keep on growing in the coming three years; nevertheless, TFT-LCD companies in Taiwan are also suffering intensive competition from their peers in Korea, Japan, and China.

Aside from competing on cost or quality, a rising trend in environmental concern has caused TFT-LCD manufacturers to invest in green/sustainable supply chain management to meet this global trend. In addition, end users' awareness of environmental protection also pushes these TFT-LCD companies to plan for greener productions. However, green/sustainable supply chain management requires a systematic and integrated approach for companies to maintain their sustainability and competitiveness; this complex but important issue involves multiple interrelated criteria, which is suitable to be analyzed using the multiple attribute decision making (MADM) framework.

134 ■ *Trends of Hybrid Multiple Criteria Decision Making*

Supplier selection is a critical issue (or problem) in green/sustainable supply chain management, which often requires multiple departments to participate in the evaluations. Different perspectives from these departments might yield conflicting opinions on selecting qualified green suppliers for a company. In addition, once suppliers have been included in its supply chain, a TFT-LCD company will often need to improve its suppliers to strengthen the whole supply chain. In this chapter, a case involving green supplier selection/improvement in a TFT-LCD company is illustrated, and we hope to show how to devise a systematic improvement plan based on the proposed hybrid MADM approach.

10.1 Background of the Case

This case study is a joint undertaking with several departments of a TFT-LCD company, named TL Company here for confidentiality. TL Company is a well-known and public-listed optoelectronics company that is among the top-five TFT-LCD manufactures in the world in terms of market share. The key components of a TFT-LCD panel are the *color filter*, the *backlight unit*, the *glass substrate*, and the *polarizer*. The cost of the key components represents more than 60% of the total cost of manufacturing a single panel, and TL Company is preparing to expand its capacity on polarizers by establishing new subsidiaries, acquisitioning, or setting up strategic alliances. Most of the well-known polarizer manufacturers are in Asia, mainly Korea, Japan, and Taiwan; for TL Company, the major portion of its materials and technical support on polarizers is from two Japanese polarizer manufacturers. To compete with its biggest rival (a company from Korea), TL Company is planning to extend its supply chain by evaluating four potential polarizer suppliers (T_1, T_2, J_1, and J_2), two from Taiwan and two from Japan. In addition, T_2 is a subsidiary of TL Company; therefore, TL Company also intends to improve the performance of T_2 to enhance its green supply chain.

Currently, all of TL Company's suppliers have to pass a two-stage assessment. At the first stage, a preliminary survey is jointly conducted by the supplier quality management (SQM), environment, safety, and health (ESH), and human resources (HR) departments. In this preliminary survey, there is a long list of audit items, and candidates have to satisfy at least 70% of the requirements to pass it. At the second stage, a formal review and performance evaluation are conducted by the SQM, material management (MM), research and development (R&D), and green product management (GPM) departments. It is obvious that the performance of a candidate in the environmental aspect is addressed at both stages. The aforementioned evaluation is a routine check on the addressed aspects for all qualified suppliers, and needs to be conducted on quarterly and yearly bases. In other words, the ranking of a qualified supplier might change based on the result of its latest assessment.

10.2 Research Framework and the Selected Criteria

Traditional supplier selection and environmental evaluation have been examined intensively, but only a few practical methods have been developed for combining supplier selection and environment evaluation in the context of green/sustainable supply chain management (Ho et al. 2010, Büyüközkan and Çifçi 2011). To solve a practical green supplier selection problem, one must inevitably consider multiple aspects/dimensions and criteria when assessing the performance of a candidate; therefore, the MADM model is a promising approach. For example, Lee et al. (2009) applied a fuzzy extended analytic hierarchy process to the evaluation of green suppliers in the high-tech industry. Kuo et al. (2010) integrated artificial neural networks and an analytic network process (ANP) with data envelopment analysis for green supplier selection. The research group of Büyüközkan proposed a hybrid fuzzy MCDM approach to evaluate green/sustainable suppliers (Büyüközkan and Çifçi 2011; Büyüközkan and Çifçi 2012). The aforementioned studies were all concerned with developing an appropriate decision support tool for green/sustainable supplier selection. Nevertheless, in the case of TL Company, it also intends to improve its subsidiary T_2. How to conduct systematic improvement planning is still an underexplored research topic in green/sustainable supply chain management.

The MADM approach aims to support DMs in the presence of multiple conflicting attributes/criteria and alternatives to make an optimal decision. Using the proposed hybrid MADM framework in Chapter 5 (Figure 5.1), the DEMATEL-based ANP (DANP) method was used to obtain the influential weight of each criterion, and this study adopted the preference ranking organization method for enrichment evaluation (PROMETHEE) to replace the VIKOR method as an aggregator for synthesizing the final performance scores of alternatives. PROMETHEE is a well-known outranking method, and PROMETHEE I and II were developed by Brans and presented for the first time in 1982 (Figueira et al. 2005). PROMETHEE I can provide a partial ranking of alternatives, and the complete ranking can be derived from PROMETHEE II. Therefore, this case study adopted PROMETHEE II in the hybrid MADM model.

Although the aforementioned two-stage assessment comprises a preliminary survey and an evaluation, TL Company pays much more attention to the second stage, to rank and select the qualified suppliers; thus, this case study was devised to support TL Company in evaluating the four potential polarizer manufacturers during the second-stage assessment. More than 10 experts from the GPM, MM, RD, and SQM departments were interviewed for this case. Referring to the opinions from those experts and the audit items of the second-stage assessment, six dimensions/aspects of the proposed decision framework are suggested: (1) *Environment* (ESH department), (2) *Purchasing* (MM department), (3) *Quality* (SQM department), (4) *Green management* (GPM department), (5) *Technology* (R&D department), and (6) *Organization* (SQM department). To select suitable criteria for each dimension, certain recent studies about green/sustainable supplier evaluation were reviewed (Awasthi et al. 2010; Kuo et al. 2010; Büyüközkan and Çifçi 2011; Amindoust et al. 2012; Tseng and Chiu

136 ■ *Trends of Hybrid Multiple Criteria Decision Making*

Table 10.1 Green Suppliers' Evaluation Criteria and Dimensions

Dimensions	Criteria	Ranges of Scores
Environment (D_1)	C_1: Health and safety	0 ~ 24
	C_2: Environmental control	0 ~ 10
	C_3: Environmental management system	0 ~ 23
Purchasing (D_2)	C_4: Delivery	0 ~ 17
	C_5: Service	0 ~ 7
	C_6: Price	0 ~ 14
Quality (D_3)	C_7: Quality system	0 ~ 305
	C_8: Quality control	0 ~ 229
	C_9: Out-of-control management	0 ~ 72
Green management (D_4)	C_{10}: Green product	0 ~ 27
	C_{11}: Recycle	0 ~ 5
Technology (D_5)	C_{12}: Current capability	0 ~ 9
	C_{13}: R&D capability	0 ~ 45
Organization (D_6)	C_{14}: Compatibility across levels	0 ~ 7
	C_{15}: Information share	0 ~ 10

Source: Data from Tsui, C.W., Tzeng, G.H., and Wen, U.P., *International Journal of Production Research*, 53(21), 6436–6454, 2015.

2013); the criteria for each dimension were decided by referring to this research and the existing audit items. Finally, 15 criteria in five dimensions were adopted to form a hybrid MADM model for TL Company, as shown in Table 10.1.

The hybrid MADM model requires two types of inputs from the experts: (1) the relative influence of one criterion over the other criteria and (2) the performance score of each alternative on each criterion. Both were collected by questionnaires. The performance data of the four candidates (i.e., J_1, J_2, T_1, and T_2) from 2012 to 2013 were evaluated by managers from four different departments. As shown in Table 10.1, the ranges of scores are not the same for those criteria; normalization was conducted to transform the scores (on those criteria) of the four alternatives into the interval [0, 1].

10.3 DANP for Finding Influential Weights of Criteria

After forming the decision framework (Table 10.1) of this research, the next step requires determining the influential weights of the criteria using the DANP method. The initial average matrix A was obtained by averaging the opinions of the experts, collected by questionnaires, regarding the influence of a criterion i on another criterion j; the influence ranges from 0 (no influence) to 4 (extreme influence), and the result of A is shown in Table 10.2.

Referring to Equation 3.2 (Chapter 3), $\alpha = 36.75$ in this initial average matrix; therefore, the normalized initial direct-relation matrix D can be obtained by dividing A by α (i.e., $D = A/\alpha$). The result of normalization is shown in Table 10.3.

According to Equation 3.3 (Chapter 3), the total relation matrix T was obtained by multiplying D with the inverse of $(I - D)$; in addition, after reorganizing the total relation matrix T based on the six defined dimensions (Table 10.1), T_C^G indicates the grouping among dimensions, as shown in Table 10.4.

Referring to Equation 4.3 (Chapter 4), the dimensional matrix T^D (Table 10.5) can be obtained by averaging all the elements within each cell in Table 10.4, and the corresponding r_i^D and c_i^D (for $i = 1,\ldots,6$) are also shown in Table 10.5.

With T_C^G and T^D, the cause–effect analyses of the dimensions and criteria can be calculated and summarized, as shown in Table 10.6.

After normalizing T^D (refer to Equations 4.7 through 4.9 in Chapter 4), T_N^D can be obtained, as shown in Table 10.7.

Since the degrees of importance and the net influential degree for each criterion and dimension are known (Table 10.6), the cause–effect analysis can thus be illustrated as an influential network relations map (INRM), as shown in Figure 10.1. The cause–effect analysis among the six dimensions shows that D_1 (*Environment*), D_3 (*Quality*), and D_4 (*Green management*) are the sources $\left(r_i^D - c_i^D > 0\right)$ that will influence the other three dimensions (i.e., D_2 [*Purchasing*], D_5 [*Technology*], and D_6 [*Organization*]).

In Figure 10.1, for simplicity, only the directional influences among the six dimensions are shown; the directional influences among the criteria that are within the same dimensions can be learned from Table 10.6. This intuitive guidance can be combined with the findings from the DANP influential weights and the measured performance gaps as a managerial tool for improvement planning.

To proceed with the calculations for DANP influential weights, the total relation matrix T_C^G is normalized to become T_N^G (Table 10.8), referring to Equations 4.2 through 4.4 (Chapter 4). After transposing T_N^G to become the unweighted supermatrix W (i.e., $W = (T_N^G)'$), the normalized dimensional matrix T_N^D can be multiplied by W to obtain the weighted supermatrix W^α $\left(W^\alpha = T_N^D \times W\right)$, as shown in Table 10.9.

By raising z consecutively for $\lim_{z \to \infty}(W^\alpha)^z$ by multiplying W^α with itself several times, a set of stable DANP influential weights can be obtained, shown with the performance evaluations (on the 15 criteria) of the four suppliers in Table 10.10.

In Table 10.10, the worst and the aspiration levels are the qualified scores and full marks of each supplier audit form, respectively. The qualified scores for the four suppliers were set by the managers of the involved departments in TL Company.

10.4 Modified PROMETHEE

The modified PROMETHEE is adopted in this hybrid MADM model, which needs additional explanation here. First, the outranking preference function should

Table 10.2 Initial Average Matrix A for Green Supplier Selection

	C_1	C_2	C_3	C_4	C_5	C_6	C_7	C_8	C_9	C_{10}	C_{11}	C_{12}	C_{13}	C_{14}	C_{15}
C_1	0.000	3.100	2.750	1.600	1.900	2.000	1.300	1.800	1.800	3.050	2.200	2.950	2.300	2.650	1.650
C_2	3.650	0.000	3.650	1.000	1.300	2.200	1.400	2.100	2.300	3.650	3.300	1.500	1.650	2.300	1.500
C_3	3.500	3.650	0.000	1.400	1.750	2.350	2.200	1.950	1.700	2.350	1.850	2.100	1.950	2.050	1.950
C_4	0.700	0.750	0.650	0.000	2.900	3.100	2.150	2.750	2.900	2.150	2.500	2.900	3.200	2.800	2.500
C_5	0.800	1.100	0.900	3.650	0.000	2.550	3.100	3.300	3.200	2.350	2.000	3.250	3.200	2.900	2.400
C_6	1.700	1.600	1.500	2.500	2.850	0.000	2.750	2.900	2.550	2.450	2.050	2.350	2.300	1.650	1.450
C_7	1.400	1.550	1.700	3.500	2.750	3.000	0.000	3.000	3.850	2.400	2.650	3.050	2.850	2.850	1.800
C_8	1.600	1.800	2.150	3.500	2.300	3.050	3.750	0.000	3.750	2.650	1.750	2.800	2.150	2.100	1.150
C_9	1.700	2.000	2.150	3.600	2.400	3.150	3.100	3.600	0.000	2.300	1.400	2.550	2.400	2.850	2.150
C_{10}	3.200	3.350	3.450	1.650	1.900	1.950	2.700	1.800	1.950	0.000	3.050	2.850	2.900	2.850	2.700
C_{11}	2.800	3.500	2.850	2.200	1.600	2.750	2.150	1.200	1.850	3.600	0.000	2.450	2.400	2.950	2.550
C_{12}	2.050	1.350	1.200	3.300	2.650	3.400	2.900	3.300	2.450	2.650	2.000	0.000	2.950	2.900	2.650
C_{13}	1.500	1.450	1.400	2.350	2.200	3.400	2.250	2.450	1.850	2.700	2.800	3.500	0.000	2.150	3.050
C_{14}	1.900	1.350	1.350	2.950	2.200	2.050	2.050	2.050	1.900	2.000	2.550	2.550	2.150	0.000	3.000
C_{15}	0.350	0.650	0.600	1.900	1.550	1.350	1.100	1.500	1.200	1.700	1.750	1.950	2.650	2.100	0.000

Source: Data from Tsui, C.W., Tzeng, G.H., and Wen, U.P., *International Journal of Production Research*, 53(21), 6436–6454, 2015.

Table 10.3 Normalized Initial Direct-Relation Matrix D

	C_1	C_2	C_3	C_4	C_5	C_6	C_7	C_8	C_9	C_{10}	C_{11}	C_{12}	C_{13}	C_{14}	C_{15}
C_1	0.000	0.084	0.075	0.044	0.052	0.054	0.035	0.049	0.049	0.083	0.060	0.080	0.063	0.072	0.045
C_2	0.099	0.000	0.099	0.027	0.035	0.060	0.038	0.057	0.063	0.099	0.090	0.041	0.045	0.063	0.041
C_3	0.095	0.099	0.000	0.038	0.048	0.064	0.060	0.053	0.046	0.064	0.050	0.057	0.053	0.056	0.053
C_4	0.019	0.020	0.018	0.000	0.079	0.084	0.059	0.075	0.079	0.059	0.068	0.079	0.087	0.076	0.068
C_5	0.022	0.030	0.024	0.099	0.000	0.069	0.084	0.090	0.087	0.064	0.054	0.088	0.087	0.079	0.065
C_6	0.046	0.044	0.041	0.068	0.078	0.000	0.075	0.079	0.069	0.067	0.056	0.064	0.063	0.045	0.039
C_7	0.038	0.042	0.046	0.095	0.075	0.082	0.000	0.082	0.105	0.065	0.072	0.083	0.078	0.078	0.049
C_8	0.044	0.049	0.059	0.095	0.063	0.083	0.102	0.000	0.102	0.072	0.048	0.076	0.059	0.057	0.031
C_9	0.046	0.054	0.059	0.098	0.065	0.086	0.084	0.098	0.000	0.063	0.038	0.069	0.065	0.078	0.059
C_{10}	0.087	0.091	0.094	0.045	0.052	0.053	0.073	0.049	0.053	0.000	0.083	0.078	0.079	0.078	0.073
C_{11}	0.076	0.095	0.078	0.060	0.044	0.075	0.059	0.033	0.050	0.098	0.000	0.067	0.065	0.080	0.069
C_{12}	0.056	0.037	0.033	0.090	0.072	0.093	0.079	0.090	0.067	0.072	0.054	0.000	0.080	0.079	0.072
C_{13}	0.041	0.039	0.038	0.064	0.060	0.093	0.061	0.067	0.050	0.073	0.076	0.095	0.000	0.059	0.083
C_{14}	0.052	0.037	0.037	0.080	0.060	0.056	0.056	0.056	0.052	0.054	0.069	0.069	0.059	0.000	0.082
C_{15}	0.010	0.018	0.016	0.052	0.042	0.037	0.030	0.041	0.033	0.046	0.048	0.053	0.072	0.057	0.000

Source: Data from Tsui, C.W., Tzeng, G.H., and Wen, U.P., *International Journal of Production Research*, 53(21), 6436–6454, 2015.

Table 10.4　Total Relation Matrix T (Termed T_C^G after Grouping Each Dimension)

| | | D_1 | | | D_2 | | | D_3 | | | D_4 | | D_5 | | D_6 | |
		C_1	C_2	C_3	C_4	C_5	C_6	C_7	C_8	C_9	C_{10}	C_{11}	C_{12}	C_{13}	C_{14}	C_{15}
D_1	C_1	0.369	0.449	0.431	0.526	0.470	0.549	0.491	0.511	0.503	0.565	0.494	0.576	0.539	0.546	0.467
	C_2	0.469	0.382	0.461	0.516	0.460	0.560	0.499	0.522	0.520	0.587	0.526	0.548	0.529	0.544	0.468
	C_3	0.451	0.458	0.357	0.514	0.460	0.550	0.505	0.508	0.495	0.542	0.480	0.549	0.524	0.525	0.466
D_2	C_4	0.391	0.396	0.383	0.508	0.511	0.595	0.530	0.553	0.548	0.555	0.513	0.594	0.579	0.565	0.502
	C_5	0.425	0.436	0.420	0.642	0.475	0.626	0.593	0.607	0.597	0.602	0.539	0.647	0.621	0.610	0.536
	C_6	0.407	0.409	0.397	0.554	0.496	0.501	0.530	0.541	0.527	0.549	0.489	0.565	0.542	0.523	0.462
D_3	C_7	0.461	0.469	0.461	0.662	0.567	0.662	0.539	0.624	0.635	0.630	0.578	0.667	0.637	0.634	0.544
	C_8	0.450	0.458	0.456	0.638	0.536	0.639	0.610	0.527	0.612	0.612	0.536	0.637	0.597	0.593	0.506
	C_9	0.454	0.465	0.458	0.645	0.542	0.645	0.598	0.620	0.522	0.608	0.531	0.635	0.607	0.614	0.534
D_4	C_{10}	0.504	0.511	0.501	0.600	0.533	0.622	0.591	0.580	0.575	0.561	0.580	0.649	0.625	0.623	0.555
	C_{11}	0.478	0.498	0.471	0.590	0.507	0.618	0.557	0.544	0.551	0.628	0.484	0.617	0.592	0.603	0.533
D_5	C_{12}	0.465	0.452	0.437	0.642	0.552	0.656	0.597	0.616	0.588	0.621	0.550	0.576	0.626	0.620	0.551
	C_{13}	0.423	0.425	0.413	0.576	0.504	0.613	0.542	0.555	0.533	0.582	0.532	0.620	0.510	0.561	0.525
D_6	C_{14}	0.397	0.387	0.377	0.544	0.463	0.532	0.492	0.500	0.490	0.518	0.484	0.549	0.520	0.461	0.484
	C_{15}	0.253	0.262	0.254	0.377	0.324	0.371	0.337	0.352	0.339	0.369	0.338	0.389	0.391	0.375	0.286

Source: Data from Tsui, C.W., Tzeng, G.H., and Wen, U.P., *International Journal of Production Research*, 53(21), 6436–6454, 2015.

Improving the Performance of Green Suppliers ■ 141

Table 10.5   Total Dimensional Matrix T^D

	D_1	D_2	D_3	D_4	D_5	D_6	r_i^D
D_1(Environment)	0.425	0.512	0.506	0.532	0.544	0.503	3.022
D_2(Purchasing)	0.407	0.545	0.558	0.541	0.591	0.533	3.175
D_3(Quality)	0.459	0.615	0.587	0.582	0.630	0.571	3.444
D_4(Green management)	0.494	0.578	0.566	0.563	0.621	0.579	3.401
D_5(Technology)	0.436	0.590	0.572	0.571	0.583	0.564	3.316
D_6(Organization)	0.322	0.435	0.418	0.428	0.462	0.402	2.467
c_i^D	2.543	3.275	3.207	3.217	3.431	3.152	

Source: Data from Tsui, C.W., Tzeng, G.H., and Wen, U.P., *International Journal of Production Research*, 53(21), 6436–6454, 2015.

be defined. If alternative q outranks alternative k for each criterion j, it can be calculated by the following equation:

$$p_j(a_q, a_k) = \begin{cases} (x_{qj} - x_{kj}) / (x_j^* - x_j^-), & \text{if } x_{qj} - x_{kj} > 0, \ j \in \{1, 2, ..., n\} \\ 0, & \text{otherwise}, \quad q, k \in \{1, 2, ..., K\} \end{cases}. \quad (10.1)$$

In the traditional PROMETHEE, $x_j^* = \max\{x_{kj} \mid k = 1, 2, ..., K\}$ and $x_j^- = \min_k\{x_{kj} \mid k = 1, 2, ..., K\}$ (for $j = 1, 2, ..., n$), representing the best and worst levels (also known as the *positive ideal point* and *negative ideal point*, respectively, in the traditional approach). In the modified PROMETHEE, $x_j^{aspired}$ is set as the aspiration level of each criterion and x_j^{worst} as the worst (least qualified) level of a criterion. The aforementioned performance scores of each alternative are listed with the aspiration and worst levels of each criterion in Table 10.10.

By setting the aspiration level for each criterion, the kth alternative's performance gap in each criterion can be calculated by the following equation: $\text{gap}_{kj} = \left[(x_j^{aspired} - x_{kj}) / (x_j^{aspired} - x_j^{worst}) \right] \times 100\%$, for $j \in \{1, 2, ..., n\}$. Then, the outranking integration index can be obtained by summing up the DANP influential weights $(w_1, ..., w_j, ..., w_n)$ with the performance gaps for each alternative, referring to Equation 10.2:

$$\pi(a_q, a_k) = \sum_{j=1}^{n} w_j P_j(a_q, a_k). \quad (10.2)$$

Table 10.6 Cause–Effect Analysis using the DEMATEL Technique

Dimensions	t_i^D	c_i^D	$t_i^D + c_i^D$	$t_i^D - c_i^D$	Criteria	r_i	c_i	r_i+c_i	r_i-c_i
D_1 (Environment)	3.022	2.543	5.565	0.480	C_1	1.249	1.289	2.539	−0.040
					C_2	1.312	1.289	2.601	0.023
					C_3	1.266	1.249	2.515	0.017
D_2 (Purchasing)	3.176	3.276	6.453	−0.100	C_4	1.613	1.704	3.317	−0.090
					C_5	1.743	1.483	3.226	0.261
					C_6	1.551	1.721	3.272	−0.170
D_3 (Quality)	3.445	3.208	6.653	0.237	C_7	1.798	1.746	3.544	0.052
					C_8	1.749	1.770	3.519	−0.022
					C_9	1.740	1.770	3.509	−0.030
D_4 (Green management)	3.401	3.218	6.691	0.183	C_{10}	1.141	1.189	2.330	−0.048
					C_{11}	1.113	1.064	2.177	0.048
D_5 (Technology)	3.317	3.432	6.749	−0.116	C_{12}	1.202	1.196	2.398	0.006
					C_{13}	1.130	1.136	2.266	−0.006
D_6 (Organization)	2.467	3.151	5.618	−0.684	C_{14}	0.945	0.836	1.781	0.109
					C_{15}	0.661	0.770	1.431	−0.109

Source: Data from Tsui, C.W., Tzeng, G.H., and Wen, U.P., *International Journal of Production Research*, 53(21), 6436–6454, 2015.

Table 10.7 Normalized Dimensional Matrix T_N^D

	D_1	D_2	D_3	D_4	D_5	D_6
D_1 (Environment)	0.141	0.169	0.167	0.176	0.180	0.166
D_2 (Purchasing)	0.128	0.172	0.176	0.170	0.186	0.168
D_3 (Quality)	0.133	0.179	0.170	0.169	0.183	0.166
D_4 (Green management)	0.145	0.170	0.167	0.166	0.183	0.170
D_5 (Technology)	0.131	0.178	0.172	0.172	0.176	0.170
D_6 (Organization)	0.130	0.177	0.170	0.173	0.187	0.163

Source: Data from Tsui, C.W., Tzeng, G.H., and Wen, U.P., *International Journal of Production Research*, 53(21), 6436–6454, 2015.

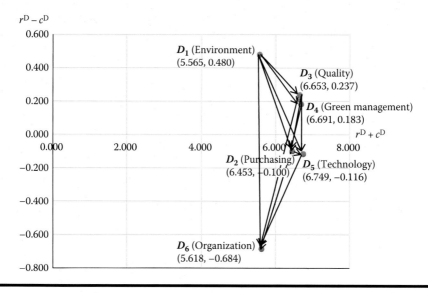

Figure 10.1 Influential network relationships map of the five dimensions.

The following step involves calculating the leaving and entering flows. The leaving flow indicates the degree to which alternative q outranks the other alternatives, and the entering flow can be defined as the degree to which other alternatives outrank alternative q. The leaving (ϕ^+) and entering (ϕ^-) flows are defined in Equations 10.3 and 10.4:

$$\phi^+(a_q) = \sum_{k=1}^{m} \pi(a_q, a_k), \quad q \neq k. \quad (10.3)$$

Table 10.8 Normalized Total Relation Matrix T_N^G

| | | D_1 | | | D_2 | | | D_3 | | | D_4 | | D_5 | | D_6 | |
		C_1	C_2	C_3	C_4	C_5	C_6	C_7	C_8	C_9	C_{10}	C_{11}	C_{12}	C_{13}	C_{14}	C_{15}
D_1	C_1	0.296	0.360	0.345	0.340	0.304	0.355	0.326	0.340	0.334	0.533	0.467	0.517	0.483	0.539	0.461
	C_2	0.357	0.291	0.352	0.336	0.299	0.364	0.324	0.339	0.338	0.527	0.473	0.509	0.491	0.538	0.462
	C_3	0.357	0.361	0.282	0.337	0.302	0.361	0.335	0.337	0.328	0.530	0.470	0.512	0.488	0.530	0.470
D_2	C_4	0.334	0.338	0.328	0.315	0.317	0.369	0.325	0.339	0.336	0.520	0.480	0.506	0.494	0.529	0.471
	C_5	0.332	0.340	0.328	0.368	0.273	0.359	0.330	0.338	0.332	0.527	0.473	0.510	0.490	0.532	0.468
	C_6	0.336	0.337	0.327	0.357	0.320	0.323	0.332	0.339	0.330	0.529	0.471	0.510	0.490	0.531	0.469
D_3	C_7	0.332	0.337	0.331	0.350	0.300	0.350	0.300	0.347	0.353	0.522	0.478	0.511	0.489	0.538	0.462
	C_8	0.330	0.336	0.334	0.352	0.295	0.353	0.349	0.301	0.350	0.533	0.467	0.516	0.484	0.539	0.461
	C_9	0.330	0.338	0.332	0.352	0.296	0.352	0.344	0.356	0.300	0.534	0.466	0.511	0.489	0.535	0.465
D_4	C_{10}	0.332	0.337	0.330	0.342	0.303	0.355	0.339	0.332	0.329	0.492	0.508	0.509	0.491	0.529	0.471
	C_{11}	0.331	0.344	0.325	0.344	0.295	0.361	0.337	0.329	0.334	0.565	0.435	0.510	0.490	0.531	0.469
D_5	C_{12}	0.343	0.334	0.323	0.347	0.298	0.355	0.331	0.342	0.327	0.530	0.470	0.479	0.521	0.529	0.471
	C_{13}	0.335	0.337	0.328	0.341	0.298	0.362	0.332	0.341	0.327	0.522	0.478	0.548	0.452	0.517	0.483
D_6	C_{14}	0.342	0.334	0.325	0.353	0.301	0.346	0.332	0.338	0.330	0.517	0.483	0.514	0.486	0.488	0.512
	C_{15}	0.329	0.341	0.330	0.352	0.302	0.346	0.328	0.342	0.330	0.522	0.478	0.498	0.502	0.567	0.433

Source: Data from Tsui, C.W., Tzeng, G.H., and Wen, U.P., *International Journal of Production Research*, 53(21), 6436–6454, 2015.

Table 10.9 Weighted Supermatrix W^α

	C_1	C_2	C_3	C_4	C_5	C_6	C_7	C_8	C_9	C_{10}	C_{11}	C_{12}	C_{13}	C_{14}	C_{15}
C_1	0.042	0.050	0.050	0.043	0.043	0.043	0.044	0.044	0.044	0.048	0.048	0.045	0.044	0.045	0.043
C_2	0.051	0.041	0.051	0.043	0.044	0.043	0.045	0.045	0.045	0.049	0.050	0.044	0.044	0.043	0.044
C_3	0.049	0.049	0.040	0.042	0.042	0.042	0.044	0.045	0.044	0.048	0.047	0.042	0.043	0.042	0.043
C_4	0.058	0.057	0.057	0.054	0.063	0.061	0.063	0.063	0.063	0.058	0.058	0.062	0.061	0.062	0.062
C_5	0.051	0.051	0.051	0.054	0.047	0.055	0.053	0.053	0.053	0.052	0.050	0.053	0.053	0.053	0.053
C_6	0.060	0.062	0.061	0.063	0.062	0.055	0.063	0.063	0.063	0.060	0.061	0.063	0.064	0.061	0.061
C_7	0.055	0.054	0.056	0.057	0.058	0.058	0.051	0.059	0.059	0.056	0.056	0.057	0.057	0.056	0.056
C_8	0.057	0.057	0.056	0.060	0.059	0.060	0.059	0.051	0.061	0.055	0.055	0.059	0.059	0.057	0.058
C_9	0.056	0.057	0.055	0.059	0.058	0.058	0.060	0.060	0.051	0.055	0.056	0.056	0.056	0.056	0.056
C_{10}	0.094	0.093	0.093	0.089	0.090	0.090	0.088	0.090	0.090	0.081	0.094	0.091	0.090	0.090	0.090
C_{11}	0.082	0.083	0.083	0.082	0.081	0.080	0.081	0.079	0.079	0.084	0.072	0.081	0.082	0.084	0.083
C_{12}	0.093	0.092	0.092	0.094	0.095	0.095	0.094	0.094	0.094	0.093	0.093	0.084	0.096	0.096	0.093
C_{13}	0.087	0.088	0.088	0.092	0.091	0.091	0.089	0.089	0.089	0.090	0.089	0.092	0.079	0.091	0.094
C_{14}	0.090	0.089	0.088	0.089	0.089	0.089	0.089	0.089	0.089	0.090	0.090	0.090	0.088	0.079	0.092
C_{15}	0.077	0.077	0.078	0.079	0.079	0.079	0.076	0.076	0.077	0.080	0.080	0.080	0.082	0.083	0.070

Source: Data from Tsui, C.W., Tzeng, G.H., and Wen, U.P., *International Journal of Production Research*, 53(21), 6436–6454, 2015.

146 ■ *Trends of Hybrid Multiple Criteria Decision Making*

Table 10.10 DANP Influential Weights and the Performance Scores of the Four Suppliers

Criteria	DANP Influential Weights	Suppliers				Aspiration Levels	Worst Levels
		T_1	T_2	J_1	J_2		
C_1	0.045	19.0	17.0	20.0	18.0	24.0	16.8
C_2	0.045	7.5	8.5	9.5	9.8	10.0	7.0
C_3	0.044	17.0	22.5	21.0	22.5	23.0	16.1
C_4	0.060	12.5	15.0	13.3	13.5	17.0	11.9
C_5	0.052	5.5	6.8	6.0	5.0	7.0	4.9
C_6	0.062	13.0	11.0	10.0	13.5	14.0	9.8
C_7	0.057	288.8	293.1	290.0	291.2	305.0	243.6
C_8	0.058	220.0	225.0	224.0	221.0	229.0	217.0
C_9	0.057	67.5	67.2	68.6	65.5	72.0	50.4
C_{10}	0.090	22.0	24.0	26.5	26.0	27.0	18.9
C_{11}	0.081	4.0	4.5	4.5	4.5	5.0	3.5
C_{12}	0.093	6.5	7.5	7.5	8.5	9.0	6.3
C_{13}	0.089	37.0	37.7	41.2	38.5	45.0	31.5
C_{14}	0.089	5.0	6.5	5.5	6.0	7.0	4.9
C_{15}	0.079	9.0	9.8	7.5	9.5	10.0	7.0

Source: Data from Tsui, C.W., Tzeng, G.H., and Wen, U.P., *International Journal of Production Research*, 53(21), 6436–6454, 2015.

$$\phi^-(a_q) = \sum_{k=1}^{m} \pi(a_k, a_q), \ q \neq k. \tag{10.4}$$

If alternative a_q outranks alternative a_k, it can be denoted as $a_q P a_k$. Generally speaking, $a_q P a_k$ will hold under two conditions: (1) alternative a_q has a greater leaving flow than that of alternative a_k, and (2) alternative a_q has a smaller entering flow than that of alternative a_k. This can be defined more precisely as follows:

Improving the Performance of Green Suppliers ■ **147**

Table 10.11 Ranking of the Four Suppliers Using the Hybrid DANP-PROMETHEE II Model

	T_1	T_2	J_1	J_2	Leaving Flow	Entering Flow	Net Flow	Rank
T_1	0	**0.043**	0.083	0.024	0.15	0.882	−0.732	4
T_2	**0.313**	0	0.174	0.117	0.604	0.256	0.348	2
J_1	0.268	0.088	0	0.083	0.439	0.435	0.005	3
J_2	0.302	0.125	0.177	0	0.603	0.224	0.379	1

Source: Data from Tsui, C.W., Tzeng, G.H., and Wen, U.P., *International Journal of Production Research*, 53(21), 6436–6454, 2015.

$$a_q P a_k \text{ iff } \begin{cases} \phi^+(a_q) > \phi^+(a_k) \text{ and } \phi^-(a_q) < \phi^-(a_k), \\ \phi^+(a_q) > \phi^+(a_k) \text{ and } \phi^-(a_q) = \phi^-(a_k), \\ \phi^+(a_q) = \phi^+(a_k) \text{ and } \phi^-(a_q) < \phi^-(a_k). \end{cases} \quad (10.5)$$

In addition, alternative a_q and alternative a_k are regarded as indifference if they have identical leaving and entering flows. With PROMETHEE II, after calculating the net flow of each alternative using Equation 10.6, the net flow score of each alternative can be obtained, which indicates the overall degree to which alternative q outranks the other alternatives.

$$\phi(a_q) = \phi^+(a_q) - \phi^-(a_q). \quad (10.6)$$

The modified PROMETHEE II adopts the full mark on each criterion as the aspiration level, which uses the DANP influential weights to aggregate the net flow for each alternative. The results for the ranking of each alternative and the associated leaving/entering flows are summarized in Table 10.11.

The outranking integration indices of any two suppliers are also shown in Table 10.11; for example, the outranking integration index of supplier T_1 to supplier T_2 is 0.043, and the outranking integration index of supplier T_2 to supplier T_1 is 0.313. Supplier T_2 is obviously superior to supplier T_1, and the same result can be concluded by comparing the net flow scores of those two alternatives.

10.5 Discussion

The new hybrid MADM, combining DEMATEL, DANP, and modified PROMETHEE II, suggests the following ranking: $J_2 > T_2 > J_1 > T_1$. Compared

148 ■ *Trends of Hybrid Multiple Criteria Decision Making*

Table 10.12 Four Suppliers' Weighted Performance Gaps in the Dimensions and Criteria

	T_1	T_2	J_1	J_2
Environment (D_1)	**78.95%**	**52.63%**	**38.01%**	**39.18%**
Health and safety (C_1)	69.44%	97.22%	55.56%	83.33%
Environmental control (C_2)	83.33%	50.00%	16.67%	6.67%
Environmental management system (C_3)	86.96%	7.25%	28.99%	7.25%
Purchasing (D_2)	**61.40%**	**45.61%**	**76.32%**	**52.63%**
Delivery (C_4)	88.24%	39.22%	72.55%	68.63%
Service (C_5)	71.43%	9.52%	47.62%	95.24%
Price (C_6)	23.81%	71.43%	95.24%	11.90%
Quality (D_3)	**31.32%**	**21.79%**	**24.63%**	**29.79%**
Quality system (C_7)	26.47%	19.38%	24.43%	22.48%
Quality control (C_8)	75.00%	33.33%	41.67%	66.67%
Out-of-control management (C_9)	20.83%	22.22%	15.74%	30.09%
Green management (D_4)	**62.50%**	**36.46%**	**10.42%**	**15.63%**
Green product (C_{10})	61.73%	37.04%	6.17%	12.35%
Recycle (C_{11})	66.67%	33.33%	33.33%	33.33%
Technology (D_5)	**64.81%**	**54.32%**	**32.72%**	**43.21%**
Current capability (C_{12})	92.59%	55.56%	55.56%	18.52%
R&D capability (C_{13})	59.26%	54.07%	28.15%	48.15%
Organization (D_6)	**58.82%**	**13.73%**	**78.43%**	**29.41%**
Compatibility across levels (C_{14})	95.24%	23.81%	71.43%	47.62%
Information share (C_{15})	33.33%	6.67%	83.33%	16.67%

Source: Data from Tsui, C.W., Tzeng, G.H., and Wen, U.P., *International Journal of Production Research*, 53(21), 6436–6454, 2015.

with the traditional MADM methods, this hybrid approach has additional implications for management; it can be adopted for improvement planning. The four alternatives' weighted performance gaps in each dimension/criterion are shown in Table 10.12, which indicates the gap between the aspiration level and the current performance level of each alternative on each criterion.

Take T_2 (a subsidiary of TL Company), for example. TL Company intends to improve T_2's competitiveness considering its limited resources. In this case, T_2 can rank its weighted performance gap in each dimension to form an improvement priority: $D_5 > D_1 > D_2 > D_4 > D_3 > D_6$. It can be observed from Table 10.12 that the gap between supplier T_2 and the aspiration level for *Technology* (D_5) is 54.32%, followed by 52.63% for *Environment* (D_1); therefore, the subsidiary supplier should regard its technological and environmental dimensions as the top-two priorities for improvement. If T_2 plans to focus on improving its top-priority *Technology* (D_5), it can learn that D_1, D_3, D_4, and D_2 all have influences on D_5, as shown in Figure 10.1. In addition, considering its performance gaps in all the criteria, its performance gap in *Health and safety* (C_1) is the highest (97.22%), which requires an immediate action plan to improve the working environment.

10.6 Conclusion

A case of polarizer suppliers' evaluations is discussed in this chapter, using the existing audit items from a well-known TFT-LCD manufacturer. Although the company's current two-stage evaluations can be used to select qualified suppliers, a detailed analysis of the weighting of each dimension/criterion in green supplier evaluation is still needed to rank those suppliers. Furthermore, the measured performance gaps and aspiration levels can guide those suppliers to plan for improvements. In a competitive and fast-changing business environment, all of the suppliers in a green supply chain are required to improve their underperforming dimensions/criteria consecutively; the proposed hybrid MADM approach provides a suitable managerial tool to support this challenging task.

Chapter 11

Exploring Smartphone Improvements Based on a Hybrid MADM Model

The smartphone is regarded as a personal information hub in the Internet era, integrating multiple functionalities, such as instant communications, web browsing, photographing, e-mail access, online shopping, and audio and video entertainment, all in a single portable device. Smartphones also collect a variety of personal information, from health statistics (e.g., heart rate and daily exercise records) to geographic information system (GIS) data, regarding the daily behaviors of the user; this personal information facilitates the development of various customized content and online services.

Based on a recent Gartner survey, the total sales of smartphones in the worldwide market were 1.24 and 1.42 billion units in 2014 and 2015, respectively, which is still moving upward in the coming years. Furthermore, the top-five smartphone vendors (Samsung, Apple, Huawei, Lenovo, and Xiaomi) had a 55.4% market share in 2015, and those vendors are pushed to keep on launching competitive products to meet consumers' needs and maintain or enlarge their market shares. In other words, how to get closer to smartphone users' expectations/needs is a critical task for vendors, which has enormous business value in practice.

11.1 Background of the Case

The smartphone penetration rate in Taiwan has surpassed 65% since 2014, which puts it in the top three in the Asia region; competition among smartphone vendors in Taiwan is very intense. Excepting the top-five brand names in the global market, certain domestic vendors also have significant market shares, such as the two well-known vendors HTC and ASUS. As a result, how to survive and thrive is a challenging task for smartphone vendors in Taiwan.

In fact, the technological advancements of semiconductors, wireless communication, multiprocessors, optical lenses, high-performance memory, and long-life batteries have all enhanced the usability of smartphones. Although the hardware's specs are important, smartphone venders also need to consider other aspects, such as marketing, to persuade consumers to select their products. Therefore, a framework based on a hybrid multiple attributes decision making (MADM) approach that considers multiple aspects is proposed to examine this complex issue. In this chapter, we examine knowledge retrieved from domain experts to evaluate five domestic vendors in Taiwan, to get closer to consumers' needs and their expectations of smartphone development.

11.2 Research Framework

Smartphones have become a personal system intimately associated with areas such as individual needs, commercial applications, and Internet access. Smartphone users in Taiwan rely on their phones to conduct various tasks; according to a survey from Statista in 2015, the top-five activities of Taiwan's smartphone users are (1) visiting social networks, (2) using search engines, (3) watching online videos, (4) checking e-mails, and (5) looking for product information. Previously, a computer would have been required to perform these activities (Pooters 2010), but smartphones carry them out with additional mobility and convenience. As smartphones become more powerful by combining the advantages of mobile phones and computers, the selection of such a personal device requires an effective evaluation method that can identify the relative importance of the considered criteria. In this regard, a hybrid MADM model combining DANP and modified VIKOR is proposed; the modified VIKOR method can aggregate the weighted performance gap in each criterion, which can be further integrated with the DEMATEL technique as a managerial tool for improvement planning. The research model is illustrated in Figure 11.1.

To develop an effective decision model for this smartphone vendor evaluation problem, three aspects were considered after literature reviews and discussions with experienced smartphone users/experts: (1) *Customer equity* (D_1), (2) *Product function* (D_2), and (3) *Mobile convenience* (D_3). The associated criteria in each dimension are also discussed in the following paragraphs and summarized in Table 11.1.

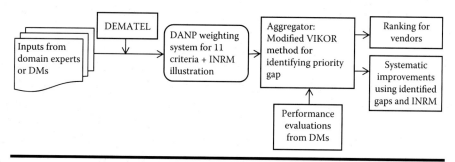

Figure 11.1 Research framework for smartphone improvement.

The idea of *Customer equity* (D_1) might not be well known in all fields; therefore, additional discussions on this dimension are provided here. Customer equity, proposed by Rust et al. (2000), regards existing customers as valuable assets from the marketing perspective. There are three main criteria in this dimension: (1) *Value equity* (C_1), (2) *Brand equity* (C_2), and *Retention equity* (C_3), briefly defined in Table 11.1. The ultimate goal of customer equity is to maximize the profits of a company by satisfying its customers' needs; to reach this goal, a company should optimize its resource allocation from the viewpoint of its customers.

Following the framework by Rust et al. (2000), the first criterion in *Customer equity* (D_1) is *Value equity* (C_1), which aims to increase the perceived value of customers while considering the benefits and pertinent costs derived from a company's products or services. The main factors affecting value equity are *quality*, *price*, and *convenience*. Quality refers to the measure of a state of being free from defects and deficiencies, which requires committed effort from a manufacturer. Price often influences consumers' choices when evaluating products, and consumers tend to select a product once they feel that its perceived value exceeds its price. Convenience relates to how manufacturers enable fast inquiries from and responses to their customers. These three factors are categorized in *Customer equity* (D_1) as those that influence consumers' evaluations of the products they are purchasing.

The second criterion in *Customer equity* (D_1) is *Brand equity* (C_2), which stems from a brand's awareness, ethics, and the attitude of its customers; it can be used to distinguish a brand name vendor from its competitors. Although brand equity involves subjective and invisible evaluations, it often plays a crucial role in consumers making product selections among a group of similar products (Buil et al. 2013). Aaker (1991) observed its importance, and found that consumers are willing to pay a higher price for a product with superior brand equity; Apple's iPhone is one of the most apparent examples. Although there are various ways for a company to promote or increase its brand equity, public relations (PR), advertisements, and certain marketing communications might be the most commonly observed ones.

Retention equity (C_3), the third criterion, relates to how customers are *attached* to a brand. In the presence of brand equity and value equity, companies are inclined to strengthen the intimacy of its relations with consumers to foster repetitive purchase

154 ■ *Trends of Hybrid Multiple Criteria Decision Making*

Table 11.1 Dimensions and Criteria for the Smartphone Evaluation Problem

Dimensions/Criteria	Descriptions
Customer equity (D_1)	
Value equity (C_1)	The objective evaluation of benefit derived from products and cost based on the customer
Brand equity (C_2)	The subjective and invisible evaluation of the brand based on the brand awareness, the brand ethics, and the brand attitude of the customer
Retention equity (C_3)	Relates to how the customer insists on the brand
Product function (D_2)	
Memory (C_4)	An electronic device manufactured using semiconductor technology that stores data during the use of the smartphone
Processor (C_5)	A programmable integrated circuit (i.e., a CPU)
Touch panel (C_6)	A reaction-type liquid crystal device (i.e., touch screen)
Operatingsystem (C_7)	Manages the hardware and software program resources
Mobile convenience (D_3)	
Remote control services (C_8)	Dominates the computer from distant locations and provides improved convenience in daily life
Location-based services (C_9)	An information service for mobile users that can accurately obtain the positional information of mobile users via a wireless communication network
Mobile wallet services (C_{10})	An application of mobile payment that replaces the physical wallet
Mobile multimedia services (C_{11})	Provides multimedia services to consumers

Source: Data from Hu, S.K., Lu, M.T., and Tzeng, G.H., *Expert Systems with Applications*, 41(9), 4401–4413, 2014.

behavior. Loyalty programs, special recognition/treatment, and affinity programs have all been adopted to enhance retention equity. For example, affinity programs establish a virtual community to promote the relationship between the manufacturer and the customers. The manufacturer becomes incorporated into the consumer's daily life to consolidate the relationship. The importance of retention equity has become apparent to branded smartphone vendors recently, and most vendors have developed their own mechanisms in this direction.

The criteria in the second and the third dimensions are relatively straightforward. The second dimension, *Product function* (D_2), includes the key hardware components and operating system of a smartphone, which influence the expected user's experience of a product. After discussions with the domain experts, three hardware components were included in the criteria: *Memory* (C_4), *Processor* (C_5), and *Touch panel* (C_6). The embedded memory in smartphones enables high-speed access and high storage capacities. The processor, or the central processing unit (CPU), is a programmable integrated circuit whose major function is sharing tasks for communication and applications with two or more processors; the memory and processor are crucial to the speed of a smartphone when processing multiple tasks simultaneously. The touch panel is the direct interface between the smartphone and the user, and offers instant feedback between fingertips and visual images. On the software side, the *Operating system* (C_7) manages the software and hardware resources of a smartphone. An operating system controls the essential functions of a smartphone, and the major functions of all smartphone operating systems (e.g., Apple iOS and Android-based operating systems) are very similar. Therefore, fine-tuning stability and intuitiveness to enhance the user's experience is a key task for the development of operating systems.

The third dimension, *Mobile convenience* (D_3), explores the convenience offered by the mobility of smartphones, and four criteria are explored in this dimension: *Remote control services* (C_8), *Location-based services* (LBSs) (C_9), *Mobile wallet services* (C_{10}), and *Mobile multimedia services* (C_{11}). Remote control services enable smartphone users to control remote devices, such as computers and home appliances, for various tasks, and have experienced rising interest recently. Location-based services are information services that can accurately obtain the location of mobile users via wireless communication networks and offer various personalized services, ranging from navigation to recording exercises. Electronic wallets or mobile wallet services provide convenient physical and online shopping functions through encrypted information processing, which replaces the wallet, includes membership cards and loyalty cards, and can be used to store personal and sensitive information. The last criterion, mobile multimedia services, also attracts users' attention, as social media (e.g., Facebook) and numerous online video/audio vendors offer rich content on the Internet. The third dimension highlights the convenience that users care about when using a smartphone, which can be referred to by smartphone vendors to develop their own niche services. Other than literature reviews and discussions with experts, the importance and validity of the aforementioned dimensions and criteria were confirmed by pretest questionnaires from experienced users.

156 ■ *Trends of Hybrid Multiple Criteria Decision Making*

11.3 DEMATEL Analysis and DANP Influential Weights for Criteria

Referring to Figure 11.1 and Table 11.1, the authors devised DEMATEL question-naires to collect the opinions of experienced users and experts on the directional influences among the criteria. Each expert was asked to indicate the degree to which he or she feels that an attribute i affects the attribute j ($i \neq j$); the degree of influence ranges from 0 (no influence) to 4 (very high influence). After averaging those opinions, the initial average matrix A can be obtained, as shown in Table 11.2.

The averaged opinions from 24 experts were examined by checking the average gap, which was defined and examined by calculating the following:

$$\frac{1}{n(n-1)}\sum\nolimits_{i=1}^{n}\sum\nolimits_{j=1}^{n}(|\,a_{ij}^{s}-a_{ij}^{s-1}\,|\,/\,a_{ij}^{s})\times100\%,$$

where:

n = 11 (the number of criteria)
s = 24 (the number of experts)

The averaged gap equals 1.32%, which suggests that the significant confidence is 98.68% in this case. Using Equation 3.2, $\alpha = 29.71$, and the initial average matrix A was normalized using α to get the initial direct-relation matrix D, as shown in Table 11.3.

Then, referring to Equation 3.3 (i.e., $T = D(I - D)^{-1}$), the total relation matrix T (termed T_C^G in DANP) can be obtained, as shown in Table 11.4; the grouping of criteria in a dimension is also illustrated by T_C^G. The averaged figure from each cell in Table 11.4 denotes the directional influences between two dimensions, which is further shown as the following dimensional matrix:

$$T^{D} = \begin{bmatrix} 0.479 & 0.513 & 0.410 \\ 0.541 & 0.530 & 0.455 \\ 0.455 & 0.465 & 0.375 \end{bmatrix}.$$

The dimensional matrix T^D and total relation matrix T can be used to calculate W^α and $r_i^C - c_i^C$ in order to identify the cause–effect relations among the three dimensions and the 11 criteria, summarized in Table 11.5.

From Table 11.5, the dimension of *Customer equity* (D_1) is influenced by the other two dimensions; in other words, the users' experience and the perceived mobile convenience offered by a smartphone will influence the customers' recognition of a brand name vendor. Next, referring to Equations 4.3 and 4.4, the transposition of

Table 11.2 Initial Average Matrix A

	C_1	C_2	C_3	C_4	C_5	C_6	C_7	C_8	C_9	C_{10}	C_{11}
C_1	0.000	2.958	2.625	2.625	3.250	3.250	3.208	1.792	2.208	1.708	2.667
C_2	3.250	0.000	2.667	2.417	2.750	2.917	2.958	1.583	1.833	1.500	2.000
C_3	2.833	3.042	0.000	2.250	2.667	2.625	2.625	1.833	1.792	1.625	1.958
C_4	2.833	2.167	2.292	0.000	3.458	2.458	2.958	1.875	2.125	1.542	2.833
C_5	3.375	2.875	2.417	3.083	0.000	2.458	3.375	2.250	2.417	1.833	3.042
C_6	3.250	3.000	2.875	2.292	2.375	0.000	2.917	2.250	2.292	1.667	3.000
C_7	3.250	2.875	2.667	3.333	3.500	3.083	0.000	2.833	2.625	2.333	3.208
C_8	1.875	1.792	1.667	2.417	2.875	2.167	2.708	0.000	2.375	2.083	2.083
C_9	2.333	2.250	2.333	2.000	2.333	1.792	2.208	2.333	0.000	2.167	2.167
C_{10}	1.958	2.167	2.125	1.667	1.917	1.833	2.375	2.167	2.583	0.000	1.958
C_{11}	2.833	2.833	2.667	2.750	2.958	2.583	2.750	2.292	2.167	2.000	0.000

Source: Data from Hu, S.K., Lu, M.T., and Tzeng, G.H., *Expert Systems with Applications*, 41(9), 4401–4413, 2014.

Table 11.3 Initial Direct-Relation Matrix D

	C_1	C_2	C_3	C_4	C_5	C_6	C_7	C_8	C_9	C_{10}	C_{11}
C_1	0.000	0.100	0.088	0.088	0.109	0.109	0.108	0.060	0.074	0.057	0.090
C_2	0.109	0.000	0.090	0.081	0.093	0.098	0.100	0.053	0.062	0.050	0.067
C_3	0.095	0.102	0.000	0.076	0.090	0.088	0.088	0.062	0.060	0.055	0.066
C_4	0.095	0.073	0.077	0.000	0.116	0.083	0.100	0.063	0.072	0.052	0.095
C_5	0.114	0.097	0.081	0.104	0.000	0.083	0.114	0.076	0.081	0.062	0.102
C_6	0.109	0.101	0.097	0.077	0.080	0.000	0.098	0.076	0.077	0.056	0.101
C_7	0.109	0.097	0.090	0.112	0.118	0.104	0.000	0.095	0.088	0.079	0.108
C_8	0.063	0.060	0.056	0.081	0.097	0.073	0.091	0.000	0.080	0.070	0.070
C_9	0.079	0.076	0.079	0.067	0.079	0.060	0.074	0.079	0.000	0.073	0.073
C_{10}	0.066	0.073	0.072	0.056	0.065	0.062	0.080	0.073	0.087	0.000	0.066
C_{11}	0.095	0.095	0.090	0.093	0.100	0.087	0.093	0.077	0.073	0.067	0.000

Source: Data from Hu, S.K., Lu, M.T., and Tzeng, G.H., *Expert Systems with Applications*, 41(9), 4401–4413, 2014.

Table 11.4 Grouped Total Relation Matrix of Criteria T_C^C

		D_1 (Customer Equity)			D_2 (Product Function)				D_3 (Mobile Convenience)			
		C_1	C_2	C_3	C_4	C_5	C_6	C_7	C_8	C_9	C_{10}	C_{11}
D_1	C_1	0.000	0.100	0.088	0.088	0.109	0.109	0.108	0.060	0.074	0.057	0.090
	C_2	0.109	0.000	0.090	0.081	0.093	0.098	0.100	0.053	0.062	0.050	0.067
	C_3	0.095	0.102	0.000	0.076	0.090	0.088	0.088	0.062	0.060	0.055	0.066
D_2	C_4	0.095	0.073	0.077	0.000	0.116	0.083	0.100	0.063	0.072	0.052	0.095
	C_5	0.114	0.097	0.081	0.104	0.000	0.083	0.114	0.076	0.081	0.062	0.102
	C_6	0.109	0.101	0.097	0.077	0.080	0.000	0.098	0.076	0.077	0.056	0.101
	C_7	0.109	0.097	0.090	0.112	0.118	0.104	0.000	0.095	0.088	0.079	0.108
D_3	C_8	0.063	0.060	0.056	0.081	0.097	0.073	0.091	0.000	0.080	0.070	0.070
	C_9	0.079	0.076	0.079	0.067	0.079	0.060	0.074	0.079	0.000	0.073	0.073
	C_{10}	0.066	0.073	0.072	0.056	0.065	0.062	0.080	0.073	0.087	0.000	0.066
	C_{11}	0.095	0.095	0.090	0.093	0.100	0.087	0.093	0.077	0.073	0.067	0.000

Source: Data from Hu, S.K., Lu, M.T., and Tzeng, G.H., *Expert Systems with Applications*, 41(9), 4401–4413, 2014.

Table 11.5 DEMATEL Cause–Effect Analysis for Dimensions and Criteria

Dimensions	r_i^D	c_i^D	$r_i^D + c_i^D$	$r_i^D - c_i^D$	Criteria	r_i^C	c_i^C	$r_i^C + c_i^C$	$r_i^C - c_i^C$
D_1 (Customer equity)	1.402	1.475	2.877	−0.073	C_1	5.483	5.777	11.260	−0.294
					C_2	5.025	5.404	10.429	−0.380
					C_3	4.881	5.079	9.960	−0.199
D_2 (Product function)	1.526	1.508	3.034	0.018	C_4	5.154	5.205	10.359	−0.051
					C_5	5.629	5.808	11.437	−0.179
					C_6	5.384	5.269	10.653	0.115
					C_7	6.088	5.795	11.884	0.293
D_3 (Mobile convenience)	1.294	1.240	2.534	0.054	C_8	4.622	4.435	9.057	0.186
					C_9	4.569	4.660	9.228	−0.091
					C_{10}	4.334	3.885	8.220	0.449
					C_{11}	5.361	5.212	10.573	0.149

Source: Data from Hu, S.K., Lu, M.T., and Tzeng, G.H., *Expert Systems with Applications*, 41(9), 4401–4413, 2014.

T_N^G can form the unweighted supermatrix W, as shown in Table 11.6. Also, referring to Equations 4.8 through 4.9, the dimensional matrix T^D was normalized to become

$$T_N^D = \begin{bmatrix} 0.342 & 0.366 & 0.292 \\ 0.355 & 0.347 & 0.298 \\ 0.351 & 0.359 & 0.290 \end{bmatrix},$$

which was used to adjust the dimensional weightings in the unweighted supermatrix W. The T_N^D-adjusted (weighted) supermatrix is shown in Table 11.7, which was calculated based on Equation 4.10 in Chapter 4.

Finally, the weighted supermatrix W^α was multiplied with itself consecutively several times (i.e., raising z for $\lim_{z \to \infty} (W^\alpha)^z$ until a set of stable DANP influential weights was reached). The DANP influential weight for each criterion will be integrated with the modified VIKOR method to aggregate the performance gap for each alternative in the next section.

11.4 Modified VIKOR for Performance Gap Aggregation

After exploring the relative importance of each criterion using the DANP method, the next stage comprises a performance evaluation of the five alternatives (all brand name smartphone vendors in the Taiwanese market) and a performance gap aggregation using the modified VIKOR method. The performance score for each alternative in each criterion ranges from 0 (the worst level) to 10 (the best or aspired level), provided by the same group of experienced smartphone users and experts.

The modified VIKOR method measures the performance gap for each alternative against the aspired level in each criterion; therefore, referring to Equation 5.5 in Chapter 5, f_j^* and f_j^- are set as 10 and 0, respectively, for the jth alternative to identify the performance gap in each criterion. For example, the performance score of alternative A on C_1 is 7.69; then, its performance gap is calculated as $gap_A^{C_1}$ = (10 – 7.69) / (10 – 0) = 0.231. The DANP influential weights of the 11 criteria and the performance gaps of the five smartphone vendors are summarized in Table 11.8.

Referring to Equations 5.12 and 5.13, the performance gap indices S_k and R_k for the five vendors can be calculated; setting δ = 0.9, δ = 0.8, and δ = 0.7, the associated V_k for the five vendors are shown and compared in Table 11.9.

Based on Table 11.9, all the performance gap indices suggest that alternative C should be the top-ranked smartphone vendor, which implies the robustness of this decision model.

Table 11.6 Unweighted Supermatrix W

		D_1 (Customer Equity)			D_2 (Product Function)				D_3 (Mobile Convenience)			
		C_1	C_2	C_3	C_4	C_5	C_6	C_7	C_8	C_9	C_{10}	C_{11}
D_1	C_1	0.315	0.380	0.372	0.360	0.360	0.355	0.357	0.355	0.352	0.349	0.353
	C_2	0.354	0.290	0.356	0.327	0.332	0.332	0.332	0.333	0.331	0.334	0.333
	C_3	0.330	0.330	0.271	0.313	0.308	0.313	0.311	0.312	0.316	0.317	0.314
D_2	C_4	0.232	0.233	0.234	0.204	0.250	0.244	0.249	0.237	0.237	0.233	0.238
	C_5	0.263	0.261	0.263	0.280	0.230	0.269	0.275	0.266	0.265	0.261	0.264
	C_6	0.243	0.243	0.242	0.244	0.244	0.211	0.248	0.234	0.236	0.238	0.238
	C_7	0.262	0.263	0.262	0.272	0.276	0.276	0.228	0.263	0.263	0.268	0.260
D_3	C_8	0.240	0.242	0.245	0.241	0.243	0.244	0.246	0.210	0.259	0.256	0.257
	C_9	0.257	0.257	0.255	0.255	0.255	0.255	0.253	0.271	0.219	0.275	0.266
	C_{10}	0.212	0.214	0.215	0.209	0.211	0.209	0.214	0.228	0.230	0.182	0.225
	C_{11}	0.291	0.287	0.285	0.294	0.291	0.292	0.287	0.291	0.292	0.287	0.252

Source: Data from Hu, S.K., Lu, M.T., and Tzeng, G.H., *Expert Systems with Applications*, 41(9), 4401–4413, 2014.

Table 11.7 Weighted Supermatrix W^α

	C_1	C_2	C_3	C_4	C_5	C_6	C_7	C_8	C_9	C_{10}	C_{11}
C_1	0.108	0.130	0.127	0.128	0.128	0.126	0.127	0.125	0.124	0.123	0.124
C_2	0.121	0.099	0.122	0.116	0.118	0.118	0.118	0.117	0.116	0.118	0.117
C_3	0.113	0.113	0.093	0.111	0.109	0.111	0.110	0.110	0.111	0.111	0.110
C_4	0.085	0.085	0.086	0.071	0.087	0.085	0.086	0.085	0.085	0.084	0.085
C_5	0.096	0.095	0.096	0.097	0.080	0.094	0.095	0.096	0.095	0.094	0.095
C_6	0.089	0.089	0.089	0.085	0.085	0.073	0.086	0.084	0.085	0.086	0.086
C_7	0.096	0.096	0.096	0.095	0.096	0.096	0.079	0.094	0.094	0.096	0.093
C_8	0.070	0.071	0.072	0.072	0.072	0.073	0.073	0.061	0.075	0.074	0.074
C_9	0.075	0.075	0.075	0.076	0.076	0.076	0.075	0.078	0.064	0.080	0.077
C_{10}	0.062	0.063	0.063	0.062	0.063	0.062	0.064	0.066	0.066	0.053	0.065
C_{11}	0.085	0.084	0.083	0.088	0.087	0.087	0.086	0.084	0.085	0.083	0.073

Source: Data from Hu, S.K., Lu, M.I., & Tzeng, G.H. *Expert Systems with Applications*, 41(9), 4401–4413, 2014.

Table 11.8 Performance Gaps of the Five Vendors and DANP Influential Weights of the Criteria

DANP Influential Weights		Brand Name Smartphone Vendors (Alternatives)				
		A	B	C	D	E
Customer equity (D_1)	**0.349**	**0.230**	**0.355**	**0.186**	**0.211**	**0.415**
Value equity (C_1)	0.124 (1)	0.231	0.375	0.202	0.250	0.375
Brand equity (C_2)	0.116 (2)	0.192	0.344	0.135	0.250	0.438
Retention equity (C_3)	0.109 (3)	0.269	0.344	0.221	0.125	0.438
Product function (D_2)	**0.357**	**0.225**	**0.456**	**0.189**	**0.424**	**0.390**
Memory (C_4)	0.084 (8)	0.298	0.375	0.260	0.375	0.313
Processor (C_5)	0.094 (4)	0.250	0.500	0.212	0.500	0.375
Touch panel (C_6)	0.085 (6)	0.163	0.438	0.154	0.375	0.500
Operating system (C_7)	0.094 (5)	0.192	0.500	0.135	0.438	0.375
Mobile convenience (D_3)	**0.294**	**0.293**	**0.470**	**0.259**	**0.511**	**0.397**
Remote control services (C_8)	0.072 (10)	0.337	0.563	0.346	0.750	0.375
Location-based services (C_9)	0.075 (9)	0.298	0.500	0.163	0.375	0.375
Mobile wallet services (C_{10})	0.063 (11)	0.385	0.500	0.375	0.750	0.563
Mobile multimedia services (C_{11})	0.084 (7)	0.183	0.344	0.183	0.250	0.313

Source: Data from Hu, S.K., Lu, M.T., and Tzeng, G.H., *Expert Systems with Applications*, 41(9), 4401–4413, 2014.

Table 11.9 Performance Gap Indices and Ranking of the Five Smartphone Vendors

Gap Indices	A	B	C	D	E
S_k	0.247	0.425	0.208	0.375	0.401
R_k	0.385	0.563	0.375	0.750	0.563
V_k while $\delta = 0.9$ (Ranking)	0.261 (2)	0.439 (5)	0.225 (1)	0.413 (3)	0.417 (4)
V_k while $\delta = 0.8$ (Ranking)	0.275 (2)	0.453 (5)	0.241 (1)	0.450 (4)	0.433 (3)
V_k while $\delta = 0.7$ (Ranking)	0.288 (2)	0.466 (4)	0.258 (1)	0.488 (5)	0.450 (3)

Source: Data from Hu, S.K., Lu, M.T., and Tzeng, G.H., *Expert Systems with Applications*, 41(9), 4401–4413, 2014.

11.5 Discussion on the Improvements of Smartphone Vendors

The findings in the previous two sections provide three kinds of analytics: (1) cause–effect analysis by DEMATEL, (2) the DANP influential weights of the criteria for smartphone vendor evaluation, and (3) a hybrid MADM model that measures and aggregates the performance gaps for the five alternatives. These three analytics can not only be used to provide ranking or selection decisions but are also capable of yielding improvement strategies for the five smartphone vendors, based on their strengths and weaknesses.

Take vendor A, for example, which is ranked as second among the five alternatives: if vendor A plans to catch up or surpass the top-ranked vendor, C, it can see that its weighted performance gap in the *Mobile convenience* (D_3) dimension is the highest, and the *Product function* (D_2) dimension is the lowest (refer Table 11.8); vendor A should invest more in R&D to enhance its functions or services in *Mobile convenience* (D_3). In addition, based on the cause–effect analysis in Table 11.5, the source dimension *Mobile convenience* (D_3) influences the other two dimensions (D_1 and D_2); as a result, vendor A may further delve into D_3, to explore the improvement priority and the cause–effect relations among the criteria (i.e., C_8, C_9, C_{10}, and C_{11}) in this dimension. A sub-INRM within the *Mobile convenience* (D_3) dimension is illustrated in Figure 11.2; also, based on the reported performance gaps in those criteria in Table 11.8, the improvement priority of vendor A in this dimension is: C_{10} (0.385) > C_8 (0.337) > C_9 (0.298) > C_{11} (0.183).

In Figure 11.2, the top priority, *Mobile wallet services* (C_{10}), is also the source factor that influences the other criteria in D_3, which holds the highest $r_i^C - c_i^C$ in

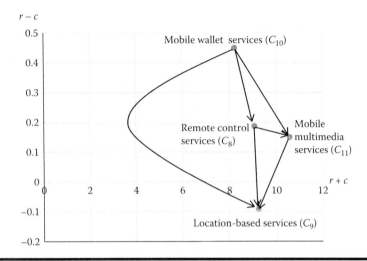

Figure 11.2 Sub-INRM of the criteria within the *Mobile convenience* (D_3) dimension. (Revised and reprinted with permission from Hu, S.K., Lu, M.T., and Tzeng, G.H., *Expert Systems with Applications*, 41(9), 4401–4413, 2014.)

this dimension. The aforementioned analyses suggest that vendor *A* should enhance its *Mobile wallet services* (C_{10}), and this improvement will influence the other mobile services. In addition, the enhancement in *Mobile convenience* (D_3) will create superior *Product function* (D_2) and higher perceived value in the *Customer equity* (D_1) dimension.

Except for improvement planning for an individual manufacturer, the result from the DANP method also indicates the influential weights of the criteria for smartphone vendors, from the experts' perspectives. It is interesting to find that the top-three influential criteria—*Value equity* (C_1), *Brand equity* (C_2), and *Retention equity* (C_3)—all belong to the same dimension: *Customer equity* (D_1). The top-three influential criteria remind vendors that the development strategy of smartphones should accumulate positive customer equity; whenever a smartphone vendor faces a dilemma on product development, the consideration of customer equity should be emphasized when selecting key components or devising new functions. This hybrid MADM model provides a useful managerial tool for smartphone vendors to refer to.

11.6 Conclusion

The growth of the smartphone market has attracted numerous companies to this industry; however, only a few brand name vendors can survive. Since the smartphone is a system product that combines the functions of computers and mobile phones, the development of such a product is complicated as well as risky and challenging. Therefore, clear guidance on smartphone development is needed in practice. This chapter paves a road to understanding the critical dimensions/criteria and their relative importance to developing successful smartphones.

In this case study, the most influential dimension is *Customer equity* (D_1), which offers an important message to smartphone vendors: the selection of key components, the development of interfaces or new functions, and the adoption of new material all need to be evaluated from the customers' perspective and to yield positive customer equity. Sometimes, R&D managers or engineers might not be aware of this point, and tend to bet on new technology without prudent cost control; the 3D-capable smartphone by Amazon might be a good example.

Aside from understanding the relative importance of the dimensions/criteria in smartphone development, the hybrid MADM model—based on DEMATEL, DANP and the modified VIKOR—shows how to identify the performance gaps closing zero toward achieving the aspiration levels on each dimension/criterion for those smartphone vendors. Using the case of vendor A, its improvement priority can be identified as a strategy for new smartphone development. Furthermore, the cause–effect analysis of the DEMATEL technique helps decision makers to identify the source factors for those highlighted criteria that require immediate attention, which indicates the effectiveness of this hybrid MADM model in practical problem solving.

Chapter 12

Evaluating the Development of Business-to-Business M-Commerce of SMEs

Small and medium-sized enterprises (SMEs) are the driving force for the economic growth of Taiwan, contributing nearly 30% of the total revenue of Taiwan's business activities in 2014, based on an official report by the SME Administration of the Ministry of Economic Affairs of Taiwan. According to Zwass (1996), "the sharing of business information, maintaining business relationships, and conducting business transactions by means of telecommunications networks" are the main functions of e-commerce (electronic commerce, or EC), and the present chapter narrows this down—considering the advantages brought by the mobile Internet—to m-commerce (mobile commerce) for SMEs. The number of registered SMEs in Taiwan exceeded 1.3 million in 2014, most of which are still lacking significant progress in adopting business-to-business (B2B) m-commerce. A practical evaluation tool to support SMEs in developing their business networks through m-commerce is urgently required, and the goal of this chapter is to construct a systematic evaluation model to explore this problem.

12.1 Research Background

M-commerce provides the convenience of doing business through the mobile Internet; with the rapid growth of wireless and smart devices (e.g., tablets and smartphones), it seems inevitable that business will catch up with this digital tide for expanding business networks. Although plenty of discussions on B2B m-commerce in the literature have mentioned the benefits information-systems/information-technology (IS/IT) can bring to businesses, such as those advantages at the operational and strategic levels, the adoption rate of B2B m-commerce is not as expected. This suggests that more effort is required to facilitate the implementation of the technology and to identify factors affecting the decision to implement B2B m-commerce. Compared with giant companies or consortiums, SMEs have only limited resources to adopt new technology such as B2B m-commerce; as a result, clear guidance on allocating scarce resources to influential factors in order to adopt B2B m-commerce is critical for SMEs.

Since the adoption of B2B m-commerce not only involves technological issues, a comprehensive framework that can reasonably consider multiple aspects is required; in this regard, the technology-organization-environment (TOE) framework is referred in this chapter, which is a prevailing theory on the implementation of technology at firm level. TOE has been applied to the adoption of various technologies for business, ranging from the operation of online e-stores for retailers (Zhu et al. 2004; Zhu and Kraemer 2005) to the supply chain management of the communications industry (Pan and Jiang 2008). Therefore, the TOE theoretical framework is applied to the construction of a MADM evaluation model for the problem addressed in this chapter.

12.2 Research Framework

The implementation of B2B m-commerce requires the consideration of multiple aspects, and TOE theory suggests that three critical dimensions should be examined: the technological, organizational, and external environments. Thus, the TOE framework is leveraged to construct a hybrid MADM model with the extended criteria in each dimension; the three dimensions and the pertaining criteria are discussed in the following subsections.

12.2.1 Technological Environment Aspect

In the technological environment aspect (or *dimension*), SMEs should consider both the new technology solutions in the market and the current status of their in-house systems or facilities; the success of adopting a new technology depends on evaluating how fit it is for the technologies and procedures that an organization already possesses (Tornatzky and Fleischer 1990; Chau and Tam 1997; Jeyaraj et al. 2006). In this aspect, previous studies have suggested that data security, network reliability, and technological complexity are the three most important factors that influence implementation decisions (Rogers 1995; Soliman et al. 2003; Soliman and Janz 2004; Narayanasamy et al. 2011). B2B relations through m-commerce involve numerous confidential and valuable information exchanges on the mobile Internet; consequently, data security and network reliability are often emphasized by decision makers when adopting innovative technologies. The third factor, technological complexity, refers to the degree to which a new technology is perceived as relatively difficult to understand and use (Rogers 1995). If the complexity of an innovative technology is too complicated for an SME's members to understand, it will be problematic to replace it or fit it into the existing system and business procedures.

12.2.2 Organizational Environment Aspect

The organizational environment aspect represents the internal criteria of an SME affecting the implementation of innovative technologies (Sila 2013). Aside from technological issues, the success of implementing new technologies often cannot be sustained without the support of top management (Sabherwal et al. 2006; Sila 2013). It is generally agreed that employees tend to resist unfamiliar and overwhelming changes in their ways of working, and the adoption of B2B m-commerce will certainly involve new procedures and operations; therefore, an emphasis on top management is necessary.

The second criterion that relates to the implementation of innovative technologies is employees' IS/IT knowledge (Mirchandani and Motwani 2001; Sabherwal et al. 2006; Pillania 2008). The higher the employees' IS/IT knowledge, the easier it will be for an organization to embrace new technologies when constructing or enhancing its B2B m-commerce, not only because of the required background knowledge on the technological aspects but also because of the mental attitude of the employees. The last criterion is firm size (Iacovou et al. 1995; Lu et al. 2013). Although the research targets of this chapter are SMEs, the size of SMEs varies. According to the definition of an SME by the Organisation for Economic Co-operation and Development (OECD), the upper ceiling of the number of employers is 500; on the other hand, plenty of small SMEs comprise only three to five employees. Large SMEs usually have dedicated IT teams evaluating new technologies, and the required resources for training and experimenting with new technologies are far less available in small SMEs; the influence of firm size cannot be overlooked in this dimension.

12.2.3 External Environment Aspect

In this case, the external environment aspect denotes the arena in which a firm conducts its business, which mainly concerns the factors that put pressure on SMEs to implement B2B m-commerce; competitive pressure and partner support are the two widely discussed sources of external pressure (Iacovou et al. 1995; Jeyaraj et al. 2006). For example, the IT manufacturing industry in Taiwan plays a crucial role in the global supply chain; in the presence of intensive global competition, SMEs are forced to comply with business protocols to be a part of those supply chains. The last factor influencing the diffusion and implementation of m-commerce for SMEs is the national regulator; its importance has been confirmed in previous studies (Papazafeiropoulou 2004; Al-Qirim 2006).

Referring to the TOE framework, a MADM conceptual model can thus be formed to explore the dimensions and criteria that influence the implementation/adoption of B2B m-commerce for SMEs; Table 12.1 summarizes the three main dimensions with the pertinent criteria of each aspect.

Furthermore, from the discussions in Sections 12.2.1 through 12.2.3, it can be seen that certain dimensions/criteria influence the others. For instance, the technological complexity factor (in the technological environment dimension) should have relations with the criteria in the organizational environment dimension; the higher the complexity of a new technology, the more important it is that the top management support the existing workers of an SME in their adoption (or learning) of the new technology. Therefore, it would be unreasonable to assume that those criteria are independent; the DEMATEL-based ANP model offers flexibility in modeling such a problem by reserving the interrelations among dimensions/criteria. The details are discussed in the following section.

12.3 DEMATEL Analysis and DANP Influential Weights for Criteria

In this case study, the implementation of IS/IT for SMEs to adopt B2B m-commerce was analyzed, and two groups of experts provided their opinions on forming the hybrid MADM model. The first group comprises 30 experts who have worked in SMEs, with at least 10 years' administrative experience. The questionnaire was devised based on the TOE framework and the pertinent criteria in Table 12.1, and a five-point scale from 0 (absolutely no influence) to 4 (very high influence) was used to determine the influence of each criterion on the other criteria. The second group only involves five experts, all of whom work for the same company, an SME that manufactures housing products. The opinions from the second group will be used and discussed in Section 12.4.

The averaged opinions from the first group were used as the initial average matrix A for the DEMATEL analysis. After normalizing A to get the initial direct-relation matrix D, the total relation matrix T (also termed T_C^G in DANP) can be obtained using Equation 3.3 (Chapter 3):

$$T = D(I - D)^{-1} \tag{3.3}$$

Evaluating the Development of SMEs ■ 173

Table 12.1 Brief Descriptions of the Dimensions and Criteria in the MADM Model

Dimensions/Criteria	Descriptions
Technological environment (D_1)	
Data security (C_1)	Data security refers to security issues associated with transactions conducted over the Internet.
Network reliability (C_2)	Network reliability addresses the ability of a firm to successfully transfer critical business applications to and from its supply chain partners over the Internet.
Technology complexity (C_3)	The degree to which an innovation is perceived as relatively difficult to understand and use.
Organizational environment (D_2)	
Top management emphasis (C_4)	Top management can provide vision, support, and a commitment to create a positive effect on the implementation of B2B m-commerce processes.
Employees' IS knowledge (C_5)	Refers to employees' attitude toward m-commerce, their experience with m-commerce, and their formal and informal m-commerce training.
Firm size (C_6)	Large firms typically have the resources necessary to experiment, pilot, and decide what technology and standards they require.
External environment (D_3)	
Competitive pressure (C_7)	Referring to peer pressure to use new technology, it has long been recognized as a driving force for the use of new technology because of its tendency to push firms to seek competitive edges through innovation.
Partner support (C_8)	Refers to the degree to which a firm's customers and suppliers are willing and ready to conduct business activities using B2B m-commerce.
Regulatory support (C_9)	This concept is similar to government policies that affect IT diffusion.

Source: Data from Lu, M.T., Hu, S.K., Huang, L.H., and Tzeng, G.H., *Management Decision*, 53(2), 290–317, 2015.

174 ■ *Trends of Hybrid Multiple Criteria Decision Making*

The sums of each row and column were used to calculate $r_i^C + c_i^C$ and $r_i^C - c_i^C$, respectively. In addition, the averaged figure from each cell in T forms the dimensional influence matrix T^D. The total relation matrix T is shown in Table 12.2. The dimensional influence matrix T^D was then obtained, as follows:

$$T^D = \begin{bmatrix} 0.404 & 0.372 & 0.376 \\ 0.472 & 0.379 & 0.406 \\ 0.389 & 0.344 & 0.325 \end{bmatrix}.$$

Using T and T^D, the DEMATEL cause–effect analyses of the dimensions and criteria can be determined, as shown in Table 12.3, which indicates that the *Organizational environment* (D_2) dimension is the source for the other two dimensions; in the meantime, the *Technological environment* (D_1) is influenced by both the D_2 and D_3 dimensions. The influential network relationships map (INRM) for this analysis is illustrated in Figure 12.1.

The grouped total-influence matrix T_C^G was normalized to become T_N^G, referring to Equations 4.2 through 4.4, shown in Table 12.4. The normalized T_N^G was then transposed to become the unweighted supermatrix W in DANP. In addition, the dimensional influence matrix T^D is normalized (refer to Equations 4.8 through 4.9) to become

$$T_N^D = \begin{bmatrix} 0.351 & 0.323 & 0.327 \\ 0.375 & 0.302 & 0.323 \\ 0.367 & 0.325 & 0.307 \end{bmatrix}.$$

Then, T_N^D was applied to adjust the dimensional weight of the unweighted supermatrix by multiplying T_N^D with W (i.e., $W^\alpha = T_N^D W$). The adjusted/weighted supermatrix W^α is shown in Table 12.5.

The weighted supermatrix can be multiplied by itself consecutively several times (raising z for $\lim_{z \to \infty}(W^\alpha)^z$) until stable; the DANP influential weights for the nine criteria can thus be applied to weight the performance gap in each criterion for the modified VIKOR aggregation method.

12.4 Modified VIKOR for Performance Gap Aggregation

As stated in Section 12.3, the second group of experts work in different departments of the same SME; the performance evaluation in this case was conducted

Table 12.2 Total Relation Matrix T (or T_C^o)

		Technological Environment (D_1)			Organizational Environment (D_2)			External Environment (D_3)		
		C_1	C_2	C_3	C_4	C_5	C_6	C_7	C_8	C_9
D_1	C_1	0.344	0.426	0.405	0.411	0.369	0.310	0.397	0.376	0.339
	C_2	0.496	0.365	0.438	0.456	0.403	0.345	0.434	0.424	0.349
	C_3	0.439	0.423	0.299	0.398	0.361	0.292	0.404	0.348	0.317
D_2	C_4	0.551	0.547	0.495	0.417	0.463	0.436	0.527	0.473	0.444
	C_5	0.512	0.504	0.460	0.481	0.320	0.333	0.441	0.416	0.357
	C_6	0.414	0.400	0.365	0.410	0.321	0.230	0.371	0.326	0.301
D_3	C_7	0.407	0.391	0.371	0.400	0.332	0.298	0.299	0.367	0.335
	C_8	0.452	0.433	0.383	0.423	0.364	0.328	0.412	0.291	0.318
	C_9	0.371	0.352	0.338	0.359	0.315	0.279	0.369	0.313	0.223

Source: Data from Lu, M.T., Hu, S.K., Huang, L.H., and Tzeng, G.H., *Management Decision*, 53(2), 290–317, 2015.

Table 12.3 DEMATEL Cause–Effect Influence Analysis of the Dimensions and Criteria

Dimensions	r_i^D	c_i^D	$r_i^D + c_i^D$	$r_i^D - c_i^D$	Criteria	r_i^C	c_i^C	$r_i^C + c_i^C$	$r_i^C - c_i^C$
D_1	1.152	1.264	2.416	-0.112	C_1	3.377	3.986	7.363	-0.609
					C_2	3.710	3.841	7.551	-0.131
					C_3	3.281	3.554	6.835	-0.273
D_2	1.257	1.095	2.352	0.162	C_4	4.353	3.755	8.108	0.598
					C_5	3.824	3.248	7.072	0.576
					C_6	3.138	2.851	5.989	0.287
D_3	1.058	1.108	2.166	-0.050	C_7	3.200	3.654	6.854	-0.454
					C_8	3.404	3.334	6.738	0.070
					C_9	2.919	2.983	5.902	-0.064

Source: Data from Lu, M.T., Hu, S.K., Huang, L.H., and Tzeng, G.H., *Management Decision*, 53(2), 290–317, 2015.

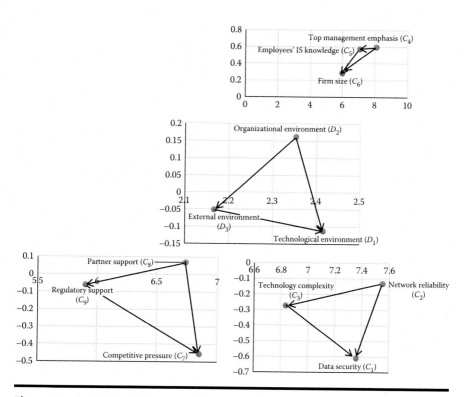

Figure 12.1   INRM of the dimensions and criteria.

for a single company in its two functional aspects of B2B m-commerce: (1) supply chain management (SCM) and (2) customer relationship management (CRM) systems. The SME has make certain initial investments in those two functions, and this evaluation is supposed to provide guidance for the company to plan for improvement.

The five experts rated this SME based on the nine criteria in the three dimensions; the score ranges from 0 (the worst) to 10 (the aspired level/ideal), indicating the satisfaction level of an expert in each criterion for this SME implementing B2B m-commerce. The average scores in those criteria were transformed into performance gaps measured against the aspired levels (i.e., 10 as the aspired level); the DANP influential weights for the nine criteria and the associated performance gaps are shown in Table 12.6.

12.5 Discussion

The aggregated indices S_i and V_i (for $\delta = 0.90$ and $\delta = 0.85$) all show that this SME performs relatively well in its CRM system (i.e., CRM ≻ SCM); based on this evaluation result, the SME can decide which system should be

Table 12.4 Normalized Total-Influence Matrix T_N^C

		Technological Environment (D_1)			Organizational Environment (D_2)			External Environment (D_3)		
		C_1	C_2	C_3	C_4	C_5	C_6	C_7	C_8	C_9
D_1	C_1	0.293	0.281	0.337	0.379	0.335	0.287	0.360	0.351	0.289
	C_2	0.382	0.281	0.337	0.379	0.335	0.287	0.360	0.351	0.289
	C_3	0.378	0.364	0.257	0.379	0.344	0.278	0.378	0.325	0.297
D_2	C_4	0.346	0.343	0.311	0.317	0.352	0.331	0.365	0.328	0.308
	C_5	0.347	0.342	0.311	0.425	0.282	0.293	0.363	0.343	0.294
	C_6	0.351	0.340	0.309	0.426	0.334	0.239	0.372	0.327	0.301
D_3	C_7	0.348	0.334	0.318	0.388	0.323	0.289	0.299	0.367	0.334
	C_8	0.356	0.342	0.302	0.379	0.327	0.294	0.403	0.285	0.311
	C_9	0.350	0.332	0.318	0.377	0.331	0.293	0.408	0.346	0.246

Source: Data from Lu, M.T., Hu, S.K., Huang, L.H., and Tzeng, G.H., *Management Decision*, 53(2), 290–317, 2015.

Table 12.5　Weighted Supermatrix W^α

		Technological Environment (D_1)			Organizational Environment (D_2)			External Environment (D_3)		
		C_1	C_2	C_3	C_4	C_5	C_6	C_7	C_8	C_9
D_1	C_1	0.103	0.134	0.133	0.130	0.130	0.132	0.128	0.131	0.129
	C_2	0.127	0.098	0.128	0.129	0.128	0.127	0.123	0.125	0.122
	C_3	0.121	0.118	0.090	0.117	0.117	0.116	0.117	0.111	0.117
D_2	C_4	0.122	0.122	0.122	0.096	0.128	0.129	0.126	0.123	0.123
	C_5	0.109	0.108	0.111	0.106	0.085	0.101	0.105	0.106	0.108
	C_6	0.092	0.092	0.090	0.100	0.088	0.072	0.094	0.096	0.095
D_3	C_7	0.117	0.118	0.124	0.118	0.117	0.120	0.092	0.124	0.125
	C_8	0.111	0.115	0.106	0.106	0.111	0.106	0.113	0.088	0.106
	C_9	0.100	0.095	0.097	0.099	0.095	0.097	0.103	0.096	0.076

Source: Data from Lu, M.T., Hu, S.K., Huang, L.H., and Tzeng, G.H., *Management Decision*, 53(2), 290–317, 2015.

Table 12.6   DANP Influential Weights and the Performance Gaps for the Modified VIKOR Method

Dimensions	Criteria	DANP weights	Supply Chain Management			Customer Relationship Management		
			Scores	N_Gaps[a]	W_Gaps[b]	Scores	N_Gaps	W_Gaps
D_1	C_1	0.127 (1)	0.127	5.600	0.440	3.600	0.640	0.081
	C_2	0.123 (2)	0.123	4.800	**0.520**	6.600	0.340	0.042
	C_3	0.114 (5)	0.114	3.600	**0.640**	3.800	0.620	0.071
D_2	C_4	0.121 (3)	0.121	5.600	0.440	5.800	0.420	0.051
	C_5	0.105 (7)	0.105	6.200	0.380	5.800	0.420	0.044
	C_6	0.091 (9)	0.091	6.200	0.380	6.600	0.340	0.031
D_3	C_7	0.117 (4)	0.117	3.800	**0.620**	6.200	0.380	0.044
	C_8	0.107 (6)	0.107	6.600	0.340	5.800	0.420	0.045
	C_9	0.096 (8)	0.096	6.200	0.380	5.600	0.440	0.042
	S_i		0.466			0.451		
	R_i		0.073			0.081		
	$V_i \, (\delta = 0.90)$		0.427			0.414		
	$V_i \, (\delta = 0.85)$		0.407			0.396		

[a] "N_Gaps" denotes the normalized performance gaps by using 0 and 10 as the worst and aspired levels, respectively.
[b] "W_Gaps" denotes the DANP weighted performance gaps

addressed first. For example, if the company feels that it should improve its laggard SCM, it can further identify its improvement priority in the criteria by observing its weighted performance gaps. According to Table 12.6, its top-three weighted performance gaps are: *Technology complexity* (C_3) > *Competitive pressure* (C_7) > *Network reliability* (C_2); and C_3 and C_2 both belong to the *Technological environment* (D_1) dimension. The SME may consider improving its weighted gap in *Technological complexity* (C_3) in two ways: for its existing system, it may request the solution provider devise a simplified version, including the functions and interface, for its employees; or it may redesign its existing operational flows to reduce the unnecessary complexities of the existing procedures. Furthermore, referring to Figure 12.1 and the cause–effect analysis in Table 12.3, the SME can see that the *Organizational environment* (D_2) has a direct influence on the *Technological environment* (D_1), which can be resolved by devoting adequate resources to this dimension, such as *Top management emphasis* (C_4) and *Employees' IS knowledge* (C_5); the support from top management may involve more training to advance employees' knowledge of SCM. The directional influences among the dimensions/criteria offer a new perspective for SMEs to plan for improvements; thus, SMEs can devise systematic action plans to address the source factors for those priority gaps.

12.6 Conclusion

The adoption or implementation of B2B m-commerce for SMEs is examined in this chapter, and the findings provide guidance for SMEs and practical insights for system providers/operators. On the one hand, the DANP influential weights highlight the relative importance of each factor to implementing B2B m-commerce, and SMEs can use it to identify its priorities for improvement. On the other hand, once an SME plans to focus on improving a specific criterion, the DEMATEL cause–effect analysis, illustrated as an INRM, shows the source factors or dimensions for that criterion, which provides a systematic overview of the whole problem. Overall, the data security and the network reliability are the two highest concerns that SMEs have when adopting or implementing B2B m-commerce; system providers/operators may strengthen their R&D on those two factors to promote their solutions. The management teams of SMEs should also be aware of the importance of the organizational environment aspect, which is the source that influences the other two dimensions. This hybrid MADM model can be used as a managerial tool for SMEs to move toward the aspired levels in all aspects.

Chapter 13

Evaluation and Selection of Glamour Stocks by a Hybrid MADM Model

Stock or equity evaluation involves complicated analyses of key financial ratios and operational figures; in practice, financial analysts tend to compare a group of stocks from the same industry, and observe the changing trends of key financial variables (also termed *criteria* or *factors*). Before making investment decisions, a decision maker (DM) will rely on his/her investment strategy or model to select targets with superior prospects. In the literature of empirical finance, several prevailing investment strategies are widely examined: value, growth stock (also termed *glamour stock*), and momentum investment strategies (Chan and Lakonishok 2004). This chapter focuses on the evaluation of glamour stock investment.

Although the glamour stock investment strategy has been examined in the previous study by Mohanram (2005) by forming a statistics-based model in the presence of multiple factors—various financial and operational ratios—statistical models will encounter obstacles when dealing with the interrelationships among variables. In addition, investors often have to compromise when target stocks indicate conflicting performance outcomes in different aspects. Therefore, a hybrid multiple attribute decision making (MADM) model that can measure the influences among factors is proposed to solve this problem; the model aims to identify the relative importance of each variable (criterion) in glamour stock evaluation, considering the complex interrelations among factors.

184 ■ *Trends of Hybrid Multiple Criteria Decision Making*

13.1 Research Background and Investment Strategy

The last two decades have seen growing emphasis placed on research examining the efficiency of stock markets, owing to the famous efficient market hypothesis (EMH) proposed by Fama (1970). Numerous empirical studies have tried to argue that investors can outperform the market either by analyzing the financial and operational data from public-traded companies, or implementing effective investment strategies; among those strategies, value stock and glamour stock investing are the two most widely examined (Chan and Lakonishok 2004; Fama and French 2006). These two strategies separately use simple valuation criteria to form different investment portfolios.

On the one hand, value stock portfolios have been found to outperform diversified portfolios or market indexes in the long term (Piotroski 2000; Fama and French 2006) and are prevailing in the financial industry. On the other hand, glamour stock investing also has attracted rising interest recently. Investors expect glamour (or growth) stocks to thrive with rising profits in the future; therefore, their present book-to-market-value (B/M) or earning-to-price (E/P) ratios are often relatively low (Hung et al. 2015). Previous studies (Brav and Heaton 2002; Mohanram 2005) have provided theoretical foundations for growth stock investing; nevertheless, investors still require clear guidance and decision aids when selecting investment targets among those low B/M or E/P stocks. The main concern of glamour stocks is that they are relatively expensive (Piotroski and So 2012); if the glamour stocks cannot meet the market's expectations, they will suffer from significant price corrections. Thus, glamour stock selection is more challenging than generic stock selection. To explore this problem, this chapter adopts the well-known G-score model as the research framework, which not only considers the financial fundamentals but also addresses the psychological aspects, to develop a hybrid MADM model for evaluating glamour stocks.

13.2 Research Framework for the G-Score and Hybrid MADM Models

According to Lakonishok et al. (1994), while investors get overly excited about stocks that have performed very well and shown growth tendency, enthusiastic buy-in forces might cause these stocks to become overpriced; these stocks are named *growth stocks* or *glamour stocks*. If investors focus merely on examining their fundamentals, such as return on equity (ROE) or free cash flow (FCF), they might overlook a critical aspect that causes the high prices of glamour stocks: psychological bias or concern. As a result, a theoretical framework that considers both the financial and psychological aspects is required. In this context, Mohanram (2005) developed the G-score model, covering three dimensions of growth stock investing: (1) *Earnings and cash flow profitability* (D_1), (2) *Naïve extrapolation* (D_2), and

Evaluation and Selection of Glamour Stocks ■ 185

(3) *Accounting conservatism* (D_3). The first and the third dimensions measure the financial and accounting aspects, respectively, and the second addresses the psychological or behavioral bias of investors. The three dimensions and their criteria are briefly defined in Table 13.1 and discussed in the following paragraphs.

The first dimension of the G-score model starts from examining profitability, which denotes the financial fundamentals of glamour stocks. Since companies operate on different scales, it is more reasonable to use financial ratios to assess and compare those stocks; three commonly examined profitability ratios are included in this dimension: return on assets (ROA), cash flow ROA, and cash flow from operations (CFO) exceed net income, all scaled by average assets. The profitability aspect accounts for a certain portion of the high pricing of glamour stocks;

Table 13.1 Influential Dimensions and Criteria of the Glamour Stock Selection (G-Score) Model

Dimensions	Criteria		Definitions
Earnings and cash flow profitability (D_1)	ROA	C_1	Net income scaled by average assets
	Cash flow ROA	C_2	Cash from operations scaled by average assets
	CFO exceeding net income	C_3	Cash flow from operations minus net income and scaled by average assets
Naïve extrapolation (D_2)	Less earning variability	C_4	Standard deviation of quarterly earnings in the previous 4 years (i.e., 16 quarters)
	Less sales growth Variability	C_5	Standard deviation of monthly sales growth in the previous 4 years (i.e., 48 months)
Accounting conservatism (D_3)	R&D expenditure	C_6	R&D expense scaled by total assets
	Capital expenditure	C_7	Capital expenditure scaled by total assets
	Marketing expenditure	C_8	Marketing expenses scaled by total assets

Source: Data from Shen, K.Y., Yan, M.R., and Tzeng, G.H., *Knowledge-Based Systems*, 58, 86–97, 2014.

186 ■ Trends of Hybrid Multiple Criteria Decision Making

however, it might not be enough to fully explain their expensive valuations. Some previous studies (Mohanram 2005) have attributed this relatively high valuation to the cognitive biases of investors, as stock markets tend to naïvely extrapolate recent fundamentals for glamour stocks (La Porta et al. 1997). In the context of cognitive biases, the G-score model includes *Naïve extrapolation* (D_2) as the second influential dimension. Thus, the variability of the two critical operational fundamentals—earnings and sales growth—are listed as the criteria in this dimension. The third dimension, *Accounting conservatism* (D_3) (Penman and Zhang 2002), highlights the implications of conservative accounting for future earnings, such as the capital expenditure, research and development (R&D), and marketing expenses. The relevance of R&D expenses and future earnings performance was confirmed in the study by Chan et al. (2001), which also found that advertising expenses are associated with the subsequent excess returns of glamour stocks. The influences of current or past expenses that might yield the high potential of future growth are thus included in the third dimension.

Although the G-score model contributes to the identification of the relevant aspects of growth/glamour stock investing, it applies a simplified approach, transforming the relevant financial information into binary signals, and forms a logistic regression model to separate gains from losses. As mentioned at the beginning of this chapter, statistics-based models have certain obvious limitations; therefore, this chapter attempts to enhance the G-score model using a hybrid MADM model, which needs not assume the independence of dimensions or criteria and the probabilistic distributions of data. Furthermore, the hybrid MADM model allows DMs to rate glamour stocks in all criteria using a 10-point performance scale, and thereby offers more precise performance measurements compared with the binary signals in the original G-score model.

To examine the proposed MADM model of glamour stock selection, a group of low B/M ratio stocks from the semiconductor industry in Taiwan were examined in this case. Since Taiwan holds the leading position in the global semiconductor industry, lots of public-listed semiconductor stocks have enjoyed relatively high valuations. The top 25% low B/M ratio stocks (in 2009) from the semiconductor industry were categorized as glamour stocks in this case, and five sample stocks from this group were selected as the target stocks—namely, MediaTek (A), Sonix Technology (B), Sitronix Technology (C), ITE Tech (D), and VIA Technologies (E).

13.3 DEMATEL Analysis and DANP Influential Weights of the G-Score Model

Eight domain experts, with more than 10 years' working experience in the financial industry, were invited to provide their opinions for this case study. Referring to Table 13.1, a DEMATEL questionnaire based on the G-score's framework was devised for the participants. The initial influence matrix A is the result of the averaged opinions from those experts.

$$
A = \begin{bmatrix}
- & 1.375 & 2.000 & 0.500 & 0.750 & 0.750 & 1.875 & 1.250 \\
2.750 & - & 3.000 & 1.750 & 3.250 & 3.125 & 1.750 & 3.125 \\
2.000 & 2.876 & - & 1.000 & 3.000 & 1.125 & 0.750 & 2.000 \\
1.000 & 0.500 & 3.125 & - & 2.875 & 1.125 & 3.000 & 0.875 \\
3.125 & 3.375 & 3.000 & 2.625 & - & 2.875 & 2.000 & 3.250 \\
2.875 & 1.250 & 1.750 & 1.875 & 0.625 & - & 3.000 & 0.750 \\
1.875 & 1.750 & 0.875 & 0.625 & 0.750 & 3.500 & - & 0.625 \\
1.375 & 1.000 & 1.875 & 1.750 & 3.250 & 0.750 & 0.625 & -
\end{bmatrix}.
$$

In this initial influence matrix A, the average gap is 3.34%, smaller than 5%; that is to say, the significant confidence is higher than 95% (100% – 3.34% = 96.66%). The averaged gap is defined by Equation 13.1:

$$
AveragedGap = \frac{1}{n(n-1)} \sum_{i=1}^{n} \sum_{j=1}^{n} \frac{\left| a_{ij}^{p} - a_{ij}^{p-1} \right|}{a_{ij}^{p}} \times 100\%, \tag{13.1}
$$

The averaged gap is defined by Equation 13.1, where a_{ij}^{p} and a_{ij}^{p-1} denote the average influence of criterion i on criterion j by experts p and $p - 1$, respectively; n denotes the number of criteria (p and n both equal 8 here).

The initial average matrix A is normalized to be the direct-influence matrix D, referring to Equation 3.2 (Chapter 3), shown in Table 13.2. Next, by multiplying D with $(I - D)^{-1}$, the total-influence matrix T can be obtained (Table 13.3).

After averaging the grouped figures in each cell, the dimensional influence matrix T^{D} can be obtained.

$$
T^{D} = \begin{bmatrix}
0.248 & 0.230 & 0.237 \\
0.316 & 0.241 & 0.285 \\
0.217 & 0.186 & 0.174
\end{bmatrix}.
$$

Using T^{D} (T), the cause–effect influential relationships between dimensions (criteria) are explored, as summarized in Table 13.4.

The total-influence matrix T, after being normalized and transposed, is turned into the unweighted supermatrix W (Table 13.5) for DANP. In addition, referring

Table 13.2 Normalized Direct-Influence Matrix *D*

Criteria	C_1	C_2	C_3	C_4	C_5	C_6	C_7	C_8
ROA (C_1)	0.000	0.068	0.099	0.025	0.037	0.037	0.093	0.062
Cash flow ROA (C_2)	0.136	0.000	0.148	0.086	0.161	0.154	0.086	0.154
CFO exceeding net income (C_3)	0.099	0.142	0.000	0.049	0.148	0.056	0.037	0.099
Less earning variability (C_4)	0.049	0.025	0.154	0.000	0.142	0.056	0.148	0.043
Less sales growth variability (C_5)	0.154	0.167	0.148	0.130	0.000	0.142	0.099	0.161
R&D expenditure (C_6)	0.142	0.062	0.086	0.092	0.031	0.000	0.148	0.037
Capital expenditure (C_7)	0.093	0.086	0.043	0.031	0.037	0.173	0.000	0.031
Marketing expenditure (C_8)	0.068	0.049	0.093	0.086	0.161	0.037	0.031	0.000

Source: Data from Shen, K.Y., Yan, M.R., and Tzeng, G.H., *Knowledge-Based Systems*, 58, 86–97, 2014.

Table 13.3 Total-Influence Matrix T (T_C^α)

Dimensions / Criteria		D_1			D_2			D_3		r_i
		C_1	C_2	C_3	C_4	C_5	C_6	C_7	C_8	
D_1	C_1	0.119	0.165	0.207	0.104	0.149	0.142	0.181	0.154	1.222
	C_2	0.374	0.214	0.387	0.254	0.379	0.354	.0292	0.342	2.595
	C_3	0.280	0.288	0.195	0.179	0.313	0.218	0.193	0.252	1.919
D_2	C_4	0.224	0.181	0.312	0.118	0.289	0.212	0.279	0.184	1.798
	C_5	0.403	0.369	0.405	0.299	0.257	0.357	0.317	0.359	2.766
D_3	C_6	0.281	0.186	0.233	0.184	0.171	0.139	0.268	0.155	1.618
	C_7	0.225	0.190	0.177	0.123	0.154	0.275	0.122	0.136	1.402
	C_8	0.222	0.185	0.250	0.191	0.296	0.174	0.164	0.134	1.616
d_i		2.128	1.778	2.166	1.451	2.008	1.871	1.816	1.716	

Source: Data from Shen, K.Y., Yan, M.R., and Tzeng, G.H., *Knowledge-Based Systems*, 58, 86–97, 2014.

Table 13.4 Cause–Effect Analysis of Dimensions and Criteria by DEMATEL

Dimensions	r_i^D	d_i^D	$r_i^D + c_i^D$	$r_i^D - c_i^D$	Criteria	r_i	d_i	$r_i + d_i$	$r_i - d_i$
D_1	0.714	0.780	1.494	−0.066	C_1	1.222	2.128	3.350	−0.906
D_2	0.841	0.657	1.497	0.184	C_2	2.595	1.778	4.374	0.817
D_3	0.577	0.695	1.272	−0.118	C_3	1.919	2.166	4.084	−0.247
					C_4	1.798	1.451	3.249	0.347
					C_5	2.766	2.008	4.773	0.758
					C_6	1.618	1.871	3.489	−0.253
					C_7	1.402	1.816	3.218	−0.415
					C_8	1.616	1.716	3.332	−0.100

Source: Data from Shen, K.Y., Yan, M.R., and Tzeng, G.H., *Knowledge-Based Systems*, 58, 86–97, 2014.

Table 13.5 Unweighted Supermatrix _W_

Criteria	C_1	C_2	C_3	C_4	C_5	C_6	C_7	C_8
ROA (C_1)	0.241	0.384	0.367	0.312	0.343	0.401	0.381	0.338
Cash flow ROA (C_2)	0.336	0.219	0.377	0.253	0.313	0.266	0.321	0.282
CFO exceeding net income (C_3)	0.423	0.397	0.255	0.435	0.344	0.333	0.299	0.380
Less earning variability (C_4)	0.410	0.401	0.364	0.290	0.538	0.518	0.443	0.393
Less sales growth variability (C_5)	0.590	0.599	0.636	0.710	0.462	0.482	0.557	0.607
R&D expenditure (C_6)	0.298	0.358	0.329	0.314	0.346	0.247	0.516	0.368
Capital expenditure (C_7)	0.379	0.295	0.291	0.414	0.306	0.477	0.229	0.348
Marketing expenditure (C_8)	0.323	0.347	0.380	0.272	0.348	0.276	0.254	0.284

Source: Data from Shen, K.Y., Yan, M.R., and Tzeng, G.H., _Knowledge-Based Systems_, 58, 86–97, 2014.

192 ■ Trends of Hybrid Multiple Criteria Decision Making

to Equations 4.8 through 4.9 (Chapter 4), the dimensional influence matrix T^D is normalized as follows:

$$T_N^D = \begin{bmatrix} 0.347 & 0.322 & 0.331 \\ 0.375 & 0.286 & 0.339 \\ 0.375 & 0.323 & 0.302 \end{bmatrix},$$

which is used to adjust the dimensional weights in DANP $\left(W^\alpha = T_N^D \times W \right)$. The adjusted supermatrix W^α (Table 13.6) needs to be multiplied by itself consecutively several times until stable, and the finalized DANP influential weights can be adopted to form a weighting system.

13.4 Modified VIKOR for Performance Gap Aggregation and Evaluation

Owing to the nature of making investment decisions, the modified VIKOR aggregation method is adopted for this hybrid MADM model. When facing a complex investment decision involving multiple criteria, alternatives often reveal inconsistent performances in different criteria. Investors often have to compromise to reach the final decision. In most circumstances, investors need to compromise concerning the complexity and interrelationships of the variables/criteria under evaluation, and the modified VIKOR method offers a reasonable approach to making compromise decisions.

The idea of the compromise decision method (VIKOR) is to select the best solution and measure the distances for all alternatives to find out the alternative with the shortest distance to the aspired level, beginning by assessing the relative performance of the five target stocks in the eight criteria. Instead of asking the domain experts to rate the target stocks through questionnaires, as in previous VIKOR research (Liou and Chuang 2010), this approach has a more objective way of obtaining the performance scores for the target stocks.

To measure the relative performance of each stock, all the semiconductor stocks in Taiwan's stock market were included for comparison. After excluding stocks with incomplete financial data for the end of 2009, there were 47 semiconductor stocks left for further analysis. All of the 47 stocks' financial performances were considered for the construction of the performance scores for the five sample stocks. The raw financial figures of the five target stocks in the eight criteria were transformed into the range [0, 10], as in Equation 13.2:

$$f_{kj} = ((g_{kj} - g_j^-) / (g_j^* - g_j^-)) \times 10, \tag{13.2}$$

Table 13.6 Weighted/Adjusted Supermatrix W^{α} by DANP Method

Criteria	C_1	C_2	C_3	C_4	C_5	C_6	C_7	C_8
ROA (C_1)	0.084	0.133	0.127	0.117	0.129	0.150	0.143	0.127
Cash flow ROA (C_2)	0.117	0.076	0.131	0.095	0.118	0.100	0.120	0.106
CFO exceeding net income (C_3)	0.147	0.138	0.089	0.163	0.129	0.125	0.112	0.143
Less earning variability (C_4)	0.132	0.129	0.117	0.083	0.154	0.167	0.143	0.127
Less sales growth variability (C_5)	0.190	0.193	0.205	0.203	0.132	0.156	0.180	0.196
R&D expenditure (C_6)	0.099	0.119	0.109	0.106	0.117	0.074	0.156	0.111
Capital expenditure (C_7)	0.126	0.098	0.096	0.140	0.104	0.144	0.069	0.105
Marketing expenditure (C_8)	0.107	0.115	0.126	0.092	0.118	0.083	0.077	0.086

Source: Data from Shen, K.Y., Yan, M.R., and Tzeng, G.H., *Knowledge-Based Systems*, 58, 86–97, 2014.

194 ■ Trends of Hybrid Multiple Criteria Decision Making

where:

f_{kj} denotes the transformed performance score of the kth alternative on the jth criterion

g_j^* and g_j^- indicate the highest/best and lowest/worst raw financial figures from the 47 semiconductor stocks on the jth criterion

The transformed performance scores of the five target stocks and the DANP influential weights (Section 13.3) are summarized in Table 13.7.

Using the performance scores of the five stocks, the performance gaps can be calculated for the modified VIKOR method; the performance gaps were calculated by setting 10 and 0 as the aspired and the worst levels, respectively. The synthesized performance gaps (Table 13.8) for each stock can be turned into indices E_k, Q_k, and U_k (refer to Equations 5.7 through 5.9 in Chapter 5; k indicates the kth alternative) for the ranking or selection of glamour stocks, and the DANP weighted performance gap in each criterion further supports or guides a glamour stock to improve its stock performance (return) by meeting investors' expectations.

The index Q_k implies the maximal regret (gap) that a stock has in a specific criterion; the combination of E_k and Q_k is one of the nonconventional linear approach on performance aggregation (non-additive type in the real world). Referring to Table 13.8, the three indices E_k, U_k ($\delta = 0.9$), and U_k ($\delta = 0.8$) all show the same ranking: $A > B > C > E > D$, which suggests the consistency and robustness of this analysis.

Table 13.7 Performance Scores of the Five Glamour Stocks and the DANP Influential Weights

Criteria	DANP Weights	Glamour Stocks				
		A	B	C	D	E
C_1	0.126	8.826	8.280	7.726	6.435	2.694
C_2	0.109	9.515	8.671	5.459	4.788	0.000
C_3	0.131	2.835	2.092	0.888	1.286	2.398
C_4	0.132	9.033	7.502	6.370	5.336	8.741
C_5	0.179	0.967	2.498	3.630	4.668	1.259
C_6	0.111	6.611	5.537	6.011	1.282	6.192
C_7	0.110	0.729	0.654	2.653	0.360	1.199
C_8	0.102	2.028	2.241	2.974	1.799	0.000

Source: Data from Shen, K.Y., Yan, M.R., and Tzeng, G.H., *Knowledge-Based Systems*, 58, 86–97, 2014.

Evaluation and Selection of Glamour Stocks ■ 195

Table 13.8 Synthesized Performance Gaps and Indices by the Modified VIKOR

Criteria	DANP Weights	Target Stock's Gaps against Aspiration Levels				
		A	B	C	D	E
C_1	0.126	0.117	0.172	0.227	0.357	0.731
C_2	0.109	0.048	0.133	0.454	0.530	1.000
C_3	0.131	0.716	0.791	0.911	0.871	0.760
C_4	0.132	0.097	0.250	0.363	0.978	0.126
C_5	0.179	0.903	0.750	0.637	0.934	0.874
C_6	0.111	0.339	0.446	0.399	0.872	0.381
C_7	0.110	0.927	0.935	0.735	0.771	0.880
C_8	0.102	0.797	0.776	0.703	0.820	1.000
Indices		Synthesized Scores (Ranking)				
E_k		0.510(1)	0.539(2)	0.556(3)	0.779(5)	0.715(4)
Q_k		0.927	0.935	0.911	0.978	1.000
U_k	$(\delta = 0.9)$	0.552(1)	0.579(2)	0.592(3)	0.799(5)	0.744(4)
U_k	$(\delta = 0.8)$	0.593(1)	0.618(2)	0.627(3)	0.819(5)	0.772(4)

Source: Data from Shen, K.Y., Yan, M.R., and Tzeng, G.H., *Knowledge-Based Systems*, 58, 86–97, 2014.

13.5 Discussion and Examination of Stock Returns

To examine the effectiveness of the proposed VIKOR-DANP method, this chapter will further explore the relationship between the ranking of target stocks and the corresponding stock's subsequent holding period return (*HPR*) from the end of May 2009 to December 2012. *HPR* is defined as follows:

$$HPR = (Price_{end} - Price_{initial} + Dividend_{period}) / Price_{initial}. \qquad (13.3)$$

Furthermore, the stock performance of the monthly average *HPR*s of the five stocks were compared with the ranking result in two periods: from May 2009 to December 2011 and from May 2009 to December 2012. The Friedman test was used to examine the ranking result compared with the model's suggested output, shown in Table 13.9. The top-three stocks' performances were consistent with the

196 ■ *Trends of Hybrid Multiple Criteria Decision Making*

Table 13.9 Friedman Test of Ranking Order in Two Time Periods

	End of May 2009~Dec 2011			End of May 2009~Dec 2012		
	$\overline{HPR}$	SD of HPR	Mean Rank[a]	$\overline{HPR}$	SD of HPR	Mean Rank[a]
Stock *A*	0.281	0.238	4.41	0.209	0.241	4.36
Stock *B*	0.209	0.205	3.97	0.144	0.207	4.02
Stock *C*	0.149	0.189	3.50	0.097	0.185	3.52
Stock *D*	−0.050	0.215	2.03	−0.175	0.278	2.02
Stock *E*	−0.395	0.146	1.09	−0.473	0.187	1.07
N (months)		32			44	
Chi-square χ^2		99.050			138.436	
df		4			4	
Sig.[b]		0.000**			0.000**	

Source: Data from Shen, K.Y., Yan, M.R., and Tzeng, G.H., *Knowledge-Based Systems*, 58, 86–97, 2014.

[a] The result was generated by SPSS 17 (a higher score represents a better ranking position; e.g., 4 is better than 3).
[b] The result was significant with a 99% confidence level, indicated by 0.000**.

ranking order in the model's output; both stocks D and E ranked at the bottom of the five target stocks, which suggests the effectiveness of this hybrid model.

Using the cause–effect analysis result from the DEMATEL technique, shown in Table 13.4, the directional influences among the dimensions/criteria are illustrated in Figure 13.1. Two interesting findings can be observed: (1) the profit-related criteria (i.e., C_1 through C_3) are not the most influential factors, and (2) the *Naïve extrapolation* (D_2) dimension has a direct influence on the other two dimensions. In other words, from the perspective of investment experts, the psychological aspect would dominate the evaluation and selection of glamour stocks in practice.

These findings have implications for the management teams of glamour stocks. To increase the shareholders' value, all management teams attempt to increase the market value of their companies using various strategic plans for performance improvements. Owing to limited resources, management teams are asked to prioritize their action plans for improvements. Take stock B, for example: its top-three weighted performance gaps are 0.134 (C_5) > 0.104 (C_3) > 0.103 (C_7), so it can see that its sales growth variability requires immediate attention. The findings of this research suggest that the stability of sales growth and earnings should be kept as a first-tier objective in preparing action plans. From the perspective of external investors, the evaluation

Evaluation and Selection of Glamour Stocks ■ 197

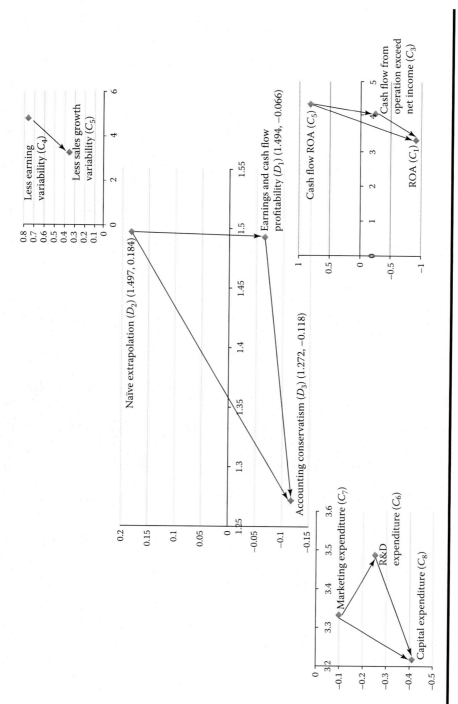

Figure 13.1 INRM of the three dimensions and pertinent criteria. (Reprinted with permission from Shen, K.Y., Yan, M.R., and Tzeng, G.H., *Knowledge-Based Systems*, 58, 86–97, 2014.)

of glamour stocks should consider the stability of sales and earnings growth as crucial indicators. In other words, the sustainability and persistence of glamour stocks should be further examined to ensure their long-term prospects.

13.6 Conclusion

This chapter has demonstrated that investment experts' implicit knowledge can be retrieved by the proposed VIKOR-DANP method when selecting glamour stocks. This hybrid approach comprises the DEMATEL, DANP, and modified VIKOR methods and a performance transformation process; the overall research flow is illustrated in Figure 13.2. The G-score model provides a theoretical framework to

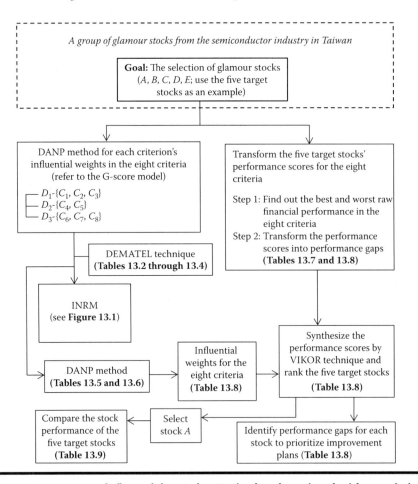

Figure 13.2 Research flow of the study. (Revised and reprinted with permission from Shen, K.Y., Yan, M.R., & Tzeng, G.H., *Knowledge-Based Systems*, 58, 86–97, 2014.)

analyze the complex decision processes involved. The findings in this chapter are twofold: (1) the importance of the psychological or behavioral aspects was highlighted, and (2) a practical decision model was constructed for glamour stock selection with positive results. The selected top-ranking stock A outperformed the other four stocks in the 44-month *HPRs* with statistical significance, which suggests the effectiveness of the proposed model. The model not only takes the interrelations among variables into consideration but also provides relative importance for each criterion regarding growth stock selection.

Despite the aforementioned findings, this work adopted an objective approach to measuring performance scores, comparing the relative performance of each stock within a specific industry in the same time frame. In practice, relative performance indicates the competitiveness of a stock compared with its peer group, and a transformation system was proposed to obtain the performance scores for target stocks in this chapter. Aside from the ranking or selection problem, this model further provides improvement guidance—such as improvement priorities and cause–effect analyses—for the management teams of glamour stocks to refer to, which is underexplored in conventional MADM research.

Chapter 14

Nonadditive Hybrid MADM Model for Selecting and Improving Suppliers

Considering the intensity of global competition, three critical challenges—cost, quality, and flexibility—are confronted by international businesses, both manufacturers and service providers. However, it is unlikely for a company to be outstanding without having strong support teams, including internal departments and external suppliers; therefore, the importance of evaluation, selection, and the continuous improvement of suppliers is gaining interest. The traditional approach in inclined to rank and select suppliers solely on price concerns. However, ranking or selection problems in the current supply chain network have become much more complicated, as potential options for evaluating decisions have to consider multiple criteria (or *factors*). Consequently, supplier selection or improvement has become a multiple criteria decision making (MCDM) problem, including multiple tangible and intangible factors (Bertolini et al. 2006). Recently, the relationships among these factors have become increasingly complex, interdependent, and dynamic, as environmental, social, political, and customer satisfaction concerns have emerged.

In addition, traditional MCDM methods only employ an additive aggregator to evaluate, rank, or select alternatives; the nonadditive effect (i.e., $1 + 1 \neq 2$) cannot be measured adequately by the these methods. More importantly, after making evaluation or selection decisions, qualified suppliers still need to improve their performance to make the whole supply chain stay ahead. In a dynamic and competitive environment,

201

to stay at the same level of performance does not guarantee the sustainability of a business. Therefore, a nonadditive hybrid approach is introduced in this chapter, revised from the previous work of Liou et al. (2014), to enhance decision making on the ranking, selection, and improvement of suppliers in the context of MCDM research.

14.1 Research Background and Literature Review

The complexity of the selection or evaluation of suppliers is increasing; despite the cost concerns, various factors have to be evaluated. Furthermore, there is no universal standard for a supplier evaluation system, which has to consider the specific strategy or requirements of a business. In other words, an effective supplier selection/evaluation system demands robust analytical methods that are capable of analyzing multiple subjective and objective criteria based on the preference—yielded from its strategy—of a business. In addition, the plausible interrelationships among variables need to be measured and preserved in an evaluation system, which cannot be overlooked in practice.

A series of literature reviews have summarized the criteria and decision methods that have been adopted to resolve the supplier selection problem; for example, in an exhaustive review of 76 articles (Weber et al. 1991), 47 articles addressed the involvement of more than one criterion. Another study (Ho et al. 2010) extended the previous literature by surveying multiple criteria supplier evaluation and improvement/selection approaches through a literature review and a classification of international journal articles from 2000 to 2008; they concluded that the MCDM approach had become the mainstream for supplier selection. The MCDM approach includes various methods and techniques, such as the analytic hierarchy process (AHP), the analytic network process (ANP), data envelopment analysis (DEA), fuzzy set theory, genetic algorithms (GA), mathematical programming, the multiple attribute rating technique (e.g., grey relations), VIKOR, the Technique for Order Preference by Similarity to an Ideal Solution (TOPSIS), and various integrations or combinations of the aforementioned methods.

Prior studies have made significant contributions to supplier selection; however, they have assumed the criteria to be independent when modeling the supplier selection problem. In the real world, the criteria are seldom independent; the relationships between the criteria are all interactive to some extent, occasionally including dependence and feedback effects (Huang et al. 2010). Others (Liao and Rittscher 2007; Hsu et al. 2013) accounted for this interdependence using ANP. Nevertheless, only the additive methods were employed to aggregate performances and weights; the additive methods might cause a certain degree of bias when handling those interdependent criteria. To resolve this issue on modeling, the nonadditive fuzzy integral technique might be a more reasonable approach to integrating the performance values of those interdependent criteria.

Over the last two decades, various decision-making methods have been proposed to address supplier evaluation and selection problems. Based on previous discussions, these methods can be roughly categorized into three groups: (1) multiple attribute decision making (MADM), (2) mathematical programming models, and (3) combined or integrated hybrid approaches.

14.1.1 Multiple Attribute Decision Making

The most popular MADM methods are AHP and ANP. A fuzzy AHP model was applied to analyze a low-carbon supply chain decision (Shaw et al. 2012); the considered factors were cost, quality, rejection percentage, late delivery percentage, greenhouse gas emissions, and demand. Another study (Bertolini et al. 2006) used AHP to select the best discount rate when defining a proposal for a public works contract. A hierarchical structure comprising 31 criteria was reported to illustrate the performance and characteristics of the proposed technique. To allow for the impreciseness of modeling, a fuzzy AHP model (Chan and Kumar 2007) was proposed to identify and discuss the development of an efficient system for global supplier selection. Since AHP assumes independent criteria, other researchers opted for ANP to consider interdependent criteria when constructing their models. For example, a study by Hsu and Hu (2009) presented an ANP approach to incorporating the issue of hazardous substance management (HSM) into supplier selection. As discussed in Chapter 4, the limitations of ANP in terms of handling the interrelationships among dimensions can be relaxed by adopting the DEMATEL-based or adjusted ANP (DANP) method; therefore, the DANP method has also been applied to the supplier selection problem (Hsu et al. 2013).

14.1.2 Mathematical Programming Models

Aside from considering multiple attributes simultaneously when evaluating suppliers, other researchers have applied multiple objective mathematical programming to this problem, in the context of multiple objective decision making (MODM). For example, a fuzzy multiobjective programming model was proposed to select a supplier while accounting for risk factors (Wu et al. 2013), using simulated historical quantitative and qualitative data for modeling. In this line of research, DEA and its derivative methods are used by many authors to address supplier selection problems by comparing the relative efficiency of each supplier (Falagario et al. 2012). Although this approach has the advantage of pursuing multiple objectives simultaneously, two critical limitations/assumptions remain: (1) linearity and (2) independence among the variables of the MODM models.

14.1.3 Combined and Integrated Hybrid Approaches

Recently, certain computational techniques (e.g., artificial neural networks [ANNs], evolutionary fuzzy systems, data-mining approaches, and expert systems) have been adopted for this supplier selection problem. For example, a hybrid intelligent model

204 ■ *Trends of Hybrid Multiple Criteria Decision Making*

that uses a fuzzy neural network and a genetic algorithm to forecast the rate of demand was devised to determine material plans and select the optimal supplier (Moghadam et al. 2008). Furthermore, owing to the fact that individual approaches often have more limitations, numerous combined or integrated approaches to supplier selection have been proposed in the last decade. One of the examples is the integrated AHP–DEA approach to supplier selection (Sevkli et al. 2007); this model used AHP to derive local weights from a given pairwise comparison matrix and aggregated the local weights to yield the overall weights. Each row and column of the matrix was assumed to be a decision-making unit (DMU) and an output. A dummy input with a value of 1 for all DMUs was deployed in DEA to calculate the efficiency scores of all suppliers. Another hybrid model (Kuo et al. 2010) combines an ANN with two multiattribute decision analysis methods (DEA and ANP) for green supplier selection; this overcomes traditional DEA drawbacks, the limitations of data accuracy, and constraints in the amount of DMUs.

Based on the preceding review, previous studies have generally assumed that the criteria are independent when establishing supplier evaluation models. A few studies have focused on the interdependence of the criteria when using ANP, but those studies still applied additive models to aggregate performance values. Unlike previous works, a nonadditive model combined with the measurement of gaps between observed aspired levels, is introduced in this chapter for the section and improvement of suppliers; it illustrates the idea of how to use nonadditive aggregator in a hybrid MADM model.

14.3 Hybrid MADM Model Using Nonadditive-Type Aggregators

To resolve the suppliers' evaluation, selection, and improvement problems, a hybrid MADM model, integrating the DEMATEL, DANP, modified VIKOR, and fuzzy integral techniques, is introduced with a numerical example. Using this new hybrid approach, the inconsistency in prior studies, that assume the interdependency of criteria but apply additive models, can be remedied.

At first, DEMATEL-based ANP is used to establish the structural relationship model and determine the criteria weights with dependence and feedback. In a complex system, all system criteria are either directly or indirectly mutually related. In such intricate systems, it is very difficult for a decision maker (DM) to obtain a specific objective/aspect and avoid interference from the rest of the system. The DEMATEL technique is adopted to determine the effect on each dimension and criterion. Next, the DANP approach, using the essential concepts of ANP (Saaty 1996), was adopted to calculate the influential weights of the criteria. The concepts of VIKOR are then applied to transform the performance values into gaps. Finally, a nonadditive-type aggregator, using fuzzy integrals, is incorporated to aggregate the weighted gaps. The framework of this hybrid model is shown in Figure 14.1.

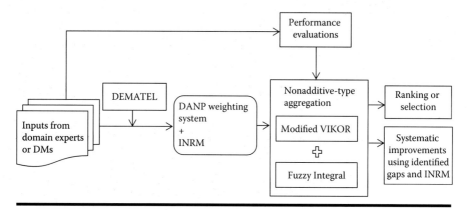

Figure 14.1 Framework of the hybrid evaluation model.

As the DEMATEL, DANP, and modified VIKOR methods have been introduced in Chapters 3 through 5, only the required procedures of how to determine the gap values and adopt the nonadditive-type aggregator (fuzzy integral) are briefly explained here.

14.3.1 Determining Gap Values Based on the New Concepts of Modified VIKOR Method

As mentioned in Chapter 5, a hybrid MADM model relies on an aggregator, either the additive or the nonadditive type, to synthesize the performance scores/gaps of an alternative in the weighted criteria. Since this hybrid model attempts to improve the performance of suppliers, gap values should be used instead of performance scores; as a result, the basic concepts of the modified VIKOR method are adopted to transform the performance scores into gap values.

Assuming that the performance score of alternative A_k on the jth criterion is denoted as f_{kj}; w_j is the relative influence weight of the jth criterion that can be obtained from DANP, where $j = 1, 2, \ldots, n$, and n is the number of criteria. The conventional VIKOR method was developed using the following traditional additive form of the L_p-metric (Equation 14.1):

$$L_k^p = \left\{ \sum_{j=1}^{n} [w_j (|f_j^* - f_{kj}|) / (|f_j^* - f_j^-|)]^p \right\}^{1/p}, \qquad (14.1)$$

where:
$1 \leq p \leq \infty$
$k = 1, 2, \ldots, m$
w_j (the influential weight) is derived from DANP

206 ■ *Trends of Hybrid Multiple Criteria Decision Making*

To formulate the ranking and gap ratio, measures $L_k^{p=1}$ and $L_k^{p=\infty}$ have been widely used in the literature of the VIKOR method, as defined in Equations 14.2 and 14.3:

$$L_k^{p=1} = \sum_{j=1}^{n} [w_j (|f_j^* - f_{kj}|)/(|f_j^* - f_j^-|)], \qquad (14.2)$$

$$L_k^{p=\infty} = \max_{j} \left\{ (|f_j^* - f_{kj}|)/(|f_j^* - f_j^-|) \big| j = 1, 2, ..., n \right\}, \qquad (14.3)$$

where $r_{kj} = (|f_j^* - f_{kj}|)/(|f_j^* - f_j^-|)$ is defined as the gap value of alternative k on criterion j. The compromise solution $\min_k L_k^p$ yields the synthesized/aggregated gap ratio that will also be minimized using Equation 14.2; while $L_k^{p=\infty}$, it indicates the improvement priority with the maximum gap value among all the criteria for alternative k. Then, the best f_j^* values are set as the aspiration levels and the worst f_j^- values are set as the tolerable levels for all criteria functions. However, the traditional VIKOR approach is constrained by the performance of the existing/available alternatives, which might force DMs to select a mediocre alternative among a group of inferior ones. Thus, the modified VIKOR asks DMs to redefine the best f_j^* and the worst f_j^- as the aspiration/ideal and the worst values, respectively. This modified approach provides reasonable guidance for DMs to find the improvement priority of an alternative in a systematic approach.

Although the modified VIKOR method is capable of guiding an alternative in systematic improvement planning, it is still based on additive-type aggregation; therefore, a nonadditive-type technique (i.e., fuzzy integrals) is introduced in the next subsection, to synthesize the final performance gap of alternatives for this hybrid model.

14.3.2 Applying λ Fuzzy Measures for Fuzzy Integrals

Based on the weight of each criterion obtained from DANP, the fuzzy measure and performance matrix can be combined to calculate the integrated performance for each alternative. Let g_λ be a λ fuzzy measure that is defined on a power set $P(x)$ for the finite set $X = \{x_1, x_2, ..., x_n\}$. The fuzzy measure has the properties shown in Equation 14.4:

$$\forall A, B \in P(X), \ A \cap B = \varnothing, \ g_\lambda(A \cup B) = g_\lambda(A) + g_\lambda(B) + \lambda g_\lambda(A) g_\lambda(B), \quad (14.4)$$

where $-1 < \lambda < \infty$. The density of the fuzzy measure $g_i = g_\lambda(\{x_i\})$ can be obtained from questionnaire responses. The local weights $(w_1, w_2, ..., w_n)$ can be obtained through DANP. Then, the fuzzy measure weights can be defined as follows:

$$(g_\lambda(\{x_1\}), g_\lambda(\{x_2\}), ..., g_\lambda(\{x_n\})) = q(w_1, w_2, ..., w_n) = (qw_1, qw_2, ..., qw_n), \quad (14.5)$$

where q is the *adjusted weight* coefficient.

In addition, the fuzzy measure has the properties shown in Equation 14.6, which will be realized for a specific case with two attributes, x_1 and x_2, with three plausible effects: multiplicative ($\lambda > 0$), additive ($\lambda = 0$), and substitutive ($\lambda < 0$).

$$g_\lambda(\{x_1, x_2, \ldots, x_n\}) = \sum_{i=1}^{n} g_\lambda(\{x_i\}) + \lambda \sum_{i=1, j>i}^{n} g_\lambda(\{x_i\})g_\lambda(\{x_j\}) + \ldots \lambda^{n-1} g_\lambda(\{x_1\})g_\lambda(\{x_2\})\ldots g_\lambda(\{x_n\}), \tag{14.6}$$

where $g_\lambda(X) = g_\lambda(\{x_1, x_2, \ldots, x_n\}) = 1$. In this hybrid model, performance scores are replaced by gap values that are equal to the aspired levels minus the evaluated values with respect to each criterion. Let h be a measurable set function (or *gap function*) defined on the fuzzy measurable space, and suppose that $h(x_1) \geq h(x_2) \geq \ldots \geq h(x_n)$; then, the fuzzy integral of fuzzy measure $g(\cdot)$ with respect to $h(\cdot)$ can be defined as follows (also shown in Figure 14.3):

$$\int h \, dg = h(x_n)g(H_n) + \left[h(x_{n-1}) - h(x_n)\right]g(H_{n-1}) + \ldots + \left[h(x_1) - h(x_2)\right]g(H_1)$$

$$= h(x_n)\left[g(H_n) - g(H_{n-1})\right] + h(x_{n-1})\left[g(H_{n-1}) - g(H_{n-2})\right] + \ldots + h(x_1)g(H_1), \tag{14.7}$$

where:

$H_1 = \{x_1\}$

$H_2 = \{x_1, x_2\}$

$\ldots$

$H_n = \{x_1, x_2, \ldots, x_n\} = X$

The fuzzy integral defined in Equation 14.7 is termed the *Choquet integral* (Sugeno et al. 1998). Using the fuzzy integral to formulate the original data, not only can fewer and more representative factors be extracted to describe the system, but the interactions between attributes are also considered.

14.4 Numerical Example

A numerical example was implemented using the historical data from a Taiwanese airline that serves over 50 international destinations. To reduce manpower costs and improve service efficiency, the company sought to contract out its ground services at foreign destinations. Data from Bangkok, are selected for this case. Currently, five major ground service companies (A_1 through A_5) are the potential alternatives to be selected as the partner. The decision is strategic because its successful completion will have a significant bearing on the company's continued competitiveness.

Several issues are important to determining the optimal collaborator in this supplier improvement/selection process, including whether there have been favorable past associations between the potential suppliers, whether the national and

208 ■ *Trends of Hybrid Multiple Criteria Decision Making*

corporate cultures of the suppliers are compatible, and whether trust exists between the suppliers' management teams. The supplier selection criteria were developed based on a literature review and a series of field discussions (Liou et al. 2014). In this case, four dimensions are concluded: compatibility, risk, quality, and cost. These dimensions are then divided into various criteria, as indicated in Table 14.1.

14.4.1 Measuring the Influential Relations and Weights by DEMATEL and DANP

The criteria in Table 14.1 were used to design the questionnaires for calculating the DEMATEL analysis, DANP weights, and fuzzy measures. The detailed steps of calculating DEMATEL and DANP are provided in Chapters 3 through 5; thereby only the initial direct-relation matrix A is shown in Table 14.2. The initial direct-relation matrix A is an 11×11 matrix, obtained by pairwise comparisons with respect to levels of influence and the direction of the relationships between criteria. The subsequent calculations for obtaining the total relation matrix T, dimensional matrix T^D, and the weighted supermatrix matrix W^∞ are omitted here; interested readers may use A as an exercise to calculate the remaining matrices.

The DEMATEL analysis shows the sum of the influence given $(r_i - c_i)$ and the difference $(r_i + c_i)$ of the influence for each dimension and criterion, and the dimensions or criteria can thus be divided into cause and effect groups, as shown in Table 14.3. The graphical illustration of the cause-effect influence analysis can be indicated as an INRM (Figure 14.2). In addition, referring to Equations 4.1 through 4.10 (Chapter 4), the DEMATEL-adjusted supermatrix (or *weighted supermatrix*) can be obtained; after raising z for $\lim_{z \to \infty}(W^\alpha)^z$ until this supermatrix is converged, the DANP influential weights are thus finalized, as shown in Table 14.4.

The results indicate that *Compatibility* (D_1) is the most important dimension in terms of influence, while *Relationship* (C_{11}) is the first priority in terms of the global weights. As previously noted, DEMATEL is combined with the ANP method to validate individual performance perspectives; the causal relationships among the criteria are shown in the INRM (Figure 14.2).

14.4.2 Integrated Weighted Gaps Using the Fuzzy Integral Technique

To rank or select the five alternatives (i.e., A_1 through A_5), the collected performance scores of the five alternatives (gathered by questionnaires) were transformed into gap values, using the basic concepts of the modified VIKOR. In the questionnaire, the performance score ranges from 0 (the worst) to 10 (the best); consequently, 10 is regarded as the aspired value, 0 the worst value, on all the criteria. Since the criteria within the same dimension have interdependent relationships, their weighted gaps should be integrated instead of treating them as individual values. Similarly, the

Table 14.1   Dimensions and Criteria of the Hybrid MADM Model

Dimensions	Criteria	Explanations
Compatibility (D_1)	Relationship (C_{11})	Includes shared risks and rewards, ensuring cooperation between the airline and ground service provider.
	Flexibility (C_{12})	Flexibility when dealing with abnormal situations, such as flight delays, overbooking, and incidents.
	Information sharing (C_{13})	Compatibility of computer systems and information sharing, such as new information/regulations at a destination airport.
Quality (D_2)	Knowledge and skills (C_{21})	The service provider's airplane maintenance facilities and their knowledge of manpower are essential.
	Customer satisfaction (C_{22})	Average customer's level of satisfaction regarding ground services, such as check-in and luggage handling.
	On-time rate (C_{23})	Ratio of airplanes delivered on time.
Cost (D_3)	Cost saving (C_{31})	Total cost of outsourcing activities.
	Flexibility in billing (C_{32})	Flexibility in billing and payment conditions, increasing goodwill between airlines and the service supplier.
Risk (D_4)	Labor union (C_{41})	Service outsourcing may be accompanied by layoffs and disturbances within the airline. Supplier employee strikes could disrupt flight schedules.
	Loss of management control (C_{42})	Poor management of the service supplier may lead to inadequate service and may cause potential flight safety problems.
	Information security (C_{43})	Mutual trust-based information sharing between the airline and the service supplier is necessary for both the continuance of the agreement and also for the security of confidential information.

210 ■ *Trends of Hybrid Multiple Criteria Decision Making*

Table 14.2   Initial Direct-Influence Matrix *A*

A	C_{11}	C_{12}	C_{13}	C_{21}	C_{22}	C_{23}	C_{31}	C_{32}	C_{41}	C_{42}	C_{43}
C_{11}	0.0	2.5	3.3	1.3	1.9	1.5	3.0	3.3	3.2	3.1	2.9
C_{12}	1.4	0.0	2.5	2.1	2.4	1.9	1.5	1.3	2.8	2.7	2.9
C_{13}	3.3	2.4	0.0	2.8	1.5	1.8	0.8	0.7	3.2	2.9	2.8
C_{21}	2.9	0.8	2.3	0.0	2.5	2.7	0.4	0.5	1.2	1.5	1.6
C_{22}	3.2	2.2	2.1	2.5	0.0	1.1	0.7	0.9	0.5	0.8	0.6
C_{23}	1.2	1.9	1.5	0.6	3.7	0.0	1.4	1.4	0.3	0.7	0.5
C_{31}	3.1	1.3	1.5	0.5	0.8	1.3	0.0	2.7	1.8	1.3	1.1
C_{32}	2.4	3.3	0.9	0.2	0.4	0.4	2.7	0.0	0.9	0.7	0.4
C_{41}	2.8	2.5	2.3	1.7	2.3	3.1	0.5	0.4	0.0	3.3	1.8
C_{42}	3.1	2.3	2.4	0.8	3.3	2.7	2.7	2.3	2.9	0.0	3.5
C_{43}	2.2	1.6	3.2	1.3	0.9	1.3	1.1	1.0	1.4	2.8	0.0

Source: Data from Liou, J.J.H. et al., *Information Sciences*, 266, 199–217, 2014.

Table 14.3   Cause–Effect Influences of Dimensions and Criteria Analyzed by DEMATEL

Dimensions					Criteria				
T^D	r_j^D	c_j^D	$r_j^D + c_j^D$	$r_j^D - c_j^D$	T^C	r_i^C	c_i^C	$r_i^C + c_i^C$	$r_i^C - c_i^C$
D_1	1.21	1.18	2.39	0.04	C_{11}	3.73	3.61	7.34	0.12
					C_{12}	3.12	3.02	6.14	0.09
					C_{13}	3.33	3.22	6.55	0.11
D_2	0.78	0.89	1.67	−0.11	C_{21}	2.43	2.11	4.54	0.33
					C_{22}	2.23	2.87	5.10	−0.65
					C_{23}	1.88	2.59	4.48	−0.71
D_3	0.76	0.79	1.54	−0.03	C_{31}	2.30	2.21	4.51	0.09
					C_{32}	1.89	2.17	4.07	−0.28
D_4	1.11	1.00	2.12	0.11	C_{41}	3.09	2.76	5.85	0.34
					C_{42}	3.68	2.96	6.64	0.72
					C_{43}	2.59	2.74	5.33	−0.16

Source: Data from Liou, J.J.H. et al., *Information Sciences*, 266, 199–217, 2014.

Nonadditive Hybrid MADM Model ■ 211

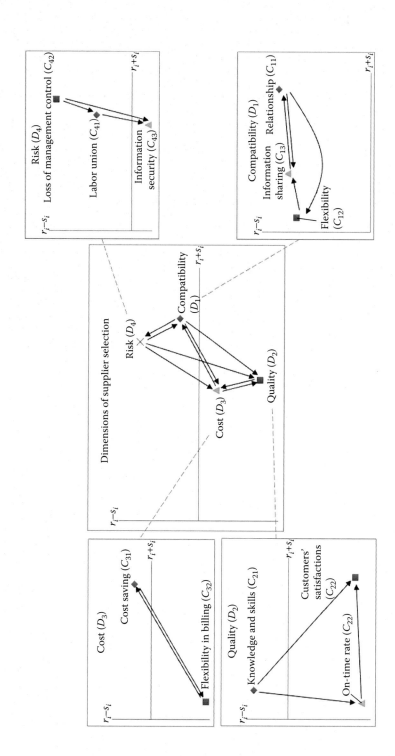

Figure 14.2 Influential network relations map. (Reprinted with permission from Liou, J.J.H. et al., *Information Sciences*, 266, 199–217, 2014.)

212 ■ *Trends of Hybrid Multiple Criteria Decision Making*

Table 14.4 Influential Weights of System Factors

Dimensions	Dimensional Weights	(Ranking)	Criteria	Local Weights	Global Weights (Ranking)
D_1	0.306	(1)	C_{11}	0.367	0.112 (1)
			C_{12}	0.310	0.095 (5)
			C_{13}	0.324	0.099 (4)
D_2	0.231	(3)	C_{21}	0.281	0.065 (11)
			C_{22}	0.379	0.088 (7)
			C_{23}	0.340	0.079 (10)
D_3	0.204	(4)	C_{31}	0.506	0.103 (2)
			C_{32}	0.494	0.101 (3)
D_4	0.259	(2)	C_{41}	0.327	0.085 (8)
			C_{42}	0.351	0.091 (6)
			C_{43}	0.322	0.083 (9)

Source: Data from Liou, J.J.H. et al., *Information Sciences*, 266, 199–217, 2014.

integrated weighted gaps of the four dimensions should be further calculated with their final synthesized values.

Through a survey questionnaire conducted by the case company's managers, fuzzy integral λ values are obtained, which range from -1 to ∞ (i.e., $-1 < \lambda < \infty$), which represent the substitutive or multiplicative properties of the relationships among the criteria. There are substitutive effects in the *Risk* (D_4) dimension and there is a multiplicative effect in the *Compatibility* (D_1), *Quality* (D_2), and *Cost* (D_3) dimensions. The λ values and the fuzzy measures $g(\cdot)$ are shown in Table 14.5. The fuzzy measures of each dimension and criterion were acquired using the questionnaire. Referring to Equation 14.5, the adjusted weight coefficient q can be obtained; then, the λ value is derived by solving the polynomial equation in Equation 14.6; the result is shown in Table 14.5. Using the obtained $g(\cdot)$ and the gap ratios transformed from their performance scores (i.e., $r_{kj} = (|10 - f_{kj}|)/(|10 - 0|)$), the synthesized final gap values for the five alternatives—using the fuzzy integral aggregator—are shown in Table 14.6.

The integrated weighted gaps of each potential supplier are then calculated as shown in Table 14.6. To illustrate the calculations, the ground service company A_1 is used as an example. Figure 14.3a indicates how the integrated weighted gap of D_1 (*Compatibility*) for company A_1 is obtained. Figure 14.3b demonstrates how

Table 14.5   Fuzzy Measure $g(\lambda)$ of Each Parameter and Parameter Combination

Fuzzy Measure $g(\cdot)$			
Supplier selection (evaluating systems), $\lambda = -0.597$, $q = 1.358$			
$g_\lambda(\{D_1\}) = 0.415$ $g_\lambda(\{D_2\}) = 0.314$ $g_\lambda(\{D_3\}) = 0.277$ $g_\lambda(\{D_4\}) = 0.352$	$g_\lambda(\{D_1, D_2\}) = 0.651$ $g_\lambda(\{D_1, D_3\}) = 0.624$ $g_\lambda(\{D_1, D_4\}) = 0.680$ $g_\lambda(\{D_2, D_3\}) = 0.539$ $g_\lambda(\{D_2, D_4\}) = 0.600$ $g_\lambda(\{D_3, D_4\}) = 0.571$	$g_\lambda(\{D_1, D_2, D_3\}) = 0.821$ $g_\lambda(\{D_1, D_2, D_4\}) = 0.866$ $g_\lambda(\{D_1, D_3, D_4\}) = 0.844$ $g_\lambda(\{D_2, D_3, D_4\}) = 0.778$	$g_\lambda(\{D_1, D_2, D_3, D_4\}) = 1$
Compatibility (D_1), $\lambda = 0.358$, $q = 0.900$			
$g_\lambda(\{C_{11}\}) = 0.330$ $g_\lambda(\{C_{12}\}) = 0.279$ $g_\lambda(\{C_{13}\}) = 0.291$	$g_\lambda(\{C_{11}, C_{12}\}) = 0.642$ $g_\lambda(\{C_{11}, C_{13}\}) = 0.656$ $g_\lambda(\{C_{12}, C_{13}\}) = 0.599$	$g_\lambda(\{C_{11}, C_{12}, C_{13}\}) = 1$	
Quality (D_2), $\lambda = 3.902$, $q = 0.539$			
$g_\lambda(\{C_{21}\}) = 0.151$ $g_\lambda(\{C_{22}\}) = 0.204$ $g_\lambda(\{C_{23}\}) = 0.183$	$g_\lambda(\{C_{21}, C_{22}\}) = 0.476$ $g_\lambda(\{C_{21}, C_{23}\}) = 0.443$ $g_\lambda(\{C_{22}, C_{23}\}) = 0.533$	$g_\lambda(\{C_{21}, C_{22}, C_{23}\}) = 1$	
Cost (D_3), $\lambda = 1.268$, $q = 0.798$			
$g_\lambda(\{C_{31}\}) = 0.403$ $g_\lambda(\{C_{33}\}) = 0.395$	$g_\lambda(\{C_{31}, C_{32}\}) = 1$		
Risk (D_4), $\lambda = -0.073$, $q = 1.025$			
$g_\lambda(\{C_{41}\}) = 0.336$ $g_\lambda(\{C_{42}\}) = 0.360$ $g_\lambda(\{C_{43}\}) = 0.330$	$g_\lambda(\{C_{41}, C_{42}\}) = 0.687$ $g_\lambda(\{C_{41}, C_{43}\}) = 0.657$ $g_\lambda(\{C_{42}, C_{43}\}) = 0.681$	$g_\lambda(\{C_{41}, C_{42}, C_{43}\}) = 1$	

Source: Data from Liou, J.J.H. et al., *Information Sciences*, 266, 199–217, 2014.

Note: The fuzzy measures for each dimension and criterion are obtained by questionnaire. The other fuzzy measures are calculated using Equation 14.6.

214 ■ *Trends of Hybrid Multiple Criteria Decision Making*

Table 14.6   Gap Ratio Values of Potential Suppliers by the Fuzzy Integral Aggregator

Criteria	Weights Local	Alternatives				
		A_1	A_2	A_3	A_4	A_5
Compatibility (D_1)	**0.306**	**0.240**	**0.197**	**0.197**	**0.182**	**0.263**
Relationship (C_{11})	0.367	0.264	0.208	0.199	0.198	0.268
Flexibility (C_{12})	0.310	0.214	0.211	0.198	0.176	0.264
Information sharing (C_{13})	0.324	0.242	0.175	0.194	0.173	0.258
Quality (D_2)	**0.231**	**0.286**	**0.224**	**0.227**	**0.227**	**0.214**
Knowledge and skills (C_{21})	0.281	0.280	0.221	0.275	0.224	0.214
Customer satisfaction (C_{22})	0.379	0.286	0.255	0.227	0.265	0.203
On-time rate (C_{23})	0.340	0.302	0.213	0.213	0.214	0.246
Cost (D_3)	**0.204**	**0.242**	**0.300**	**0.319**	**0.339**	**0.268**
Cost saving (C_{31})	0.506	0.246	0.333	0.313	0.324	0.267
Flexibility in billing (C_{32})	0.494	0.239	0.278	0.328	0.362	0.269
Risk (D_4)	**0.259**	**0.252**	**0.245**	**0.227**	**0.249**	**0.277**
Labor unions (C_{41})	0.327	0.257	0.292	0.214	0.219	0.275
Loss of management control (C_{42})	0.351	0.255	0.208	0.218	0.248	0.288
Information security (C_{43})	0.322	0.242	0.235	0.249	0.278	0.268
Total gap (Rank)		**0.258** (4)	**0.245** (1)	**0.246** (2)	**0.254** (3)	**0.262** (5)

Source: Data from Liou, J.J.H. et al., *Information Sciences*, 266, 199–217, 2014.

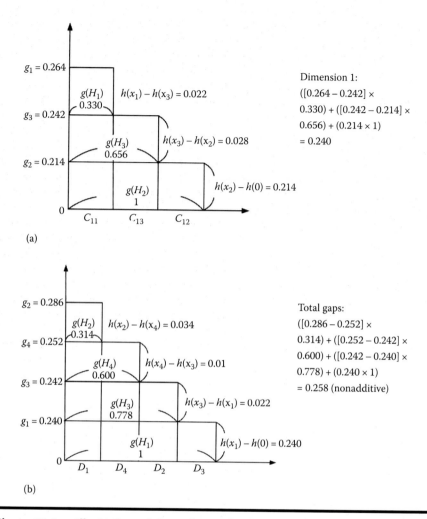

Figure 14.3 Illustration of fuzzy integral calculations: (a) fuzzy integral calculation for the gap in Dimension 1 in A_1, (b) fuzzy integral calculation for the total gap in A_1. (Reprinted with permission from Liou, J.J.H. et al., *Information Sciences*, 266, 199–217, 2014.)

the total weighted gap is aggregated from the synthesized values of the four dimensions. According to this fuzzy integral hybrid model, A_2 has the smallest weighted gap and hence should be selected.

Table 14.5 shows the outcome of the λ values in the four dimensions. When λ is equal to 0 (additive model), the gap is not affected during the synthesis/aggregation processes. However, the gap will increase after synthesis/aggregation when the dimension exhibits a substitutive effect ($\lambda < 0$). Conversely, the multiplicative effect ($\lambda > 0$) will reduce the gap after synthesis/aggregation. This phenomenon can

216 ■ *Trends of Hybrid Multiple Criteria Decision Making*

be observed in this empirical example. The multiplicative effect on *Quality* (D_2) reduces the gap of A_3, and the substitutive effect ($\lambda = -0.597$) within the dimensions increases the gap of A_3. Based on the plausible substitutive or multiplicative effects within the dimensions and the INRM, certain strategies can be derived for improvements. For example, for companies seeking to reduce the overall gap, controlling risk should be the most important task, as risk ranks first in the INRM and there is a substitutive effect among the dimensions.

14.5 Discussion on Improving toward the Aspired Levels

According to the global weights (Table 14.4) of the improvement/selection criteria, *Relationship* (C_{11}) is the most important criterion (11.2%) in supplier improvement/selection, followed by *Cost savings* (C_{31}) (10.3%) and *Flexibility in billing* (C_{32}) (10.1%). However, based on the INRM and the influential degree analysis (Table 14.3), *Cost* (D_3) has the lowest $r_j^D - c_j^D$ value. These interesting results indicate that managers do not believe that cost influences the other criteria; however, they nonetheless consider cost an important factor when evaluating a supplier. Furthermore, these results do not necessarily suggest that less attention should be paid to risk factors. In fact, Table 14.4 indicates that *Risk* (D_4) has the highest degree of influence given $r_j^D - c_j^D$, and will influence the other dimensions more than they influence it. In other words, risk considerations between the firm and its supplier will affect how the supplier fulfills the other needs of the firm, such as *Compatibility* (D_1) and *Quality* (D_2). It should again be emphasized that the proposed model is capable of handling such interdependencies. Another advantage of the proposed model is the observed directional influences between dimensions through the INRM (Figure 14.2), which provide improved supplier strategies. For example, the consideration of *Knowledge and skills* (C_{21}) has the highest value $r_i^C - c_i^C$ in the *Quality* (D_2) dimension, meaning that having employees with superior knowledge skills could lead to increased service quality and avoid the possibility of delayed flights.

In the traditional MADM approach, relative performance values are generally applied to prioritize the alternatives. However, in this new hybrid approach, DMs can set an aspiration level as the benchmark, which implies that alternatives should keep on improving until the target (i.e., zero gap) is reached. The performances are replaced by the weighted gaps that represent the direction of improvement between the alternative and the benchmark, which is more suitable in the contemporary business environment. Besides, the traditional approach can only determine the gaps between a company and its leading competitors; however, this hybrid approach not only helps companies discover the gaps between current performance and aspiration levels but also provides an opportunity for them to outperform their leading competitors. In addition, even the top leading supplier is required to keep on enhancing its performance in a dynamic and highly competitive environment. Take A_3, for example: in Table 14.6, its total gap (0.246) is very close to the top ranked A_2 (0.245); once A_3 surpasses A_2, it would be lack of a systematic guidance on continuous improvements using the traditional approach.

In the numerical results (Tables 14.3 through 14.6) of this case study, certain interesting findings can be learned. If *Cost savings* (C_{31}) is the only criterion, it is obvious that A_1 should be selected. However, when multiple criteria and network relationships are included in the evaluation system, the ranking result will change dramatically. The global weights from DANP (Table 14.4) indicate that *Compatibility* (C_{11}) is more important than *Cost savings* (C_{31}), and A_2 should be selected as the best service provider. These findings echo the discussions in Section 14.1; that is, (1) cost concern might not be the most crucial factor for supplier selection, and (2) the interrelationships among factors (including dimensions and criteria) have an impact on the overall evaluation.

14.6 Conclusion

In this chapter, a supplier evaluation model using the fuzzy integral–based nonadditive aggregation method is illustrated. Compared with the traditional approach, several advantages can be found in this new hybrid approach.

First, the traditional models assume that the criteria are independently and hierarchically structured; however, in reality, decision problems are frequently characterized by interdependent criteria and dimensions and may even exhibit feedback-like effects. To overcome this issue, this hybrid MADM model first applies the DEMATEL technique to explore the network relationships among variables; the DANP method is then used to derive the influence weights in a way that eliminates the time-consuming pairwise comparisons of the original ANP.

Second, relatively good solutions from the existing alternatives are replaced by aspiration levels to meet the demands of contemporary competitive markets. In this hybrid model, the VIKOR concepts are used to transform the performance levels into weighed gaps for each aspiration level. This enables a DM to reduce the gaps in alternatives to reach the aspiration level.

Third, the emphasis in MADM applications has shifted from ranking/selection to improving the performance of existing alternatives. INRM identifies how and in which directions the criteria influence each other, which helps managers to understand the root causes of performance issues and to devise strategies for improvement.

Fourth, nonadditive-type information fusion techniques, such as the fuzzy integral technique in this case, have been developed to aggregate performance values, which should be more reasonable than conventional additive models (Shen et al. 2017). The empirical example indicates that the effect of the interdependencies among criteria is

218 ■ *Trends of Hybrid Multiple Criteria Decision Making*

significant, which further supports the validity of adopting the nonadditive aggregator to model complicated MADM problems.

The plausible reasons for adopting this new hybrid MADM approach have been highlighted in Chapter 2; here, a real case of supplier selection has been examined and illustrated to show the advantages of this hybrid model. Interested readers are suggested to make comparisons, using the traditional methods and this approach, with their own MADM problems, and the distinctions should deepen their understanding of this new approach.

Chapter 15

New Perspectives on Modeling Strategic Alliances by De Novo Programming

While strategic alliances have been extensively discussed and explained by various economic or management theories, how to guide a strategic alliance on optimal resource allocation is rarely explored. As a result, this chapter not only aims to propose an integrated framework for modeling strategic alliances based on multi-objective programming (MOP) but also intends to guide the subsequent resource allocation decisions of alliance members. The mainstream theories on strategic alliances will be discussed, and a new perspective, termed *De Novo programming*, will be proposed that models strategic alliances considering both the plausible synergy and scale effects brought by the cooperation of alliance members. A numerical example will be used to illustrate how to adopt De Novo programming for evaluating strategic alliances.

15.1 Introduction to Strategic Alliances and Literature Review

Although there are numerous types of strategic alliances, in general, a strategic alliance may be defined as a formal agreement by at least two individual companies to share or exchange resources for long-term cooperation. Various motivations

220 ■ *Trends of Hybrid Multiple Criteria Decision Making*

for forming a strategic alliance have been discussed (Wassmer 2008), such as (1) reaching economies of scale, (2) overcoming legal and trade barriers, and (3) reducing the cost and risk of entering a new market. The case in this chapter mainly covers the first motivation. The formation of strategic alliances has been widely examined; several economic/management theories—transaction cost theory, resource-dependent theory, and the strategic behavior and organizational learning perspectives—are briefly introduced in the following subsections.

15.1.1 Transaction Cost Theory

Transaction cost theory was proposed by Coase (1937) to explain the decisions of businesses when exchanging resources from the market or internalizing within the hierarchy of a company. Transaction costs can be regarded as a threshold; once the transaction cost of exchange is too high, companies will look for internal mechanisms to accomplish their missions; if not, the exchange of resources, between two or more companies, will continue. The original transaction cost theory only explains certain extreme conditions; it was extended by the further research of Williamson (1991) to explain strategic alliances. The extension of transaction cost theory highlighted two types of costs, those of transaction and production, to determine the decision between market or hierarchy. In addition, the cost of managing relationships and the possible costs of inferior governance should also be considered.

In a sense, transaction cost theory is a reasonable approach to explaining the formation of strategic alliances; nevertheless, it is inclined to focus on minimizing the cost of a single party. In other words, if a strategic alliance is taken as a system, transaction cost theory focuses on optimizing a subpart of the system to explain the formation of an alliance; a holistic approach that can cover the whole system's costs or benefits is still needed. This insufficiency gave rise to the resource-based view, also named *resource-dependent theory*.

15.1.2 Resource-Dependent Theory

To explain strategic alliances based on the whole system's interests, resource-dependent theory shifts its attention to the *resources* that might bring synergies to its alliance members (Grant 1991; Barney et al. 2001). Roughly speaking, resources fall into two categories: tangible (e.g., financial results and patents) and intangible resources (e.g., bargaining power). Alliance members seek to acquire or leverage the complementary competences/resources from the alliance, which can be regarded as a dynamic adaption process (Teece 2009) for the alliance members to stay competitive.

Although there are multiple ways (e.g., hierarchy, market, and alliances) for a company to strengthen or accumulate its resources, the assumed heterogeneity among different companies implies that alliances often yield superior outcomes. Consequently, by forming or joining a strategic alliance, a company may plausibly gain synergies from the complementary resources that the other alliance members possess.

The resource-based view has provided a broader perspective to explain strategic alliances. However, one critical issue remains: what should a company do to optimize its objectives after joining/forming a strategic alliance? In other words, the optimization of resource allocations is key to whether a company can earn synergies and be competitive in the market; in this chapter, we adopt the De Novo perspective to explore this issue.

15.1.3 Strategic Behavior and Organizational Learning Perspectives

As well as the transaction cost and resource-dependent theories, other theories from management science also compete to explain the formation of strategic alliances, such as the strategic behavior and organizational learning perspectives. Compared with transaction cost theory, which emphasizes minimizing the total relevant costs, the strategic behavior perspective seeks to maximize the long-term benefits (Kogut 1988). As a result, by using different theories to explain strategic alliances, the motivations for alliance formation or the selection of alliance partners will not be the same. The last perspective discussed here is organizational learning theory (Contractor and Lorange 1988; Hotho et al. 2015), which focuses on learning complementary skills and knowledge from the other alliance members; the ultimate goal is for a company to form its own core competences or competitive advantages by learning from the strategic alliance. It is relatively difficult to precisely measure the two perspectives discussed in this subsection within a specific time frame; therefore, the subsequent resource allocation model of strategic alliances will mainly be grounded on transaction cost and resource-dependent theories.

15.2 Resource Allocation and De Novo Programming Perspectives

It is obvious from the previous discussions on strategic alliances that resource and cost concerns are critical factors in the formation of strategic alliances. Nevertheless, the aforementioned economic and management theories only provide explanations for the motivations of strategic alliances; how to allocate scarce/valuable resources is still unanswered. To tackle this critical issue and yield concrete results, mathematic programming technique can be used, which is a prevailing method for resolving optimization problems.

Conventional mathematic programming allocates scarce resources efficiently under certain constraints to reach an optimal goal/objective, and linear programming may well be the most prevalent method. A typical linear programming model is described in Equation 15.1:

222 ■ Trends of Hybrid Multiple Criteria Decision Making

$$\text{MAX } \boldsymbol{Cx}$$

$$s.t.$$

$$\boldsymbol{Ax} \leq \boldsymbol{l}, \tag{15.1}$$

$$\boldsymbol{X} \geq \boldsymbol{0},$$

where:

C denotes the matrix of objective functions' parameters

x denotes the vector of amount/quantity of each variable of the system to be optimized

A denotes the matrix of limitation parameters

l is the vector of limitations of the constraints

Although linear programming provides a plausible approach to the resource allocation problem, there is a potential drawback to modeling strategic alliances: the assumed additivity of a linear programming system. Linear programming (or MOP) presumes that all productive resources are independent and the total effect equals the summation of each individual effect; this presumption excludes the possibility of the synergy effect that a strategic alliance might yield, which is likely to result in a biased decision.

The side effect of resource independence can be explained using a newly observed phenomenon in the manufacturing industry: *mass customization*. Traditionally, manufacturers have two ways to enlarge profits: (1) reducing unit costs by reaching economies of scale and (2) increasing unit revenues by customization, which entails higher unit production costs. Under the assumption of resource independence, manufacturers are unlikely to reduce unit costs and increase unit revenues simultaneously. However, in mass customization, the assumption of resource independence is removed to explain this phenomenon.

Taking a different perspective, De Novo programming (Zeleny 1982, 1986) attempts to redesign or reshape a system for optimization, relaxing the assumption of resource independence; instead, it assumes that resources are not only dependent but also can be brought from the market. Therefore, the original system, including the constraints, can be redesigned to allow the existence of synergy effects (i.e., $1 + 1 > 2$). In this regard, the De Novo perspective is applied, combining both transaction cost and resource-dependent theories, to model the formation of strategic alliances.

Hereby, a simplified case of two companies (α and β) is illustrated to compare traditional linear programing with the De Novo perspective on modeling strategic alliances. Assuming a plausible alliance comprises companies α and β, P_α and P_β denote the profits in α and β, respectively. The goal of these companies is to maximize their profits, as shown in Figure 15.1. Usually, compromises are the most efficient solutions, residing on the curve $\widehat{\alpha\beta}$. Nevertheless, the ideal/aspired point γ cannot be attained in traditional linear programming solutions. On the other

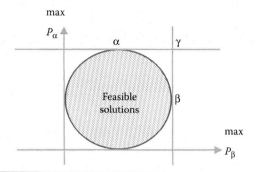

Figure 15.1 Feasible solutions using linear programming. (Revised and reprinted with permission from Huang, J.J. et al., *Mathematical and Computer Modelling*, 41(6), 711–721, 2005.)

hand, the De Novo perspective attempts to overcome the limitations to achieving the ideal/aspired solution by reshaping a MOP system based on the transaction cost and resource-dependent theories (Figure 15.2). A generalized model comprising N possible alliance members can be explained as follows:

a. Transaction cost theory
Assume that there are mainly three types of costs covered in the formation of strategic alliances of production companies: (1) transaction costs, (2) production costs, and (3) alliance costs (e.g., shared operational and information alignment costs). The logic for explaining the formation of strategic alliances can be summarized as follows:
IF (alliance cost) $\leq \sum_{i=1}^{N}$ (individual cost), THEN (the companies tend to form strategic alliances).

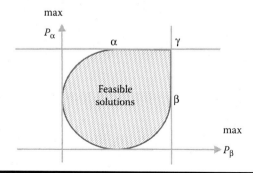

Figure 15.2 Feasible solutions from the De Novo perspective. (Revised and reprinted with permission from Huang, J.J. et al., *Mathematical and Computer Modelling*, 41(6), 711–721, 2005.)

224 ■ *Trends of Hybrid Multiple Criteria Decision Making*

b. Resource-dependent theory

The view of resource-dependent theory is that companies tend to form strategic alliances while the synergy effect exists. The logic is as follows:

IF (alliance total revenue) $\geq \sum_{i=1}^{N}$ (individual company revenue), THEN (the companies tend to form a strategic alliance).

Combining these two theories, the formation of a strategic alliance by two companies (α and β) can be expressed as

IF ($R(\alpha \cup \beta) - C(\alpha \cup \beta) > R(\alpha) + R(\beta) - C(\alpha) - C(\beta)$), THEN (companies α and β tend to form a strategic alliance),

where $R(\bullet)$ and $C(\bullet)$ denote the revenue and cost functions, respectively. One thing needs to be noted here: $C(\alpha \cup \beta)$ includes not only the production costs and transaction costs, but also the alliance costs. This logic can be easily extended for multiple (i.e., more than two) alliance members. The following set of functions (Equations 15.2a and 15.2b) more formally describe the tendency of companies to form strategic alliances (denoted by $T(\bullet)$).

$$T(\alpha) = \begin{cases} 1, \lambda \times R(\alpha \cup \beta) - \theta \times C(\alpha \cup \beta) > R(\alpha) - C(\alpha), \\ 0, \lambda \times R(\alpha \cup \beta) - \theta \times C(\alpha \cup \beta) < R(\alpha) - C(\alpha), \end{cases} \tag{15.2a}$$

and

$$T(\beta) = \begin{cases} 1, (1-\lambda) \times R(\alpha \cup \beta) - (1-\theta) \times C(\alpha \cup \beta) > R(\beta) - C(\beta), \\ 0, (1-\lambda) \times R(\alpha \cup \beta) - (1-\theta) \times C(\alpha \cup \beta) < R(\beta) - C(\beta), \end{cases} \tag{15.2b}$$

where:

λ denotes the percentage of increased alliance revenue in α
θ denotes the percentage of reduced alliance costs in α

Based on the aforementioned logical expressions yielded from transaction cost and resource-dependent theories, the De Novo perspective can be applied to support the modeling of a strategic alliance within a given budget; a numerical example is provided in the next section.

15.3 Numerical Example

A numerical case from Zeleny (1995) is used to demonstrate how to adopt the De Novo perspective on modeling strategic alliances, which is revised from the previous work by Huang et al. (2005d). Assume that there are two manufacturing companies, A and B, and they are considering forming an alliance; the two companies only produce two products (x_1 and x_2) by using five raw materials ($m_1, m_2, \ldots, m_5$), and the unit market prices for each raw material are as follows ($ per unit): $p_1 = 30$,

New Perspectives on Modeling Strategic Alliances ■ 225

$p_2 = 40$, $p_3 = 9.5$, $p_4 = 20$, and $p_5 = 10$. In addition, the cost function of B is assumed to be 90% of company A's result (i.e., $C(B) = 0.9 \times C(A)$). Objective functions f_1 and f_2 indicate the revenue functions for product 1 and product 2, which are to be maximized, and the total budget of companies A and B are both \$1300 (i.e., $B_A = B_B = 1300$). The optimization problem can be rewritten as follows:

$$\text{MAX } f_1 = 400x_1 + 300x_2$$

$$\text{MAX } f_2 = 6x_1 + 8x_2$$

s.t.

$$4x_1 \le 10,$$

$$2x_1 + 6x_2 \le 12,$$

$$12x_1 + 4x_2 \le 30,$$

$$3x_2 \le 5.25,$$

$$4x_1 + 4x_2 \le 10,$$

$$x_1, x_2 \ge 0.$$

Using traditional mathematical programming, the optimized results for (x_1, x_2) are (2.125, 1.125). The total revenue can be summed up by adding f_1 to f_2: $1187.5 + 21.75 = 1209.25$. The profit for company A can be written as $1209.25 - C(A)$; similarly, the profit for company B is $1209.25 - C(B)$ or $1209.25 - 0.9 \times C(A)$.

Suppose that the total budget for this alliance is 2600 ($B_{A \cup B} = 2600$), and the MOP can be optimized for two individual systems separately:

$$\text{MAX } f_1 = 400x_1 + 300x_2$$

s.t.

$$354x_1 + 378x_2 \le 2600,$$

$$x_1, x_2 \ge 0.$$

Then, if x_1 and x_2 are 7.34 and 0, respectively, f_1 is 2937.85, which is the optimized outcome considering merely objective f_1.

$$\text{MAX } f_2 = 6x_1 + 8x_2$$

s.t.

$$354x_1 + 378x_2 \le 2600,$$

$$x_1, x_2 \ge 0.$$

226 ■ Trends of Hybrid Multiple Criteria Decision Making

Similarly, if x_1 and x_2 are 0 and 6.88, respectively, f_1 is 55.03; the optimized revenue for product 2 can thus be reached.

The ideal (f_1, f_2) for this MOP problem is (2937.85, 55.03) if $x^* = (7.34, 6.88)$. However, this ideal outcome requires a total budget of \$5200 ($B^*_{A \cup B} = 5200$), which is not attainable under the assumed budget of \$2600 ($B_{A \cup B} = 2600$). As a result, the De Novo perspective suggests that the optimal-path ratio equals 0.5 ($B_{A \cup B} / B^*_{A \cup B} = 2600 / 5200 = 0.5$), which yields the associated $x^{**} = (0.5 \times 7.34, 0.5 \times 6.88) = (3.67, 3.44)$. The optimized revenue of this alliance using the De Novo perspective would be $(400 \times 3.67 + 300 \times 3.44) + (6 \times 3.67 + 8 \times 3.44) = 2549.54$; the profit of this alliance would then be $2549.54 - C(A \cup B)$.

Although the alliance's revenue is higher than the sum of the two individual companies ($2549.54 > 1209.25 + 1209.25$), the reasons for those two companies to form a strategic alliance are still not sufficient. Be aware of the plausible synergy effect; the requirements for forming a strategic alliance using the De Novo perspective are as follows:

$$2549.54 - C(A \cup B) > \left(1209.25 - C(A)\right) + \left(1209.25 - C(B)\right). \quad (15.3)$$

In addition, referring to Equations 15.2a and 15.2b, the high tendency of companies A and B to join this alliance is shown in Equations 15.4 and 15.5:

$$\lambda_A \times 2549.54 - \theta_A \times C(A \cup B) > 1209.25 - C(A), \quad (15.4)$$

$$\lambda_B \times 2549.54 - \theta_B \times C(A \cup B) > 1209.25 - C(B), \quad (15.5)$$

where:

λ_A or λ_B denote the percentage of increased alliance revenue in A or B

θ_A or θ_B denote the percentage of reduced alliance cost in A or B

Since $C(B) = 0.9 \times C(A)$, Equation 15.3 can be rewritten as

$$C(A \cup B) < 131.04 + 1.9 \times C(A).$$

15.4 Discussion

It can be seen that mainstream economic or management models such as the transaction cost and resource-dependent theories are able to explain the formation of strategic alliances with persuasive reasoning. Nevertheless, the subsequent resource allocation problem for alliance members is neither highlighted nor discussed with concrete solutions. Therefore, in this chapter, the De Novo perspective is adopted to resolve this critical problem.

New Perspectives on Modeling Strategic Alliances ■ 227

Table 15.1 Comparisons of Different Perspectives Explaining Strategic Alliances

	Transaction Cost Theory	Resource-Dependent Theory	De Novo Perspective
Level of analysis	Company	Company	Alliance and company
Focus of analysis	Transaction costs	Complementary resources	Both
Method for resource allocation	N/A	N/A	De Novo programming
Motive for strategic alliances	Minimize individual cost	Maximize individual profit or value by leveraging the alliance's resources	Both

In addition, it is insufficient to consider the formation of a strategic alliance from merely a single angle; the whole alliance's perspective and the motivations of each alliance member all should be addressed, but have been overlooked in previous economic/management theories. The previous mainstream theories and the De Novo perspective on strategic alliances are summarized and compared in Table 15.1.

Based on the De Novo perspective on the illustrated numerical example, Equation 15.3 is able to analyze the required conditions to form this alliance. Furthermore, the sufficient conditions for each alliance member to join the alliance are also established in Equations 15.4 and 15.5. Decision maker(s) can derive the optimal path to reach the ideal point in the presence of a budget constraint. Real cases in the modern business environment are usually much more complex, and the constraints might involve hundreds or even thousands of equations. However, the De Novo perspective has paved a way to analyzing the motivations and resource allocations of strategic alliances with a holistic approach.

15.5 Conclusion

In the competitive and dynamic business environment, companies are facing numerous challenges, and internal hierarchy is the typical approach for a company to take to adapt and survive. Once an individual company has found certain candidates in the market, the formation of a strategic alliance is an attractive alternative if the alliance can offer additional value or incentive for the company. Previous economic and management theories mainly use an individual company's angle to measure the pros and cons of joining an alliance; this is enhanced by the proposed De Novo perspective on evaluating the alliance and its members.

Furthermore, how strategic alliances allocate resources efficiently is not discussed in both transaction cost and resource-dependent theories; this important issue is suitable to be solved by mathematical programming. However, owing to the assumption of additivity, the synergy effect is not able to be modeled by traditional linear programming techniques, which might be the dominant concern when forming a strategic alliance; the De Novo perspective offers a new approach to resolving this issue. Through the restructuring of the resource allocations of the alliance members, the alliance will have chance to devised an approach to reaching its ideal/aspired goal, considering internal and external resources, within a given budget constraint. This is the major advantage of the De Novo perspective: redesign a system for optimization.

Chapter 16

Automated Factory Planning Using the New Idea of Changeable Spaces

Two types of decisions, ranking/selection and optimization, are often confronted by decision makers (DMs) when handling business/management operations. On the one hand, ranking/selection decisions require existing alternatives to be evaluated in multiple aspects, which belong to multiple attribute decision making (MADM) research. On the other hand, optimization decisions that belong to the field of MODM involve single or multiple objectives, and DMs aim to allocate limited and valuable resources (e.g., budgets) within a set of constraints to reach ideal or compromised outcomes. In this chapter, the optimization of automated factory planning—by balancing the capacities of skilled laborers and automated robots—is illustrated by adopting the new idea of *changeable spaces* from MODM.

Typically, a decision model formulated for multiobjective or single-objective programming is limited by fixed constraints that reflect the boundaries of a given system in feasible solutions. When dealing with a MODM problem, *trade-offs* are often required, which means that one cannot increase the levels of satisfaction of one objective without decreasing those of another.

Traditional optimization methods in multiobjective problems are only concerned with the *Pareto-optimal* solutions (also called *noninferior*, *nondominated* solutions, or *effective* solutions) within a system, rather than with the concept of optimality itself or with expanding the notion of true optimization (Zeleny 1990). As introduced in

229

230 ■ Trends of Hybrid Multiple Criteria Decision Making

Chapter 6, De Novo programming proposes a fixed-budget approach to calculating an optimization ratio toward an ideal point; however, as Zeleny mentioned in a previous study (1995), the notions of true optimization are still underexplored. To bridge this gap, changeable spaces are illustrated and compared in this chapter with the De Novo programming approach, followed by a discussion.

16.1 Background of the Case

More recently, there has been a shift in attention in production-oriented manufacturing from low labor costs to the quality and flexibility of automated manufacturing systems. The emphasis on low-cost production is also termed the *original equipment manufacturing* (OEM) business strategy (Hobday 1995), still prevalent among developing countries such as the Philippines, Vietnam, and China, and a part of the global information technology (IT) supply chain. However, advancements in automated robots and information systems have fostered the progress of unmanned factory production; on the business side, the cost concern of automated production is also diminishing. As a result, automated or unmanned factory production can be regarded as either a threat or an opportunity by those countries that have thrived by leveraging the low cost of labor (or land) in the OEM strategy. However, to embrace unmanned factory production is not a one-stop matter. The existing skilled laborers, facilities, manufacturing procedures, and systems all need to be redesigned and upgraded step by step, based on the constraints at each stage. During the migration process, it can be regarded as a continuing improvement process toward the final ideal/aspired level: fully automated factory production.

To support the aforementioned OEM manufacturers as they head toward the ideal of automated unmanned factory production, this chapter proposes a MODM approach to modeling this problem. The road to fully automated production can be regarded as a path between the use of laborers and robots. This is a challenging task. DMs will encounter conflicts between the allocation of the workforce to improve production standards and the allocation of robots to automate production. In addition, as well as the cost concern, other objectives such as flexibility, speed, or quality cannot be ignored. Therefore, the establishment of an unmanned factory is the ultimate goal. A guide that shows how to manage/allocate scarce resources while considering multiple objectives, and how to reach this destination effectively and efficiently, is valuable in practice.

A wafer-manufacturing OEM company is analyzed as a case study here, revised from the previous work by Tzeng et al. (2014). This company wants to optimize operational profits and manufacturing quality simultaneously; the number of industrial robots and workforce personnel are the two production resources to be allocated, under a budget concern. Two approaches from MODM are discussed and compared in this chapter, and the concept of changeable spaces, which offers a new perspective on continuous improvements toward an ideal/aspired level, will be explained with an illustration.

16.2 Research Framework and the Evolution of Optimization

This work adopts a new idea combining multiobjective programming (MOP) with the changeable space MODM model to solve an optimization problem: the resource allocation of a wafer manufacturer. Changeable space programming is based on the fundamental concept of De Novo programming (Zeleny 1982, 1986, 1995; Tzeng and Huang 2014). Here, it deals with the trade-offs involved in optimizing the use of industrial robots and a workforce of skilled laborers to achieve the aspiration level, considering multiple objectives.

In this case, the wafer manufacturer is attempting to optimize its operational profit (f_1) and manufacturing quality (f_2) simultaneously. Two manufacturing resources—the number of industrial robots (x_1) and workforce personnel (x_2)—are to be determined for the optimal allocation. The associated technical coefficients of this case are shown in Table 16.1. The average operational profits from wafer sales brought by each unit of automated robots and each unit of laborers are \$500 (x_1) and \$100 (x_2), respectively. Moreover, the second objective function (f_2) is to maximize the total wafer-manufacturing quality index: 100 points per x_1 and 90 points per x_2. The two objectives are assumed to be equally important.

With the technical coefficients in Table 16.1 and the two aforementioned objectives (f_1 and f_2), this optimization problem can be formulated as follows:

$$\text{MAX} f_1 = 500x_1 + 100x_2$$

$$\text{MAX} f_2 = 100x_1 + 90x_2$$

s.t.

$$180x_1 + 120x_2 \leq 840,$$

$$90x_1 + 60x_2 \leq 480,$$

$$90x_1 + 90x_2 \leq 540,$$

$$10x_1 \leq 50,$$

$$100x_1 + 50x_2 \leq 500,$$

$$x_1 \geq 0,$$

$$x_2 \geq 0.$$

Since the assigned number of automated robots and workforce personnel should be integers, this optimization model can be solved as an integer optimization problem; several MODM approaches to this optimization problem are illustrated in the following subsections.

232 ■ *Trends of Hybrid Multiple Criteria Decision Making*

Table 16.1 Technical Coefficients of the Unmanned Factory Planning Problem

Constraints	Decision Variables		Resource Limitations
	x_1	x_2	
Setup time	180	120	840
Maintained time	90	60	480
Running test time	90	90	540
System changing	10	0	50
Troubleshooting	100	50	500

16.2.1 *Conventional Pareto Solution*

To solve this optimization using the conventional Pareto solution approach, the two objectives should be optimized separately to find the ideal point for each objective first; then, a compromised MODM programming technique can be applied to combine the two objectives into a single objective function. Under the first objective (f_1), the current ideal result will be 2100 (i.e., $f_1 = 2100$), while $(x_1, x_2) = (4, 1)$. Similarly, the ideal result under the second objective solely will be 560, while $(x_1, x_2) = (2, 4)$. Then, using a compromised programming method, where $p = 2$ (refer to Equation 6.10), the combined compromise objective can be defined as follows under the same set of constraints within a budget of \$232,900:

$$\text{MIN} \left[\left(0.5 \times \frac{2100 - (500x_1 + 100x_2)}{2100} \right)^2 + \left(0.5 \times \frac{560 - (100x_1 + 90x_2)}{560} \right)^2 \right]^{\frac{1}{2}}.$$

Then, if there is no budget limitation, the ideal $(x_1, x_2) = (4, 1)$, using the Pareto solution approach; the associated $(f_1, f_2) = (2100, 490)$.

16.2.2 *De Novo Programming for Optimization*

As explained in Chapter 6, using information on the resource unit prices in a system, De Novo programming further eliminates trade-offs among objects to reach the optimal resource allocation (Zeleny 1990; Huang et al. 2005a, 2006c; Chen and Tzeng 2009). Assuming that the total budget was given as \$232,900 for this wafer manufacturer, the unit price/cost of each constraint is shown in Table 16.2.

According to De Novo programming, this problem can be redefined as the following model by minimizing the total cost.

Table 16.2 Unit Price for Each Constraint

Constraints	Unit Prices
Setup time	50
Maintained time	30
Running test time	100
System changing	450
Troubleshooting	200

$$\text{MIN } Cost = 45,200x_1 + 26,800x_2$$

s.t.

$$500x_1 + 100x_2 \geq 2100,$$

$$100x_1 + 90x_2 \geq 560,$$

$$x_1 \geq 0,$$

$$x_2 \geq 0,$$

where x_1 and x_2 should be integers. This model can be solved easily with a minimal *Cost* of \$234,400, where $(x_1, x_2) = (4, 2)$. Since the minimal *Cost* exceeds the assumed budget (\$232,900), an optimum-path ratio can be obtained (\$232,900 / \$234,400 = 0.9936) to approach the ideal point under the budget's constraint. Resources $(x_1, x_2) = (3.54, 1.77)$ when 0.9936 is multiplied by 4 and 2, respectively. However, owing to the requirement of x_1 and x_2 being integers, the optimized (x_1, x_2) by De Novo programming could be either (4, 1) or (3, 2); the associated (f_1, f_2) will reach (2100, 490) or (1700, 480), respectively. Since the objective functions are superior while $(x_1, x_2) = (4, 1)$, the optimal solution by De Novo programming is the same as the conventional solution while x_1 and x_2 are integers.

16.2.3 New Ideas of Changeable Spaces

The compromised results of De Novo programming are confined by a fixed-budget constraint. If we take a new perspective—termed as the changeable spaces approach—to explore this problem, such as external financing to resolve the fixed budget limitation, the obtained result could move one step closer to the ideal/aspired point. For example, assume that the wafer manufacturer can secure an additional loan of up to \$100,000 at a 4% interest rate. The optimization can be rewritten as follows:

$$\text{MIN } 1.04 \times Loan$$

s.t.

$$500x_1 + 100x_2 \geq 2100,$$

$$100x_1 + 90x_2 \geq 560,$$

$$50 \times (180x_1 + 120x_2) + 30 \times (90x_1 + 60x_2) + 100 \times (90x_1 + 90x_2) +$$

$$450x_1 + 200 \times (100x_1 + 50x_2) \leq 232900 + Loan,$$

$$x_1, x_2 \geq 0.$$

The minimal *Loan* will be $1500 by solving the above linear programming model, and the optimized (x_1, x_2) will be (4, 2); the ideal/aspired (f_1, f_2) will be (2200, 580), which are superior results compared with those of the Pareto solution and De Novo programming approaches.

Aside from raising additional funding, the objective space can also be altered to enhance this optimization problem. For example, assuming that this wafer manufacturer hopes to increase its operational profit (f_1) and manufacturing quality (f_2) targets from (2100, 560) to (2690, 700) by launching an upgraded subsystem, including reengineered software and work processes as production aids, the associated unit improvement costs and contributions (of x_1 and x_2) for each objective are shown in Table 16.3.

This model incorporates the required unit improvement cost and contributions of x_1 and x_2 for each objective, which changes the objective space of this MOP problem.

Table 16.3 Associated Unit Improvement Costs for the Two Objectives

Two Objectives	Existing Targets	New Targets (Ideal/Aspired Level)	$x_1 = 1$	$x_2 = 1$
f_1 Unit contributions	2,100	2,690	25	23
(Unit improvement costs)			($200.12)	($0.58)
f_1 Unit contributions	560	700	20	28
(Unit improvement costs)			($30.32)	($0.87)

$$\text{MIN } Cost$$

$$s.t.$$

$$(500 + 25)x_1 + (100 + 23)x_2 \geq 2690,$$

$$(100 + 20)x_1 + (90 + 28)x_2 \geq 590,$$

$$50 \times (180x_1 + 120x_2) + 30 \times (90x_1 + 60x_2) + 100 \times (90x_1 + 90x_2) +$$

$$450x_1 + 200 \times (100x_1 + 50x_2) \leq 232{,}900 + Cost,$$

$$x_1, x_2 \geq 0,$$

where $Cost = 200.12x_1 + 0.58x_2 + 30.32x_1 + 0.87x_2$. The newly obtained $(x_1, x_2) = (5, 1)$, and the required minimal $Cost = \$1154$ (x_1 and x_2 still have to be integers in this model). The optimized results from the three approaches are summarized in Table 16.4.

16.3 Discussion

The case of the wafer manufacturer has shown how to pursue resource optimization based on various MOP approaches. Conventional MODM methods, such as the goal-programming or compromised MOP techniques, search for an ideal point within a bounded system, considering the trade-offs between multiple objectives. Zeleny (1986) offers a different viewpoint on the MODM problem for redesigning an optimal system: De Novo programming. In this approach, most resources are assumed to be purchasable/exchangeable from the market at reasonable prices, and the only constraint should be the total budget. As such, the main difference

Table 16.4 Comparison of the Three Approaches for MOP Optimizations

	Conventional MODM	De Novo Programming	Changeable Spaces (Additional Funding)	Changeable Spaces (Additional Improvements)
(f_1, f_2)	(2100, 490)	(2100, 490)[a]	(2200, 580)	(2690, 700)
Budgets	\$232,900	\$232,900	\$234,400	\$234,054
(x_1, x_2)	(4, 1)	(4, 1)	(4, 2)	(5, 1)

[a] While $(x_1, x_2) = (3.54, 1.77)$, using the optimum-path ratio, the associated $(f_1, f_2) = (1947, 513.3)$.

236 ■ *Trends of Hybrid Multiple Criteria Decision Making*

between De Novo programming and traditional MODM is that De Novo programming determines the resource reallocation of the redesigned system.

In the newly devised system, its objective functions, subject to existing constraints, can eliminate trade-offs and thus achieve an ideal solution (i.e., ideal point), according to the reconstructed decision space. Take the numerical example in Subsection 16.2.2, for example: the new budget ($232,900) was less than the required ideal budget ($234,400) for optimization; the optimum-path ratio ($232,900 / $234,000 = 0.9936) could thus be applied to derive the associated manufacturing resources x_1 and x_2 accordingly. Although the notion of reforming an MOP system, within a fixed budget, indicates the potential to enhance an existing system using De Novo programming, many more possibilities and potentials can be evaluated and planned.

Based on the concept of using changeable spaces to achieve the ideal point (Tzeng and Huang 2014), the two numerical examples that change the budget limitation and the objective space in Subsection 16.2.3 have shown how to redesign a system for further optimization. Businesses often encounter constraints to reaching an ideal or aspired level in practice; however, innovative management teams should find more resourceful ways of thinking instead of using fixed mindsets to deal with bottlenecks. Financial, marketing, research and development (R&D), and human resources (HR), all can be considered to alter the current system for reaching superior outcomes. This is the key spirit of changeable spaces in practical and continuous improvement planning.

16.4 Conclusion

In this chapter, three approaches—conventional compromised MODM, De Novo programming, and changeable spaces—are illustrated regarding the production resource optimization of a wafer manufacturer. The first, the compromised MOP approach, is adopted to solve optimization problems with fixed constraints, and merely searches for feasible solutions within a given decision space. Subsequently, De Novo programming introduces the idea of resource exchanges by considering the unit cost of each resource, which assumes that existing constraints can be relaxed by purchasing certain limited resources from the market, within a specific budget. Although this move has opened a new window for conventional MODM research, it is still confined by a fixed budget. In other words, the concept of changeable spaces was proposed based on the essential idea of De Novo programming to explore broader spaces, including objective spaces, decision spaces, technical coefficients, and budget limitations; this new concept can be adopted to redesign or redefine an optimization system. This is in line with business practices in multiple aspects. Since businesses reside in a dynamic environment, management teams or DMs have to adapt to constraints to reach their objectives. During this dynamic adaption process, a company usually has to try all available means of optimizing its value. Marketing activities, financial instruments, and R&D enhancements can all be applied to meet this end. As a result, the new perspective that changeable space programming has brought is close to business practice; furthermore, it offers a managerial tool for businesses to plan for continuous or consecutive improvements, which is still underexplored in management science, in the context of MODM research.

In terms of industrial applications, the primary benefits of setting up an automated factory are the reduced production times and increased flexibility, along with optimized operational profits and production quality. It is also possible to decrease the operation times (e.g., setup, maintenance, and system change) and efficiently plan the work to be undertaken by the unmanned factory. Although current OEM manufacturers mainly care about the profitability (i.e., low cost) and quality brought about by automated productions, it can also be foreseen that other benefits or objectives (e.g., the safety and health of laborers) will be taken into consideration in the long run. Changeable space programming proposes a flexible way for DMs to plan for various improvement plans to reach the ideal/aspired results.

Further research into the proposed method is needed to develop technical innovations for intelligent robots. Parameters such as budget, resources, and technical upgrades can be changed or even expanded within the decision spaces of DMs to move beyond the current limitations. Meanwhile, the study of dynamic, changeable decision space transition modeling is needed to extend our understanding of how to redesign optimal systems effectively. In the new era of intelligent robot-based production, the problem of how to redesign an optimal system will be more important than the optimization of an existing fixed system.

Chapter 17

Fuzzy Inference-Supported MRDM for Technical Analysis: A Case of Stock Investment

This chapter aims to construct an intelligent decision support system, based on technical analysis (TA), to aid investments. TA is a prevailing investment approach in financial markets, ranging from trading foreign exchange derivatives to stock investments. Although the validity of TA has been examined extensively in the literature (e.g., Park and Irwin 2007), previous research has mainly explored the effectiveness of each technical indicator separately. Sometimes, different TA indicators might generate divergent signals; a practical system that can consider the inconsistency of various technical indicators simultaneously is still underexplored. To bridge this gap, a hybrid model is proposed that integrates the variable-consistency dominance-based rough set approach (VC-DRSA) with fuzzy inference–enhanced technical signals, and which adopts the idea of multiple rules–based decision making (MRDM) to support stock investment decisions. A numerical case using historical data from the weighted average index of the Taiwan stock market (from 2002 to 2014) is illustrated, with positive findings.

17.1 Background of Technical Analysis and Computational Intelligence Techniques

TA is widely applied by institutional and individual investors in various financial markets, and mainly adopts trading information—such as trading volumes and the high, low, opening, and closing prices of an underlying instrument—to form signals. Since applying TA does not require any financial knowledge, it has been overlooked in formal financial education for decades. However, nearly all online trading platforms and web-based financial market services (e.g., Yahoo Finance) offer various kinds of technical indicators and charts, which implies its popularity in financial markets. The importance (or influence) of TA, then, deserves more attention. According to a survey (Menkhoff 2010), nearly 87% of 692 fund managers across 5 countries, would refer to TA—with at least certain degree of importance—to support their investment decisions. Owing to the prevalence of TA among professionals, its effectiveness is affirmed. However, the ways to interpret TA indicators diverge; in addition, different TA indicators sometimes generate inconsistent suggestions. When making investment decisions, investors (decision makers [DMs]) still encounter difficulties gathering clear and useful guidance from TA.

Despite its broad usage, TA is often criticized for lacking theoretical support and economic implications (Menkhoff 2010). The rise of behavioral finance has enabled researchers to explain recurrent price patterns, which are thereby detected by TA indicators, based on the irrationality (e.g., greed, myopia, and fear) of investors (Shiller 2003; Park and Irwin 2007); the present chapter holds a similar viewpoint, assuming that some price patterns will reoccur due to certain commonly held mentalities of investors.

Aside from the debate on what makes TA useful, researchers are also interested in examining the effectiveness of TA indicators/signals. According to a comprehensive survey (Park and Irwin 2007), standard TA studies used to rely on statistical methods to examine the validity of each technical indicator (or rule) separately. This kind of research design is often constrained by prerequisite assumptions about the probabilistic distribution of variables and the linear relationships of a model, which makes it far removed from practice (Wei et al. 2011). Furthermore, it is common for investors to refer to more than one technical indicator simultaneously to make a decision; the aforementioned studies neglected this joint consideration issue.

Other than statistics, various computational intelligence techniques have competed to retrieve useful knowledge and patterns based on TA. Since TA involves analyzing numerous data patterns and transaction records, its data-centric characteristics are suitable for exploration by computational intelligence techniques, and those techniques are roughly divided into two categories: machine learning and soft computing.

The most prevalent machine learning techniques for predicting financial markets might be artificial neural network (ANN)-related algorithms. Researchers do not need to assume the linearity or the probabilistic distribution of the ANN model; instead, only the relevant input variables (indicators), the network structure, and the target output with the selected learning algorithms are assigned. According to the aforementioned survey (Park and Irwin 2007), the average number of input variables is between 4 and 10, and the most commonly adopted variables include several technical indicators and the four major trading data of an underlying alternative. However, at least two obvious drawbacks can be found in this approach: (1) the ANN learning algorithms are devised to minimize modeling errors by adjusting the weights among neurons, and the obtained connecting weights are difficult for investors to interpret; (2) ANN algorithms process all of the involved inputs simultaneously and are not able to retrieve useful trading signals from only a partial set of input variables. In trading practice, professional investors often refer to two or more indictors when making buy-in or sell-out decisions; nevertheless, it would be unlikely for investors to consider all available technical indicators simultaneously to reach a final decision.

Soft computing techniques, on the other hand, focus on obtaining understandable rules or logic by allowing for the imprecise characteristics of data or patterns. Among the various soft computing techniques, fuzzy set theory might be the most prevalent. Fuzzy set theory (Zadeh 1965) has been applied widely in engineering (Precup and Hellendoom 2011) and socioeconomics (Zavadskas and Turskis 2011; Peng and Tzeng 2013). The applications of fuzzy set theory to TA may be categorized into two representative types: fuzzy pattern recognition and fuzzy logic reasoning. The first one adopts visual pattern recognition based on fuzzy set theory (Zhou and Dong 2004), which relies on the judgments of researchers/DMs; the second one adopts fuzzy logic, which is incorporated to resolve the uncertainties of market timing and order size (Gradojevic and Gencay 2013) and is relatively easy to reuse as investment knowledge. The fuzzy inference technique is also adopted in this chapter to enhance experts' judgments, which will be explained in the next section.

Other studies have chosen an integrated approach to forecast stock markets, such as the integration of ANNs and fuzzy inference systems (FISs), also termed ANF (Boyacioglu and Acai 2010; Shen 2013). Despite its capability of learning fuzzy parameters from examples, the target output of ANFIS has to be a real number to conduct Sugeno-type fuzzy reasoning. Therefore, a previous study (Wei et al. 2011) was constrained to using the subsequent $(t + 1)$ stock price as the output for its FIS; it could only forecast the next day's closing price based on the trained FIS at time t. Investors may only have one-day foresight of stock movements, which is insufficient to support a buy-in decision that needs to last for more than two days. Aside from ANFIS, a combination of RST and GA was proposed for stock price forecasting (Cheng et al. 2010), using a rough set algorithm to derive rough decision rules; a GA was then applied to refine the extracted rules for higher classification accuracy. Although this research (Cheng et al. 2010) leveraged the machine learning capabilities of rough sets and GAs, the model has certain potential drawbacks; for

242 ■ Trends of Hybrid Multiple Criteria Decision Making

example, the discretization of the conditional attributes (i.e., technical indicators) was based on a cumulative probability distribution approach, which has to presume the probabilistic distribution of each indicator.

It can be seen from the previous discussion that although machine learning techniques can support the induction of price patterns from historical data, the in-depth knowledge of investment veterans should not be overlooked and is still the key to forming reasonable models and experiments. Considering the aforementioned issues in the previous research, this chapter attempts to devise a hybrid soft computing model based on the knowledge of investment professionals, to prevent unreasonable experiment settings. Furthermore, domain experts were consulted and their knowledge/experience transformed for the discretization of VC-DRSA, which aligns the model setting with trading practice (to be explained in the next section).

17.2 Hybrid Investment Support System Based on Fuzzy and Rough Set Techniques

The proposed model comprises two stages: (1) data preprocessing and (2) VC-DRSA modeling, shown in Figure 17.1. The basic ideas of VC-DRSA are introduced in Chapter 7, and more details on DRSA or VC-DRSA can be found in previous research (Greco et al. 1999; Błaszczyński et al. 2011, 2013) or retrieved from the online resources of the Intelligent Decision Support System (IDSS) of the Poznan University of Technology, Poland. Since VC-DRSA is adopted to induce rough decision rules from multiple TA signals, how to generate meaningful TA signals needs to be ensured at the first stage, which is also the emphasis of this model.

The involvement of investment experts helped to select commonly used TA indicators in this model, and all the TA signals were generated using the testing function from the XQ professional trading suite; using XScript, a trading language–like grammar can be applied for back-testing. The required process for forming this hybrid model can be summarized in six steps.

Step 1: Define involved TA indicators and divide them into the crisp and fuzzy signal groups.

Though VC-DRSA algorithms may retrieve patterns from various trading signals, the complexity of certain technical indicators that require imprecise judgments means that they need further processing to discern their meaning; therefore, the involved TA indicators should be divided into crisp and fuzzy groups. The adopted technical indicators that belong to the fuzzy group will be further explained in the numerical case.

Step 2: Conduct back-testing for each crisp group TA indicator.

This step uses the defined and built-in TA indicators from the XQ professional trading suite. Within a testing period, the associated investment

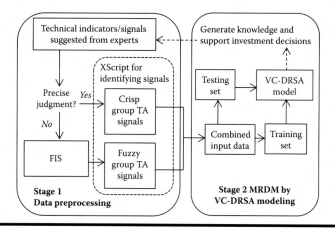

Figure 17.1 Research model of the two stages. (Reprinted with permission from Shen, K.Y., and Tzeng, G.H., *International Journal of Fuzzy Systems*, 17(3), 375–389, 2015c.)

performance of each TA signal, subject to various investment strategies, will be recorded and transformed as the inputs for VC-DRSA at the next stage.

Step 3: Form an FIS and conduct back-testing for each fuzzy group TA indicator.

The TA indicators in the fuzzy group (Figure 17.1) require imprecise judgments by experts; therefore, a FIS is proposed to imitate the reasoning processes of investment experts for those indicators. A FIS is similar to the human reasoning process, which maps given inputs to an output based on fuzzy logic. The structure of a FIS comprises three parts: fuzzy inputs, logical reasoning, and the defuzzified output. The required steps for a FIS are (1) define fuzzy intervals for the input and output variables, (2) form rules or logical reasoning, (3) apply fuzzy operators, (4) aggregate all outputs from the rules in the FIS, and (5) defuzzify the final output. Details can be found in the previous research by Shen and Tzeng (2014a, 2015f).

Step 4: Transform and combine the outcomes from Step 2 and Step 3 as inputs for VC-DRSA.

All the considered technical indicators, including the crisp and the fuzzy signal groups, will be combined as the conditional attributes for the subsequent VC-DRSA model; this step is also termed *data fusion*, and a sample format is illustrated at the bottom of Figure 17.2. The available data should also be divided into a training set and a testing set.

Step 5: Induce rough decision rules by VC-DRSA using the training set.

Step 6: Examine/validate the VC-DRSA model using the testing set.

A partial set of the available inputs from Step 4 will be used as the training set; those remaining will be used as the testing set. The VC-DRSA model formed by the training set comprises multiple rough decision rules, which should be examined and validated by the testing set.

244 ■ *Trends of Hybrid Multiple Criteria Decision Making*

17.3 Numerical Experiments

To illustrate the proposed approach, historical trading data from TAIEX, from August 2002 to May 2014 (3000 trading days), were analyzed to form a hybrid MRDM model. In addition, to retrieve the implicit TA knowledge of experts, eight domain professionals were involved to form the VC-DRSA hybrid model. The research flow of the empirical case is illustrated in Figure 17.2.

17.3.1 Data Preprocessing

The daily trading data of TAIEX were retrieved from the database of SysJust, which is a Taiwan-based leading financial trading solutions provider. In addition, since there are numerous technical indicators (considering combinations of different parameter settings) used in the market, in order to retrieve knowledge and trading experience from investment professionals, questionnaires were used to collect the most commonly used indicators and corresponding parameter settings from eight domain experts. The intersection of the technical indicators suggested by the experts were used in this case (Table 17.1). Six of the eight experts work in the same financial holding company in Taipei, and they are mainly responsible for trading foreign exchanges, stock indices, and interest rates in derivative markets; the other two experts are a senior fund manager and a manager from an investment consulting firm. All the experts have worked in financial markets for more than 15 years.

Most of the technical indicators can be judged by crisp logic, except one commonly adopted indicator: moving average (*MA*). The *MA* indicator has been widely examined in previous research (Menkhoff and Taylor 2007; Gradojevic and Gencay 2013; Taylor 2014), and the invited professionals recommended a dual *MA* combination (i.e., a short *MA* and a mid-term *MA*), which requires imprecise judgments on the trends (patterns) of two *MA* lines and a daily closing price; therefore, the dual moving average combination indicator (i.e., *MA*) was categorized into the fuzzy signal group.

The selected technical indicators and corresponding trading signals were further defined as conditional attributes (criteria) according to Table 17.1, as suggested by the eight experts. Then, the experiment executed the back-testing function—applying the XScript from SysJust—on the TAIEX for each technical indicator. In this case, the signals suggested by each indicator are *Buy-in*, *Neutral*, and *Sell-out*; once an indicator suggests a *Buy-in*, the consequence of this buy-in decision will be associated with a *Gain* or *Loss* depending on a predefined trading strategy, referring to the real price movements. The so-called trading strategy is to sell the position based on the predefined upside and downside intervals after a buy-in action. For example, if the trading strategy was set as "−5% and +10%," it means that this simulated investment would be sold when the deal reached either the downside (−5%) or the upside (+10%) return level based on the buy-in price, and marked as *Loss* or *Gain*, respectively. This experimental setting can consider the downside risk and cut losses, which is closer to investment practice. Previous studies (Park and Irwin 2007; Gradojevic and Gencay

Fuzzy Inference–Supported MRDM for Technical Analysis ■ 245

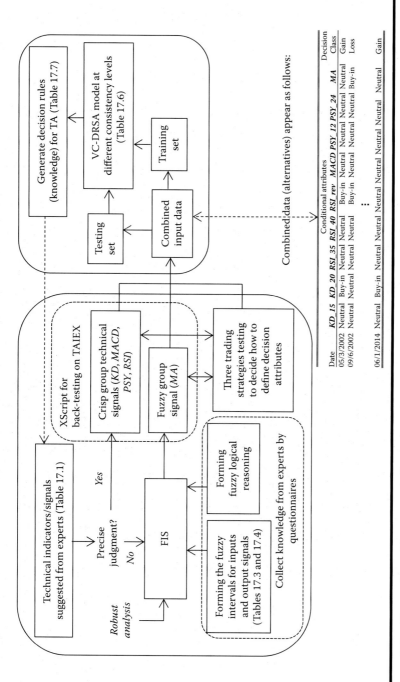

Figure 17.2 Illustration of the research flows. (Reprinted with permission from Shen, K.Y., and Tzeng, G.H., *International Journal of Fuzzy Systems*, 17(3), 375–389, 2015c.)

Table 17.1   Technical Signals Suggested by Domain Experts

Property (Crisp)	Symbols	Length (Days)	Buy-In Signals (Precise Judgments)	Neutral	Sell-Out Signals (Precise Judgments)
Stochastic oscillator	KD_15	9	[K(9)[a] and D(9) < 15] AND [K(9) cross above D(9)]	Others	[K(9) and D(9) > 85] AND [K(9) cross below D(9)]
	KD_20	9	[K(9) and D(9) < 20] AND [K(9) cross above D(9)]	Others	[K(9) and D(9) > 80] AND [K(9) cross below D(9)]
Moving average convergence/ divergence	MACD	9	DIF(9) cross above MACD	Others	DIF(9) cross below MACD
Psychology indicator	PSY_12	12	PSY(12) < 15	Others	PSY(12) > 85
	PSY_24	24	PSY(24) < 20	Others	PSY(24) > 80
Relative strength indicator	RSI_35	6/12	RSI < 35 AND [RSI(6) cross above RSI(12)]	Others	RSI > 65 AND [RSI(6) cross below RSI(12)]
	RSI_40	6/12	RSI < 40 AND [RSI(6) cross above RSI(12)]	Others	RSI > 60 AND [RSI(6) cross below RSI(12)]
	RSI_rev	6/12	RSI < 40 AND Close < LowBound AND [RSI(6) cross above RSI(12)]	Others	RSI > 60 AND Close > HighBound AND [RSI(6) cross below RSI(12)]

(Continued)

Table 17.1 (Continued)   Technical Signals Suggested by Domain Experts

Property (Fuzzy)	Symbols	Length (Days)	Buy-In Signals (Imprecise Judgments)	Neutral	Sell-Out Signals (Imprecise Judgments)
Dual combination of moving average	MA	5/20	If deviation is small AND [ClosePrice cross above MA(5) and MA(20)] If deviation is small AND short trend is up AND mid-trend is up If deviation is large AND short trend is sharp decline AND mid-trend is up	Others	If deviation is small AND [ClosePrice cross below MA(5) and MA(20)] If deviation is large AND short trend is up AND mid-trend is up If deviation is small AND short trend is sharp rebound AND mid-trend is down

Source: Data from Shen, K.Y., and Tzeng, G.H., *International Journal of Fuzzy Systems*, 17(3), 375–389, 2015c.

[a] K(9) denotes the value of 9-days K.

248 ■ *Trends of Hybrid Multiple Criteria Decision Making*

2013) mainly used a fixed time period or used the selling signal to calculate the holding period return, and measured the return volatility to denote the associated risk of a deal; they were not able to control the downside risk for investors.

Three trading strategies were examined after discussions with the experts: [–3%, +7%], [–5%, 10%], and [–7%, 20%]. Based on these settings, the number of triggered buy-ins (i.e., alternatives or instances) and the associated consequences of each technical indicator for the three trading strategies are summarized in Table 17.2. As shown in Table 17.2, the trading strategy [–7%, +20%] indicated the best outcome (i.e., 90/149 = 60.4% as *Gain*); therefore, the subsequent analyses adopted this strategy to assign a value for the decision attribute of each alternative.

Referring to Step 2 and Step 3 in Subsection 17.2, the conditional attributes comprise the crisp signal group (i.e., *KD*, *MACD*, *PSY*, and *RSI*) and the fuzzy signal group (i.e., the dual *MA* combination). On the one hand, all of the technical indicators in the crisp signal group are categorized into three values (i.e., *Buy-in*, *Neutral*, and *Sell-out*) to denote the discretization results. On the other hand, the dual *MA* combination signal requires imprecise judgments and natural language-like expressions (e.g., IF "deviation is small" AND "short trend is upward" AND "mid-trend is upward," THEN "the signal is *Buy-in*"), with more than one rule

Table 17.2 Preprocessed Instances with Decision Classes

Symbols	Triggered Instances	[–3%, 7%]		[–5%, 10%]		[–7%, +20%]	
		Gain	*Loss*	*Gain*	*Loss*	*Gain*	*Loss*
KD_15	10	5	5	5	5	6	4
KD_20	28	14	14	15	13	18	10
RSI_35	6	3	3	3	3	2	4
RSI_40	12	5	7	5	7	6	6
RSI_rev	6	3	3	3	3	5	1
MACD	56	28	28	31	25	33	23
PSY_12	11	6	5	6	5	8	3
PSY_24	4	2	2	2	2	2	2
MA	16	9	7	9	7	10	6
Total	149	75	74	79	70	90	59
Successful rate		50.3%		53.0%		60.4%	

Source: Data from Shen, K.Y., and Tzeng, G.H., *International Journal of Fuzzy Systems*, 17(3), 375–389, 2015c.

Fuzzy Inference–Supported MRDM for Technical Analysis ■ 249

associated with the *Buy-in*, *Neutral*, or *Sell-out* signals, which require imprecise judgments. Therefore, this study assumes that fuzzy inference could enhance the outcomes in this model, which is explained in the following subsection.

17.3.2 Discretization of Fuzzy TA Signals by Fuzzy Inference Systems

As suggested by the experts, the dual *MA* combination involves imprecise judgments such as "small," "large," "upward," and "downward" (Table 17.1), which can be enhanced by a (Mamdani-type) FIS to be closer to the experts' reasoning. The widely applied triangular membership function was adopted to fuzzify and defuzzify the experts' opinions. The FIS comprises three parts: the input variables, logical reasoning, and the output signal; the three input variables (i.e., Deviation, ShortTrend, and MidTrend) were all assigned to the same three triangular membership functions to denote small, middle, and large (or downward, flat, and upward). The output signal was assigned to three values (i.e., *Sell-out*, *Neutral*, and *Buy-in*). The eight experts' opinions were collected and averaged using questionnaires, and the averages of the parameters for the corresponding fuzzy membership functions were adopted for the FIS. The opinion range in the questionnaire was from 0 to 10, and the experts would give their opinions regarding each triangular membership function. For example, one expert gave his opinions as (0, 3.5), (2.5, 5, 7.25), and (6.25, 0) to denote the parameters for the three triangular membership functions. The results of the parameters are provided in Table 17.3. Similarly, the three fuzzy intervals for the output variable are

Table 17.3 Fuzzy Parameters for the Input Variables

Experts	(S_m, S_h)[a]	(M_l, M_m, M_h)	(L_m, L_h)
1	(0, 3)	(2.5, 5, 7)	(6, 10)
2	(0, 4)	(3, 5, 7)	(6.5, 10)
3	(0, 3)	(2, 5, 7)	(6.5, 10)
4	(0, 3)	(3, 5, 7.5)	(7, 10)
5	(0, 4)	(2, 4.5, 7.5)	(7.5, 10)
6	(0, 5)	(2, 4.5, 7.5)	(6, 10)
7	(0, 4)	(3, 5, 7)	(6.5, 10)
8	(0, 3)	(3, 4.5, 6.5)	(6.5, 10)
Averages	(0, 3.63)	(2.56, 4.81, 7.13)	(6.56, 10)

Source: Data from Shen, K.Y., and Tzeng, G.H., *International Journal of Fuzzy Systems*, 17(3), 375–389, 2015c.

[a] (S_m, S_h), (M_l, M_m, M_h), and (L_m, L_h) denote the three fuzzy intervals *Small, Middle,* and *Large* (or *Downward, Flat,* and *Upward*), respectively.

250 ■ *Trends of Hybrid Multiple Criteria Decision Making*

Table 17.4 Fuzzy Parameters for the Output Variable

Experts	(N_m, N_h)[a]	(NE_l, NE_m, NE_h)	(P_m, PU_h)
1	(0, 2.5)	(2.5, 5, 7.5)	(6, 10)
2	(0, 2)	(2, 5, 7)	(6.5, 10)
3	(0, 3)	(2, 5, 7.5)	(6.5, 10)
4	(0, 2.5)	(2, 5, 7.5)	(7, 10)
5	(0, 3)	(2.5, 4.5, 8)	(7.5, 10)
6	(0, 3.5)	(2, 4.5, 7)	(6, 10)
7	(0, 2)	(1.5, 5, 7)	(6.5, 10)
8	(0, 3)	(2, 4.5, 7.5)	(6.5, 10)
Averages	(0, 2.69)	(2.56, 4.81, 7.38)	(6.56, 10)

Source: Data from Shen, K.Y., and Tzeng, G.H., *International Journal of Fuzzy Systems*, 17(3), 375–389, 2015c.

[a] (N_m, N_h), (NE_l, NE_m, NE_h), and (P_m, P_h) denotes the three fuzzy intervals *Sell-out*, *Neutral*, and *Buy-in*, respectively.

shown in Table 17.4. The logical reasoning is given in Table 17.5 (i.e., R1 to R8), as recommended by the experts.

During the back-testing period, once any indicator was triggered as *Buy-in*, whether it was from the crisp or the fuzzy group, all the remaining indicators were checked and marked with the corresponding outputs as the data of the conditional attributes of an instance; in addition, the associated consequence, based on the trading strategy [−7%, +20%], was recorded as *Gain* or *Loss* as the result of the decision attribute. This step is also termed *data fusion*, which is essential to forming the training and testing sets for the subsequent VC-DRSA model.

17.3.3 VC-DRSA Model

The available data (instances), after data fusion, were divided into two sets: the training and the testing sets, comprised of 101 and 48 observations (instances), respectively. The training set covered the period August 2002–December 2009, and the testing set the period January 2010–June 2014. The VC-DRSA algorithm was implemented by using jMAF (Błaszczyński et al. 2013), jMAF is a rough set data analysis framework written in Java language; it is based on java Rough Set (jRS) library; jMAF and jRS library implement methods of analysis provided by the DRSA and VC-DRSA. At different consistency levels (CLs), the classification accuracy (CA) of each model was computed using fivefold cross-validation, as summarized in Table 17.6.

Fuzzy Inference–Supported MRDM for Technical Analysis ■ 251

Table 17.5 Logical Reasoning of the Fuzzy Inference System

Rules	Fuzzy Logical Reasoning
R1	IF (Deviation is *Small*) AND (ShortTrend is *Upward*) and (MidTrend is *Upward*) THEN (signal is *Buy-in*).
R2	IF (Deviation is *Large*) AND (ShortTrend is *Downward*) and (MidTrend is *Upward*) THEN (signal is *Buy-in*).
R3	IF (ShortTrend is *Flat*) or (MidTrend is *Flat*) THEN (signal is *Neutral*).
R4	IF (Deviation is *Middle*) THEN (Signal is *Neutral*).
R5	IF (Deviation is *Middle*) OR (ShortTrend is *Flat*) OR (MidTrend is *Flat*) THEN (signal is *Neutral*).
R6	IF (Deviation is *Large*) AND (ShortTrend is *Upward*) and (MidTrend is *Upward*) THEN (signal is *Sell-out*).
R7	IF (Deviation is *Small*) AND (ShortTrend is *Upward*) and (MidTrend is *Downward*) THEN (signal is *Sell-out*).

Table 17.6 VC-DRSA Classification Results

	VC-DRSA (%)				
	$CL^a = 95$	$CL = 90$	$CL = 85$	$CL = 80$	$CL = 75$
1st	79.41	83.17	88.12	85.15	81.19
2nd	74.51	84.16	86.14	82.18	83.17
3rd	77.45	82.18	85.15	83.17	81.19
4th	77.45	86.14	89.11	84.16	84.16
5th	76.47	83.17	87.13	83.17	82.18
AVG	76.47	83.76	87.13	83.57	82.38
SD	1.78	1.34	1.40	1.01	1.15

Source: Data from Shen, K.Y., and Tzeng, G.H., *International Journal of Fuzzy Systems*, 17(3), 375–389, 2015c.

[a] CL denotes consistency level.

While CL = 0.85, the averaged CA reached 87.13% (the highest one in Table 17.6); furthermore, the untouched testing set was examined by the trained VC-DRSA model (CL = 0.85), and the classification accuracy reached 79.17% (38/48). Furthermore, the CORE attributes comprise six technical indicators—*KD_15*, *RSI_rev*, *MACD*, *MA*, *PSY_12*, and *PSY_24*—and the fuzzy group signal was also included. The

252 ■ *Trends of Hybrid Multiple Criteria Decision Making*

Table 17.7   Decision Rules from VC-DRSA Model

Decision Rules	Rules for the At-Least Gain Decision Class	Supports
DR1	MA $\geq$ Buy-in	14
DR2	PSY_12 $\geq$ Neutral	11
DR3	PSY_24 $\geq$ Neutral	8
DR4	MACD $\geq$ Neutral AND MA $\geq$ Neutral AND PSY_12 $\geq$ Neutral	53
	Rules for the At-Most Loss Decision Class	*Supports*
DR5	MA $\leq$ Neutral	29
DR6	MACD $\leq$ Sell-out	26
DR7	RSI_rev $\leq$ Neutral AND MACD $\leq$ Neutral AND MA $\leq$ Neutral AND PSY_24 $\leq$ Neutral	44
DR8	KD_15 $\leq$ Neutral AND MA $\leq$ Neutral AND PSY_12 $\leq$ Neutral AND PSY_24 $\leq$ Neutral	45

Source: Data from Shen, K.Y., and Tzeng, G.H., *International Journal of Fuzzy Systems*, 17(3), 375–389, 2015c.

Note: CL = 0.85 in VC-DRSA.

VC-DRSA model generated 11 certain rules; the rules with more than 10 supports (i.e., DR1 to DR8) are listed in Table 17.7.

17.4 Simulated Investment Performance and Discussion

To compare the trained VC-DRSA model with each technical indicator and the simple buy-and-hold strategy during the training period, the trading strategy [−7%, +20%] was adopted for each indicator. To estimate the accumulated returns generated by each indicator, a simple additive calculation approach was applied, and a 0.6% transaction cost for each simulated transaction was adopted based on convention (transaction fees and transaction tax). For example, if the *KD_15* indicator generated five buy-in signals during the testing period with four *Gain* outcomes and one *Loss*, then the accumulated return would be $(20\% \times 4) - (7\% \times 1) - (0.6\% \times 5) = 70\%$; the calculations were based on the daily closing price, and the results are consolidated in Table 17.8.

In addition, two supplementary experiments were conducted to examine the robustness of the FIS: (1) replacing the triangular membership function with the Gaussian membership function in the FIS, and (2) inviting a new investment professional to give opinions on the *MA* indicator for the testing period. The Gaussian membership function used the average of the previous opinions (Table 17.3) for the three input variables. Take the fuzzy interval (concept) *Middle*, for example, which adopted 4.81 to denote the mean of *Middle*; 2.56 and 7.38 (the averages of M_l and M_h, respectively) were used to define the full range of *Middle*. The calculation of the fuzzy membership function for the output signal was done in the same way. The FIS outputs (generated by the Gaussian membership function) were close to the previous ones, except in only one instance. Also, the new professional's inputs for the *MA* indicator generated similar FIS outputs, which yielded the same classification accuracy for the testing set (i.e., 79.17%). These two additional experiments suggest the robustness of the FIS results.

Table 17.8 Simulated Investment Results for Testing Period

Investment Decisions by	Buy-In Times	Result as Gain	Result as Loss	Accumulated Returns (%)
KD_15	7	7	0	135.8
KD_20	10	10	0	194.0
RSI_35	15	10	5	156.0
RSI_40	24	17	7	276.6
RSI_rev	29	20	9	319.6
MACD	35	23	12	379.6
PSY_12	7	7	0	135.8
PSY_24	5	5	0	97.0
MA	5	5	0	97.0
Buy-and-hold	1	1	0	13.9
VC-DRSA	36	27	9	455.4

Source: Data from Shen, K.Y., and Tzeng, G.H., *International Journal of Fuzzy Systems*, 17(3), 375–389, 2015c.

17.5 Conclusion

How to apply MRDM to resolve a data-centric problem were illustrated with a numerical case in this chapter. The fusion of implicit and imprecise knowledge, from domain experts and historical complex patterns, is key to forming a practical investment decision support system, and the hybrid VC-DRSA model shows acceptable results when inducing trading signals from combinations of multiple technical indicators. The classification accuracy reached nearly 87% on average (by setting CL = 0.85 for VC-DRSA); in other words, the uncertainty and the volatility of the stock market is still inevitable when inducing decision rules. Also, considering the required imprecise judgments on certain TA indicators, an FIS subsystem was incorporated to enhance the VC-DRSA model; two supplementary experiments supported the robustness of the FIS.

Compared with previous studies, several merits can be observed in this work: (1) the proposed approach encodes the buy-in and sell-out signals based on expert knowledge; (2) the commonly adopted indicators are divided into two groups (i.e., crisp and fuzzy)—depending on the required essence of the judgments—to generate signals, which is more practical and reasonable; (3) the simulated trading strategy (i.e., [−7%, +20%]) is closer to investment practice and can control the downside risk by setting a bottom line for cutting losses. Previous studies are constrained to forecasting the daily profits/losses of price movements (or those of a specific period); these studies (Cheng et al. 2010; Gradojevic and Gencay 2013) are neither sufficient to support an investment that needs to last more than two days nor capable of controlling the downside risk in practice. In addition, the decision rules obtained indicate that certain technical indicators (e.g., *MACD*, *MA*, and *PSY_12* in DR4) should be considered jointly to make judgments, which has been overlooked in previous research.

Despite the positive findings in this work, certain limitations remain; future studies are suggested to work in the following directions: (1) adopting high-frequency trading data for intraday investment analysis, (2) devising a mechanism to incorporate the suggested sell-out signals with a trading strategy to decide the timing of disposition, and (3) enhancing the discretization of input signals by other soft computing or machine learning techniques. The emergence of financial technology (or FinTech) has attracted increasing attention to automated investment analysis and financial pattern detection; the MRDM approach has shed light on discovering valuable and understandable knowledge for investors.

Chapter 18

Financial Improvements of Commercial Banks Using a Hybrid MRDM Approach

Banks play a crucial role in facilitating and stabilizing the economy of a nation; their importance is unquestionable. While nearly everyone would agree that the safety or soundness of a bank hinges on its financial performance (FP), the ways to analyze and interpret complicated FP reports (or indicators) diverge. Unlike the commonly observed financial statements of general companies, certain specific indicators—devised to detect the healthiness of a bank's operations—are difficult for investors to decipher. Furthermore, the influences of financial ratios/indicators are often interrelated; the complexity of the interrelationships among multiple variables often impedes investors' comprehension or even forecasts of the FP of banks. To resolve this complicated problem, a rough set–based hybrid multiple criteria decision making (MCDM) model is proposed, and not merely for investors to refer to; the results may provide constructive suggestions for banks to improve their future performance.

18.1 Research Background

Since the financial crisis in 2008, the importance of monitoring and forecasting the future FP of banks has been the priority of central banks all over the world.

As a consequence, there has been increasing interest in exploring the relationship between historical data (mainly financial ratios and special indicators for the banking industry) and future FP. While bank performance is traditionally analyzed by using financial ratios and statistical methods, the complexity of multiple dimensions and criteria has motivated researchers to adopt advanced quantitative techniques from other fields (Fethi and Pasiouras 2010). An integrated model fusing soft computing and MCDM methods is proposed here.

The diagnosis of a bank's FP can serve multiple purposes in practice: detecting bankruptcy, evaluating the credit scores of banks, making investment decisions, and helping the management teams of banks plan for improvements. Owing to the needs of practical fields, many methods have been adopted to solve the problem; these can be roughly categorized into three approaches: statistics, decision science, and computational intelligence.

Conventional studies mainly rely on statistical methods; however, statistical models are constrained by certain unrealistic assumptions (Liou and Tzeng 2012). Take the most commonly used regression model, for example: assumptions of both the independence of variables and the linearity relationship are required to form regressions. These unrealistic assumptions create limitations to exploring the entwined relationships of complex problems in practice (Liou 2013; Tzeng and Huang 2011).

As for decision science, MCDM is a reasonable approach to solving the problem, due to its main focus on handling multiple variables. Among the MCDM methods, data envelopment analysis (DEA) might be the most commonly used technique for evaluating the performance or efficiency of banks (Fethi and Pasiouras 2010; Kao and Liu 2014). Other group decision methods have also been adopted, such as multiple-group hierarchical discrimination (Zopounidis and Doumpos 2000), the analytic network process (ANP) (Niemira and Saaty 2004), and Utilités Additives Discriminantes (UTADIS) (Kosmidou and Zopounidis 2008). Group decision methods transform the opinions of domain experts into evaluation models for ranking or selecting alternatives.

Computational intelligence (or machine learning–based) techniques, such as artificial neural networks (ANNs) (Zhao et al. 2009; Ao 2011), support vector machines (SVMs) (Luo et al. 2009; Wu et al. 2007, 2009), genetic programming (Ong et al. 2005a; Huang et al. 2006e), and decision trees (DTs) (Ravi and Pramodh 2008), all have their own advantages in handling nonlinear data. In addition, the rough set approach (RSA) is a mathematical theory (Pawlak 1982) that uses computational algorithms to induce findings from large and imprecise data. The rising computational capability of computers makes these machine learning–based techniques more efficient and effective in handling data-centric problems in the Big Data era.

Although various computational methods/techniques have been applied to predict the FP of banks, the aforementioned studies mainly depend on a single approach: either constructing models from the opinions of experts (e.g., ANP) or finding patterns in large data sets (Verikas et al. 2010). An integrated model that can leverage advantages from different approaches is still underexplored. Thus, to tackle this complicated issue, the FP diagnosis problem is decomposed into three stages using a multiple rules-based decision making (MRDM) approach. At the first stage, considering the large number of related variables used to assess FP, rough set–based machine learning is proposed to obtain the critical variables with decision rules from historical data. Next, the implicit knowledge of domain experts is retrieved to comprehend the interrelationships among variables and the influential weights of each criterion. At this stage, the DEMATEL-based ANP (DANP) (Hsu et al. 2013) is adopted by asking the experts to compare only two variables (i.e., the relative influences of criterion CA and criterion CB) each time, which makes it easier for them to give their opinions on a complex problem. At the third stage, a modified VIKOR decision method (Opricovic and Tzeng 2004) is incorporated to find the performance gap of each bank in each criterion. The VIKOR model can rank the alternatives—while facing conflicting and noncommensurable criteria—by minimizing the total performance gap. To demonstrate this three-stage model, a group of commercial banks is examined as an empirical case. The raw data come from the quarterly reports (from 2008 to 2012) of the central bank of Taiwan.

18.2 Core Attributes–Based MRDM Approach for Financial Performance Improvement

The research flow of this three-stage approach is illustrated in Figure 18.1, and the main functions of the three computational methods, DRSA, DANP, and modified VIKOR, are briefly discussed. The integrated model comprises three stages: The first stage focuses on exploring and retrieving patterns—decision rules and indispensable attributes—from the historical data. The second stage adopts the CORE attributes (by DRSA) from the first stage, induced from historical data, to form a DANP decision model, which may indicate the directional influences among the attributes and the relative importance of each criterion. And the modified VIKOR method is used to synthesize the performance gap of each bank for ranking or selection at the final stage. Since the modified VIKOR method weights and synthesizes the performance gap of each alternative in each criterion, it can yield a priority gap for each bank to refer to. An empirical case showing how to transform the analytics from this hybrid model into improvement planning will be provided in the following section.

To summarize the proposed model, the steps involved in the three stages are listed in sequence as follows:

Step 1: Discretize the raw financial figures for the conditional attributes and decision attribute for DRSA at the first stage. The three-level discretization method will be further explained in the empirical example later in the chapter.

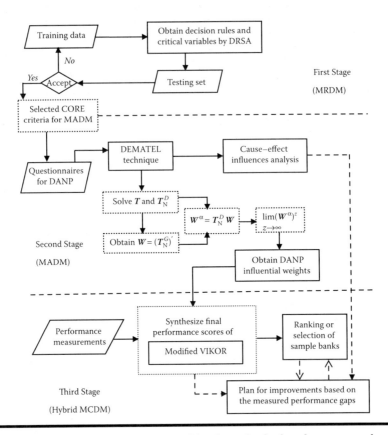

Figure 18.1 Illustration of the combined methods for the proposed model. (Revised and reprinted with permission from Shen, K.Y. and Tzeng, G.H., *Soft Computing*, 19(4), 859–874, 2015a.)

Step 2: Apply the DRSA algorithm to generate decision rules from the discretized information system table and adopt the CORE attributes for the subsequent DANP model. In this case, the data will be divided into a training set and a testing set. Once the classification result meets the expected accuracy, the strong decision rules and core attributes (obtained by setting a threshold for cutting support) will be further analyzed at the next stage.

Step 3: Calculate the initial average matrix A, referring to Equation 3.1 (Chapter 3), by collecting opinions from domain experts. Among the core attributes (obtained from Step 2), experts are asked to compare the relative influence that they feel criterion i has on criterion j. Then, the direct-influence matrix D can be calculated from A, referring to Equation 3.2 (Chapter 3).

Step 4: Obtain the influential weight of each criterion in the CORE attributes using Equations 4.1 through 4.10 (Chapter 4) until the limiting supermatrix becomes stable.

Step 5: Using questionnaires, collect opinions from experts regarding the performance score of each bank for each criterion in the core attributes. The actual financial figures of sample banks and the industrial average in each criterion are provided in the questionnaire, and domain experts are asked to rate the performance of the target banks in each criterion.

Step 6: Synthesize the final score for each bank using the modified VIKOR method. Using Equations 5.6 through 5.15 (Chapter 5), researchers may set δ to form the final compromise-ranking index for each bank.

Step 7: Plan for improvements based on the performance gap information and the cause-effect relationships obtained from the INRM.

18.3 An Empirical Case of Five Commercial Banks

In this case, the raw data were retrieved from the quarterly reports of the central bank in Taiwan, released under the title "Condition and Performance of the Domestic Banks." There were 34 commercial banks included in this analysis, and the year-end reports from 2008 to 2011 were adopted to construct the DRSA model. There were two reasons for the selection of this time period: (1) the financial crisis began in 2008, and the management teams of banks were heavily influenced by this in their operations afterward; (2) the government (i.e., the central bank) was also involved, offering additional support to and making requirements of the banking industry after 2008. The financial results and operational performances may differ from previous patterns; therefore, the research period was set to be from 2008 to 2012 for this empirical case.

The data from 2008 to 2011 were used as the training set, and the data from 2012 the testing set. The report comprises six dimensions: (1) *Capital sufficiency*, (2) *Asset quality*, (3) *Earnings and profitability*, (4) *Liquidity*, (5) the *Sensitivity of interest rates*, and (6) the *Growth rates of main business*. The six dimensions include 25 variables/attributes (financial ratios or special indicators for the banking industry). A short description and definition for each variable is shown in Table 18.1. Furthermore, the growth rate of the return on assets (ROA) in the subsequent year was used to define the *Good* or *Bad* decision classes. The 25 financial ratios of a bank in each year were ranked and transformed into categories 1, 2, and 3 to represent *low*, *middle*, and *high*, respectively; the 25 ratios represent the conditional attributes in DRSA. For example, the top third of stocks (34 / 3 = 11 stocks) in the ratio E_1 were categorized as 3 in the model. As for the decision class, we categorized the banks with more than 10% growth in *ROA* as *Good* and those with more than 10% decline as *Bad*. The banks that performed in the middle (i.e., $-10\% \leq \Delta ROA \leq 10\%$) were not used to form decision rules. As a whole, there were 84 records (the training set) used in the DRSA model.

The training set was examined by a threefold cross-validation five times, and its average and standard deviation (SD) were compared with the results of

260 ■ *Trends of Hybrid Multiple Criteria Decision Making*

Table 18.1 Description of Variables Used in the Central Bank's Report

Dimensions	Symbols	Descriptions	Definitions
Capital sufficiency (C)	C_1	Regulatory capital to risk-weighted assets	Regulatory capital/risk-weighted assets
	C_2	Tier 1 capital to risk-weighted assets	Tier 1 capital/risk-weighted assets
	C_3	Debt–equity ratio	Debt/net worth
	C_4	Net worth to total assets	Net worth/total assets
Asset quality (A)	A_1	Nonperforming Loan (NPL) ratio	NPL/loan and discount
	A_2	Loan loss reserve to NPL	Loan loss reserves/NPLs
	A_3	Possible loss of classified assets to reserve	Possible loss of classified assets/reserves
Earnings and profitability (E)	E_1	Net income before tax to equity	NIBT/average equity
	E_2	NIBT with loan loss provision to equity	NIBT with loan loss provision/equity
	E_3	NIBT to asset	NIBT/average asset
	E_4	NIBT and loan loss Provision to average assets	(NIBT + loan loss provision)/average asset
	E_5	Net interest revenues to NIBT	Net interest revenues/NIBT
	E_6	NIBT to total net revenues	NIBT/total net revenues
	E_7	NIBT per employee	NIBT/employees
Liquidity (L)	L_1	Liquidity ratio	Liquidity ratio
	L_2	Loans to deposits	Loans/deposits
	L_3	Time deposits to deposits	Time deposits/deposits
	L_4	NCDs to time deposits	NCDs / time deposits
	L_5	180 days' accumulated gap of assets and liabilities to equity	Accumulated gap of assets and liabilities (180 days)/equity

(Continued)

Financial Improvements of Commercial Banks ■ 261

Table 18.1 (Continued)   Description of Variables Used in the Central Bank's Report

Dimensions	Symbols	Descriptions	Definitions
Interest rate sensitivity (S)	S_1	Interest rate sensitivity assets to interest rate sensitivity liabilities	Interest rate sensitivity assets/interest rate sensitivity liabilities
	S_2	Interest rate sensitivity Gap to equity	Interest rate sensitivity gap/equity
Growth (G)	G_1	Deposit growth rate	Deposit growth rate
	G_2	Loan growth rate	Loan growth rate
	G_3	Investment growth rate	Investment growth rate
	G_4	Guarantee growth rate	Guarantee growth rate

Source: Data from Shen, K.Y., and Tzeng, G.H., *Soft Computing*, 19(4), 859–874, 2015a.

discriminant (DISCRIM) and decision tree (DT) analysis in Table 18.2. DRSA generated 92.38% classification accuracy on average, superior to the results of DISCRIM and DT.

The jMAF software (Błaszczyński et al. 2013) was used for DRSA modeling, and the DTREG software was used for the calculations of DISCRIM and DT; 33 decision rules were retrieved from the training set, and those decision rules successfully reclassified the 84 objects with 97.62% (82 / 84 = 97.62%) accuracy. To validate the DRSA model, the untouched testing set (21 banks) was examined by the 33 decision rules, and the accuracy of approximation reached 90.91% (20 / 21 = 90.91%), which indicated its effectiveness in modeling. To select the crucial variables for the second stage, only the CORE attributes (12 attributes, shown in Table 18.3) were included for forming the subsequent DANP model. The decision rules with more than six supports are listed in Table 18.4.

Table 18.2   Classification Accuracy of the Training Set

Threefold Cross-Validation	DRSA (%)	DISCRIM (%)	DT (%)
Average[a]	92.38	71.19	75.71
SD	2.47	6.76	4.95

Source: Data from Shen, K.Y., and Tzeng, G.H., *Soft Computing*, 19(4), 859–874, 2015a.

[a] Result of threefold cross-validation repeated five times.

262 ■ *Trends of Hybrid Multiple Criteria Decision Making*

Table 18.3 CORE Attributes (12 Criteria) in Four Dimensions

Dimensions	Criteria[a]
Capital structure (D_1)	C_1, C_2, C_4
Profitability (D_2)	E_2, E_3, E_4
Liquidity (D_3)	L_1, L_2
Growth (D_4)	G_1, G_2, G_3, G_4

Source: Data from Shen, K.Y., and Tzeng, G.H., *Soft Computing*, 19(4), 859–874, 2015a.

[a] See Table 18.1 for the definition of each criterion.

Table 18.4 DRSA Decision Rules with High Supports

Decision Rules	Supports
If ($C_2 \geq 3$) and ($E_4 \geq 2$) and ($L_1 \geq 2$) and ($L_2 \geq 2$), then decision class = at least *Good*	7
If ($E_3 \geq 2$) and ($L_1 \geq 2$) and ($L_2 \geq 2$) and ($G_2 \geq 3$), then decision class = at least *Good*.	7
If ($C_2 \leq 1$) and ($E_2 \leq 2$) and ($G_3 \leq 1$), then decision class = at most *Bad*.	7
If ($C_2 \leq 2$) and ($C_4 \leq 1$) and ($E_3 \leq 2$) and ($L_1 \leq 1$), then decision class = at most *Bad*.	8

Source: Data from Shen, K.Y., and Tzeng, G.H., *Soft Computing*, 19(4), 859–874, 2015a.

At the second stage, the 12 variables obtained from the first stage were used to collect knowledge from experts regarding the FP prediction problem. All the experts (eight experts in total) have more than 10 years' working experience in the banking or financial industries; their job titles include senior consultant, vice president, chief financial officer (CFO), senior analyst, director, associate professor (retired government official), and manager.

The initial average influence matrix A for DEMATEL analysis was normalized to form direct-influence matrix D; referring to Equation 3.3 (Chapter 3), Table 18.5 shows the total-influence matrix T, obtained by calculating $D \times (I - D)^{-1}$. The directional influences among the four dimensions were obtained by finding the normalized dimension matrix T_N^D, referring to Equations 4.7 through 4.9 (Chapter 4), as shown in Table 18.6.

The unweighted supermatrix $W = (T_N^G)$ is the transpose matrix of the normalized direct-influence matrix, as shown in Table 18.7. The weighted supermatrix, adjusted by DEMATEL ($W^\alpha = T_N^D W$), can thus be obtained, as shown in Table 18.8. The stable limiting supermatrix was arrived at by raising the power z of $\lim_{z \to \infty} (W^\alpha)^z$; the final influential weights of each criterion are shown in Table 18.8 with the evaluations of the target banks.

Table 18.5   Total-Influence Matrix T

T	C_1	C_2	C_4	E_2	E_3	E_4	L_1	L_2	G_1	G_2	G_3	G_4	r_i
C_1	0.219	0.206	0.297	0.290	0.403	0.256	0.318	0.223	0.398	0.350	0.358	0.237	3.557
C_2	0.332	0.175	0.335	0.332	0.325	0.342	0.348	0.199	0.408	0.333	0.290	0.221	3.642
C_4	0.285	0.282	0.198	0.312	0.363	0.290	0.325	0.185	0.395	0.325	0.260	0.184	3.403
E_2	0.317	0.258	0.321	0.191	0.371	0.285	0.307	0.176	0.397	0.305	0.260	0.200	3.390
E_3	0.100	0.061	0.097	0.063	0.085	0.072	0.108	0.053	0.119	0.119	0.108	0.074	1.057
E_4	0.256	0.174	0.163	0.178	0.245	0.150	0.307	0.153	0.267	0.194	0.167	0.140	2.394
L_1	0.181	0.123	0.161	0.165	0.263	0.267	0.246	0.235	0.283	0.333	0.310	0.286	2.853
L_2	0.138	0.104	0.131	0.094	0.235	0.142	0.214	0.079	0.225	0.159	0.126	0.104	1.752
G_1	0.155	0.109	0.134	0.132	0.160	0.178	0.217	0.121	0.138	0.146	0.125	0.096	1.711
G_2	0.298	0.177	0.216	0.222	0.391	0.267	0.399	0.198	0.321	0.258	0.350	0.318	3.414
G_3	0.264	0.175	0.209	0.213	0.363	0.303	0.394	0.195	0.315	0.363	0.229	0.317	3.339
G_4	0.130	0.093	0.108	0.103	0.225	0.140	0.299	0.116	0.156	0.282	0.267	0.137	2.056
c_j	2.676	1.937	2.369	2.295	3.429	2.693	3.481	1.934	3.421	3.167	2.850	2.315	

Source: Data from Shen, K.Y., and Tzeng, G.H., *Soft Computing*, 19(4), 859–874, 2015a.

264 ■ *Trends of Hybrid Multiple Criteria Decision Making*

Table 18.6 Normalized Dimension Matrix T_N^G

Dimensions	D_1	D_2	D_3	D_4
D_1	0.2227	0.2784	0.2292	0.2696
D_2	0.2567	0.2410	0.2433	0.2590
D_3	0.1849	0.2572	0.2559	0.3020
D_4	0.1962	0.2560	0.2761	0.2717

Source: Data from Shen, K.Y., and Tzeng, G.H., *Soft Computing*, 19(4), 859–874, 2015a.

Five commercial banks were chosen to illustrate this hybrid model at the final stage: (1) E. Sun Commercial Bank (*A*), (2) Standard Chartered Bank (Taiwan) (*B*), (3) Mega International Commercial Bank (*C*), (4) Taipei Fubon Commercial Bank (*D*), (5) Taishin International Bank (*E*). The experts were asked to give ratings for the five banks in the 12 criteria. To be consistent with the reasoning processes of the DRSA model, the experts were provided with the raw financial ratios and the contemporary industry averages on the 12 variables of the five banks, and were asked to give ratings of *Bad*, *Middle*, and *Good* for the five banks in each criterion.

The performance scores of each bank in each criterion were collected from the same group of domain experts (to form the DANP model). Since the highest score in each criterion is 3, the aspired level in each criterion was set as 3, and the performance gaps of each bank in each criterion were calculated as shown in Table 18.9. Taking the performance score of Bank *A* on criterion C_1 as an example, the raw score was 2.125, and the transformed performance gap was calculated as $(3 - 2.125)/(3 - 0) = 0.292$ in Table 18.9. The influential weights obtained from DANP showed that L_1, E_3, and L_2 were the top-three influential criteria in predicting future FP, and we may conclude that liquidity ratios have a dominant effect in the evaluation model.

18.4 Analytical Results by the Modified VIKOR Method

At the final stage, the modified VIKOR method formed the compromise ranking index V_k (Table 18.9), which indicates that Bank *A* is the top choice, using different weights in δ (i.e., $\delta = 1$, $\delta = 0.7$, and $\delta = 0.5$); despite the minor inconsistencies, the ranking results for the top-three banks (using different δ) are all in line with the actual ΔROA performance in 2012. In addition, to compare the results, another aggregation operator, the fuzzy simple additive weighting (FSAW) method, was further applied. FSAW considers every expert's differences in subjective judgments

Table 18.7   Unweighted Supermatrix $W^{\alpha} = (T_N^G)'$

Criteria	C_1	C_2	C_4	E_2	E_3	E_4	L_1	L_2	G_1	G_2	G_3	G_4
C_1	0.304	0.394	0.372	0.354	0.389	0.432	0.390	0.371	0.389	0.431	0.408	0.393
C_2	0.285	0.208	0.369	0.288	0.237	0.293	0.265	0.280	0.275	0.256	0.269	0.280
C_4	0.411	0.398	0.259	0.358	0.375	0.274	0.345	0.350	0.336	0.313	0.323	0.327
E_2	0.305	0.333	0.324	0.226	0.288	0.311	0.237	0.200	0.281	0.252	0.242	0.219
E_3	0.425	0.325	0.376	0.438	0.385	0.428	0.378	0.499	0.341	0.444	0.413	0.481
E_4	0.270	0.342	0.300	0.336	0.327	0.261	0.385	0.301	0.379	0.304	0.345	0.300
L_1	0.588	0.637	0.637	0.635	0.670	0.666	0.511	0.730	0.642	0.668	0.669	0.721
L_2	0.412	0.363	0.363	0.365	0.330	0.334	0.489	0.270	0.358	0.332	0.331	0.279
G_1	0.296	0.326	0.339	0.341	0.283	0.348	0.233	0.366	0.274	0.258	0.257	0.185
G_2	0.261	0.266	0.279	0.263	0.284	0.252	0.275	0.258	0.289	0.207	0.296	0.335
G_3	0.267	0.231	0.223	0.224	0.258	0.217	0.256	0.206	0.248	0.281	0.187	0.317
G_4	0.177	0.177	0.158	0.172	0.176	0.183	0.236	0.170	0.189	0.255	0.259	0.163

Source: Data from Shen, K.Y., and Tzeng, G.H., *Soft Computing*, 19(4), 859–874, 2015a.

Table 18.8   Adjusted/Weighted Supermatrix $W^a = T_N^D W$

Criteria	C_1	C_2	C_4	E_2	E_3	E_4	L_1	L_2	G_1	G_2	G_3	G_4
C_1	0.068	0.088	0.083	0.091	0.100	0.111	0.072	0.069	0.076	0.085	0.080	0.077
C_2	0.063	0.046	0.082	0.074	0.061	0.075	0.049	0.052	0.054	0.050	0.053	0.055
C_4	0.092	0.089	0.058	0.092	0.096	0.070	0.064	0.065	0.066	0.061	0.063	0.064
E_2	0.085	0.093	0.090	0.054	0.069	0.075	0.061	0.051	0.072	0.064	0.062	0.056
E_3	0.118	0.091	0.105	0.106	0.093	0.103	0.097	0.128	0.087	0.114	0.106	0.123
E_4	0.075	0.095	0.084	0.081	0.079	0.063	0.099	0.078	0.097	0.078	0.088	0.077
L_1	0.135	0.146	0.146	0.155	0.163	0.162	0.131	0.187	0.177	0.184	0.185	0.199
L_2	0.094	0.083	0.083	0.089	0.080	0.081	0.125	0.069	0.099	0.092	0.091	0.077
G_1	0.080	0.088	0.092	0.088	0.073	0.090	0.070	0.111	0.074	0.070	0.070	0.050
G_2	0.070	0.072	0.075	0.068	0.073	0.065	0.083	0.078	0.079	0.056	0.081	0.091
G_3	0.072	0.062	0.060	0.058	0.067	0.056	0.077	0.062	0.067	0.076	0.051	0.086
G_4	0.048	0.048	0.043	0.045	0.046	0.047	0.071	0.051	0.051	0.069	0.070	0.044

Source: Data from Shen, K.Y., and Tzeng, G.H., *Soft Computing*, 19(4), 859–874, 2015a.

Financial Improvements of Commercial Banks ■ 267

Table 18.9 VIKOR–DANP Evaluation Result of the Five Sample Banks

DANP Weights	Criteria	Commercial Banks				
		A	B	C	D	E
0.083	C_1	**0.292**	0.125	0.333	0.125	0.125
0.059	C_2	0.042	0.042	0.375	0.125	0.583
0.073	C_4	0.542	0.042	0.250	0.583	0.458
0.069	E_2	0.625	0.625	0.250	0.042	0.625
0.105	E_3	0.042	0.042	0.583	0.583	0.458
0.084	E_4	0.667	0.292	0.625	0.375	0.583
0.161	L_1	0.000	0.208	0.042	0.125	0.500
0.092	L_2	0.042	0.167	0.333	0.625	0.208
0.080	G_1	0.042	0.292	0.292	0.375	0.625
0.075	G_2	0.125	0.333	0.292	0.375	0.292
0.067	G_3	0.000	0.042	0.542	0.667	0.250
0.054	G_4	0.208	0.667	0.042	0.583	0.625
Actual ΔROA in 2012		209%	35%	24%	10%	−10%
S_k		0.388(1)	0.607(2)	0.665(3)	0.790(5)	0.785(4)
R_k		0.056	0.043	0.061	0.058	0.081
V_k $(v = 0.7)$		0.288(1)	0.438(2)	0.484(3)	0.570(5)	0.574(4)
V_k $(v = 0.5)$		0.222(1)	0.325(2)	0.363(3)	0.424(4)	0.433(5)

Source: Data from Shen, K.Y., and Tzeng, G.H., *Soft Computing*, 19(4), 859–874, 2015a.

regarding *Bad*, *Middle*, and *Good*; each expert was also asked to fill out their subjective fuzzy membership parameters for these ratings. The commonly adopted triangular membership function (with FSAW) was then used to transform the experts' judgments into performance scores.

The briefing of FSAW is as follows: Assume that there are s experts for a fuzzy performance measurement, and E_{kj}^h denotes the hth expert's fuzzy judgment for the kth alternative on criterion j. The average operation was selected to obtain the representative result for the kth bank in criterion j, as expressed in Equation 18.1:

$$E_{kj} = \left(E_{kj}^1 \oplus E_{kj}^2 \oplus \cdots \oplus E_{kj}^s \right) / s = \left(L_{kj}^s, M_{kj}^s, H_{kj}^s \right). \tag{18.1}$$

268 ■ *Trends of Hybrid Multiple Criteria Decision Making*

While w_j denotes the influential weights for criterion j, the fuzzy synthetic performance measurement for the kth bank can be expressed as Equation 18.2:

$$E_k = \left(\frac{\sum_{j=1}^{n} w_j \times L_{kj}^s}{n}, \frac{\sum_{j=1}^{n} w_j \times M_{kj}^s}{n}, \frac{\sum_{j=1}^{n} w_j \times H_{kj}^s}{n} \right) = \left(L_k^w, M_k^w, H_k^w \right), \quad (18.2)$$

where n is the number of total criteria under evaluation.

Then, the fuzzy synthetic performance measurement E_k can be defuzzified into performance score P_k (Opricovic and Tzeng 2004) for the ith alternative, as in Equation 18.3:

$$P_k = L_k^w + \frac{\left(H_k^w - L_k^w \right) + \left(M_k^w - L_k^w \right)}{3}. \quad (18.3)$$

The five example banks' FSAW performance scores P_k ($k \in \{A, B, C, D, E\}$) are: 5.39 (A), 4.42 (B), 4.05 (C), 3.87 (D), and 3.02 (E), respectively; the final ranking sequence is $A > B > C > D > E$. This ranking sequence is the same as the result from VIKOR, where $\delta = 0.5$ (refer to Table 18.9). Although the final ranking sequence of the five banks from the modified VIKOR method (setting different values in δ) and that of FSAW are not exactly the same, the consistent ranking for the top three banks suggests the stability of the proposed approach.

18.5 Discussion

Combining the findings from the DEMATEL and VIKOR analyses, the proposed hybrid MCDM model not only carries out rankings and selections but also helps banks plan for improvements. Take the top-ranked Bank A, for example; if Bank A attempts to improve its FP for better growth, it should take the dimension *Earnings and profitability* (D_2) as its top priority, because the aggregated performance gap in D_2 equals 0.104 ([0.069 × 0.625] + [0.105 × 0.042] + [0.084 × 0.667] = 0.104) (Table 18.9); the performance gap in D_2 is the highest among the four dimensions (the performance gaps of Bank A in dimensions D_1, D_3, and D_4 are 0.066, 0.056, and 0.024, respectively). Moreover, the relative influences and a cause–effect analysis of the dimensions can be obtained by observing $r_i^D - c_i^D$ in Table 18.10.

The weighted dimensional performance gaps of Bank A, against the aspired levels, are illustrated with the directional influences in Figure 18.2. Referring to Equation 4.7 and its associated explanation in Chapter 4, the results of $r_i^D - c_i^D$ can divide dimensions into a cause group ($r_i^D - c_i^D > 0$) and an effect group ($r_i^D - c_i^D < 0$). The dimension *Capital structure* (D_1) might cause changes in the

Financial Improvements of Commercial Banks ■ 269

Table 18.10 Influential Weights of Dimensions

Dimensions	r_i^D	c_i^D	$r_i^D + c_i^D$	$r_i^D - c_i^D$
D_1	1.162	0.765	1.927	0.397
D_2	0.756	0.925	1.681	−0.169
D_3	0.756	0.886	1.642	−0.130
D_4	0.878	0.976	1.854	−0.098

Source: Data from Shen, K.Y., and Tzeng, G.H., *Soft Computing*, 19(4), 859–874, 2015a.

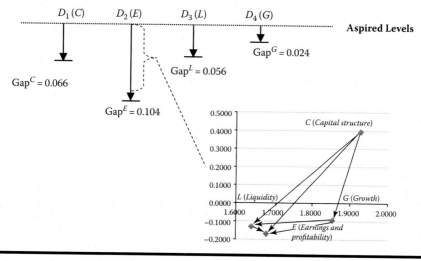

Figure 18.2 Performance gaps against aspired levels with directional influences among the dimensions. (Reprinted with permission from Shen, K.Y. and Tzeng, G.H., *Soft Computing*, 19(4), 859–874, 2015a.)

dimension *Earnings and profitability* (D_2); the dimension D_1 also has the highest influence (i.e., $r_i^D + c_i^D = 1.927$) among the four dimensions (i.e., $D_1 = 1.927$, $D_3 = 1.642$, and $D_4 = 1.854$). Therefore, a reasonable improvement plan should focus on improving the dimension D_1, which would yield the highest marginal effect. Likewise, by calculating $r_i^C - c_i^C$ (Equations 3.4 and 3.5, Chapter 3), Bank A can analyze the directional influences among the 12 criteria and identify a priority list for its performance improvements.

Aside from ranking and improvement planning, the proposed model not only formed decision rules to identify future improvements but also generated decision rules to detect deteriorating FP in the future (Table 18.4). This finding can be applied to detecting the symptoms of potential crises, acting as a warning mechanism. As commercial banks are crucial to the stability of economies, the obtained decision rules may prove useful for identifying the early symptoms of potential crises.

18.6 Conclusion

To conclude, this chapter proposed an integrated hybrid MCDM model for predicting the FP of commercial banks. The complexity of the multiple dimensions and criteria of financial reports impedes decision makers dealing with large and imprecise data sets; therefore, the rough set–based DRSA algorithm is useful for finding patterns and critical variables to predict FP. In this chapter, 12 critical indicators were successfully selected from the original 25 financial attributes with the capability of discriminating positive (or negative) FP changes, which is also the strength of MRDM in processing complex data-centric problems. In addition, several easy-to-understand decision rules that can predict future performance improvements/deterioration with strong supports were found.

With fewer variables, the domain experts were able to give opinions to form the DANP model. Since the experts were asked to compare the relative influence of one criterion with another, we are more likely to retrieve reliable knowledge from experts by making pairwise comparisons with fewer criteria. The constructed DANP model can analyze the interrelationships among the criteria, and the influential weights of each criterion were also found. Furthermore, the obtained DEMATEL analysis at this stage divided the dimensions (criteria) into a cause group and an effect group, which can be integrated with the VIKOR method to guide FP improvements.

To examine the constructed model, the raw financial data of the five sample banks and the industrial averages were provided to the experts for them to rate each criterion for those banks. At this stage, the compromised ranking VIKOR method was used to aggregate the performance gaps of each alternative for ranking. The top-three-ranked banks were in line with their actual FP changes in 2012, which indicated the effectiveness of the proposed model. Furthermore, the selected Bank A was illustrated to identify its top-priority dimension for improvement, and the way to explore the source dimension for improvements—by combining VIKOR and DEMATEL analysis—was also discussed. Thus, the present chapter contributes to the application of soft computing and MCDM methods in the banking industry.

Despite the contributions of this chapter, there are still certain limitations. First, the DRSA model only uses one period-lagged data set to predict FP (i.e., associating data on a bank's conditional attributes in period $t - 1$ with its decision class in period t). Some latent tendencies in relatively long lag periods (e.g., more than 2 years) might not be captured in the model. Second, although the model may identify the performance gaps of banks and suggest improvement priorities, it is still at the experimental stage. To cooperate with banks on evaluating the feasibility of those improvement suggestions from the analytical results would be helpful in solving practical problems; future studies are suggested to work in this direction. Lastly, this approach only integrates MRDM with MADM to unravel implicit knowledge and make ranking and selection decisions; how to allocate limited resources for improvements is not addressed. The integration of MRDM, MADM, and MODM might be a plausible approach to tackle this complicated issue.

Chapter 19

FCA-Based DANP Model Using the Rough Set Approach: A Case Study of Semiconductor Companies

In this chapter, another case involving the financial performance (FP) improvement problem is discussed, combining multiple rules-based decision making (MRDM) with multiple attribute decision making (MADM). Although various computational techniques can serve as modeling tools, a logical approach that can retrieve understandable knowledge to guide systematic improvements is still required in practice. As a result, a case study on improving the FP of semiconductor companies is illustrated here to show the proposed approach.

19.1 Research Background

The intensification of global competition and rapid changes in the economic environment have created severe challenges for companies to maintain or improve their FP, which is key to a company sustaining itself in the long run. As a result, the importance of using advanced analytical techniques to support financial decision making has become clear to both the academic and practical fields (Spronk et al. 2005).

273

Considering the complexity among the factors involved in FP, the multiple criteria decision making (MCDM) approach is suitable for tackling this issue and has been widely explored (Steuer and Na 2003; Xidonas et al. 2009). Moreover, computational intelligence techniques that deal with the uncertain and imprecise characteristics of business information are incorporated to enhance the effectiveness of the models. A rising trend of integrating multiple methods based on the strengths of each technique has emerged, and a hybrid approach based on MRDM and multiple attribute decision making (MADM) is devised and integrated here using the logical reasoning technique of formal concept analysis (FCA) (Ganter et al. 1997; Wille 2005).

The study of FP has gained increasing interest from various groups for at least three purposes: (1) supporting investment decisions, (2) detecting warning signs of insolvency, and (3) planning for performance improvements (Penman 2007). Due to its potential value and the practical needs of businesses, researchers from various fields have tried to address this research topic. Conventional studies mainly rely on statistical analyses to identify the main variables that have positive or negative influences on future FP (Piotroski 2000). However, the most frequently used statistical model, the regression model, has obvious limitations in several aspects. For example, the presumed linear relationship of the regression model is not realistic; also, the assumed independence of variables might not be valid. In the real world, factors are often interrelated in certain ways (Liou and Tzeng 2012). To extend the limitations of conventional studies, various MCDM and computational intelligence methods have been used to explore the FP prediction problem. The most commonly used MCDM methods—such as the analytic hierarchy process (AHP) (Saaty 1988), the analytic network process (ANP) (Saaty 2004), DEMATEL (Gabus and Fontela 1972) and TOPSIS (Opricovic and Tzeng 2004)—collect the opinions of domain experts to decide the relative importance of each criterion and rank or select alternatives. However, these kinds of MCDM models depend on the subjective judgments of experts, and have been criticized for lacking objective support. On the contrary, computational intelligence techniques or methods depend on advanced algorithms to induce implicit patterns from large data sets. The well-known artificial neural network (ANN) (Lam 2004; Shen 2011), the neuro-fuzzy inference technique (Boyacioglu and Avci 2010), decision tree (DT) analysis (Wang and Chan 2006), and the rough set approach (RSA) (Shyng et al. 2010a) have all been applied to solve this FP prediction problem. ANN-related techniques have strength in minimizing modeling errors through various machine learning algorithms; nevertheless, the implicit knowledge is embedded in its connections among nodes and its network structure, which makes it difficult to gain reusable and understandable knowledge. Compared with ANN-related techniques, the DT and RSA methods can induce more understandable decision rules; however, certain limitations remain in terms of exploring the interrelationships among criteria and understanding the in-depth managerial implications. Both MCDM methods and computational intelligence techniques have their own strengths and limitations; thus, a logical combination that can capture changes in FP patterns from large data sets and clarify the interrelationships among the critical criteria is needed.

Aside from the FP prediction problem, once a company has been classified as underperforming in the next period, which criterion it should focus on improving at the current stage—based on limited resources—is another important issue. Furthermore, the involved criteria are often interrelated in a real business environment (Liou and Tzeng 2012; Liou 2013), and how to consider the plausible directional influences to design or evaluate an improvement plan for a specific criterion is still underexplored. To bridge the gap, an FP diagnosis model is proposed fusing two approaches: computational intelligence and MCDM methods. To be more accurate, the following questions are to be answered: Which criteria are crucial to identifying changes in FP? Once a company has found the critical gap that it plans to improve, which are the relevant criteria that should be considered contextually? Moreover, if the company comes out with several potential improvement plans for its identified gaps, how should it select among them? To answer the aforementioned questions, a dominance-based rough set approach (DRSA) (Greco et al. 2002) is proposed to explore the patterns of FP changes. DRSA is extended from RSA (Pawlak 1982), which considers the preferential characteristics of criteria when making inductions. The evaluation of FP involves various financial attributes, and those attributes often have preferential characteristics in nature; for example, a higher gross margin ratio is generally preferred concerning FP evaluation. The decision rules obtained from DRSA may support underperforming companies to identify performance gaps for improvements; however, if a company further attempts to devise or select improvement plans for its performance gaps, a logical and systematic evaluation method is still required. As a result, an FCA-based MCDM model is formed at the next stage, to extend the findings from the DRSA decision rules.

At the second stage, if a criterion is identified as the priority, it is necessary to identify the most interrelated criteria to plan for improvements; otherwise, implementing a new plan might cause unwanted side effects. To meet this goal, a FCA-based DEMATEL-based ANP (DANP) analysis (Shen et al. 2014) is proposed for the diagnosis. FCA is a mathematical theory (Wille 2005) that can use historical data to generate implication rules and find pertinent criteria using a logical approach. Finally, the associated criteria can be analyzed (using the DANP model) to explore the directional influences and relative weights concerning the improvement of the addressed criterion. The simplified two-stage research flow is illustrated in Figure 19.1.

19.2 Reviews of MCDM and Soft Computing Methods in Financial Applications

This section briefly reviews the relevant soft computing techniques (RSA and FCA) and MCDM methods used in financial applications. In addition, the main purposes and advantages of adopting each method and technique are discussed.

The growing complexity of the business environment has increased the need to evaluate financial decisions while considering multiple aspects and criteria;

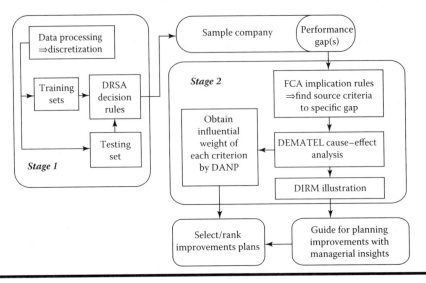

Figure 19.1 Research flow of the hybrid MCDM model for improvement planning. (Reprinted with permission from Shen, K. Y., and Tzeng, G.H., *Technological and Economic Development of Economy*, 22(5), 685–714, 2015e.)

therefore, it is apt to adopt various methods in the framework of MCDM (Spronk et al. 2005; Zopounidis and Doumpos 2013; Zopounidis et al. 2015). Two research mainstreams can be observed: investment analysis (e.g., portfolio management and investment evaluation) and corporate finance (e.g., FP prediction and evaluation, business failure prediction, capital budgeting, credit scoring, auditing, and financial planning); this work belongs to the latter.

In practice, the use of financial ratios to compare and forecast the FP of a company, also termed *fundamental analysis* (FA), is quite common. FA is generally conducted by comparing the relative FP of a company within its industry or modeling the patterns of its own FP indicators over a period of time (Penman 2007). In financial literature, FA is mainly modeled by regressions; however, the unrealistic assumptions (e.g., the independence of the variables involved, the linear relationships between the target variable and other variables) of regression models can cause unpersuasive results. Furthermore, regression represents the "average" result (Spronk et al. 2005), which has difficulty indicating the contingent performance in a specific context. On the other hand, certain MCDM methods are not constrained by assuming the probabilistic distributions of variables and the independence of the criteria, which makes them more realistic in practice. As a result, MCDM methods have strengths that complement existing financial research into solving FP problems (Liou and Tzeng 2012).

In the literature, several MCDM outranking methods have been introduced to assess the FP of companies, such as the two groups ELECTRE (Ergul and Oktem

2011) and PROMETHEE (Mareschal and Brans 1991; Mareschal and Mertens 1992; Tsui et al. 2015). Another main branch of MCDM is the utility-based approach, which includes multiattribute utility theory (MAUT) (Diakoulaki et al. 1992; Tzeng et al. 1989; Yeh et al. 2000) and the utility additive (UTA) method (Zopounidis et al. 1995). Utility-based theory uses its theoretical foundations in economics to aggregate the preferences of decision makers (DMs) on multiple attributes. The aforementioned outranking methods and the utility-based approach can conduct ranking and selection without the prerequisite assumption of the probabilistic distribution of variables. Nevertheless, the assumption of the independent relationships among criteria remains (Liou and Tzeng 2012; Liou 2013; Peng and Tzeng 2013).

Until recently, extended AHP methods were adopted to resolve the FP evaluation problem (Shen et al. 2014; Shen and Tzeng 2014a). This approach asks domain experts (or DMs) to make pairwise comparisons between two criteria, and the relative influence of each criterion may be obtained to form the overall performance evaluation model. Detailed discussions on this approach can be found in Chapter 5. Although the interdependence among criteria is allowed in this approach, the selection of criteria relies on subjective judgment. Considering the complexity of FP evaluation, how to select the minimal criteria (variables) with discernibility for modeling is still an unanswered issue. Therefore, FCA implication analysis (Shyng et al. 2010b; Ou Yang et al. 2011) is adopted in this case to find the most highly related criteria—for the identified performance gap based on DRSA decision rules—from historical data, which can form the basis for the subsequent decision model in a more objective way.

RSA is a mathematical theory (Pawlak 1982) that has been implemented in various applications to deal with the imprecise and uncertain characteristics of data sets. Several studies have attempted to apply RSA to FP predictions, which can be categorized into two types: (1) a single RSA model implemented by different induction algorithms (Dimitras et al. 1999; Tay and Shen 2002), and (2) a hybrid or a combined model that infuses two or more techniques with RSA, such as ANN (Ahn et al. 2000), data envelopment analysis (DEA) (Shuai and Li 2005), and the rough-AHP with fuzzy decision-making model (Aydogan 2011). The first type employs RSA algorithms to induce rules from data sets, and the obtained rules are used to make classifications or predictions. Though this model yields positive results, it cannot support an individual company's plans for improvements. The second type (Ahn et al. 2000) uses RSA as a preprocessor to remove redundant attributes, and an ANN is applied to increase the model's accuracy. However, the ANN technique has been criticized for its black-box processing (Ravi et al. 2008); the obtained knowledge is mainly stored in the network structure and the connections between nodes, so DMs would have difficulty gaining understandable insights. The last model discussed here—the rough-AHP with fuzzy decision-making model (Aydogan 2011)—integrates RSA with MCDM to make selections. The conditional entropy and attribute significance concepts in RSA are used to improve the judgment consistency in AHP. Nevertheless, the AHP method cannot evaluate the interdependence among attributes and is also unable to support improvement planning.

As for the DRSA (Greco et al. 2002) adopted in this model, it was extended from classical RSA. DRSA was proposed to consider the preferential characteristics of attributes when making classifications, which is more suitable to tackling multiple criteria decision problems; a discussion on solving MCDM problems with DRSA can be found in Chapter 7. Recently, the social sciences have become aware of the strengths of DRSA (Shyng et al. 2010a,b; Zaras 2011; Shen and Tzeng 2014a,b, 2015a,b,c), and the aforementioned advantages are also the reasons why this chapter adopts DRSA to explore the patterns of FP changes in semiconductor companies.

The last technique discussed here is FCA, which originates from applied mathematics (Ganter et al. 1997; Wille 2005). FCA analysis can organize objects and related attributes in a contextual way to form hierarchical concept lattices and has the capability to clarify the formalization of concepts with implication rules. The applications of FCA have recently gained increasing interest in the international research community, ranging from software engineering, knowledge discovery, marketing, and personal investment portfolios (Shyng et al. 2010a) to information retrieval (Tilley et al. 2005). The use of FCA in financial applications (e.g., investment analysis or performance evaluation) is relatively underexplored. The use of FCA can support explorations of the related source criteria of a specific financial attribute in a logical manner, and we attempt to construct a persuasive model based on FCA implications to select the critical attributes in the DANP model. The subsequent *logical* improvement planning will be hinged on the findings from FCA.

Despite the discussions on MCDM or soft computing techniques, the combination or integration of multiple methods/techniques (also termed *hybrid models*) has emerged as a rising trend. One of the hybrid MCDM approaches—including DEMATEL, DANP, and modified VIKOR—has been introduced in Chapter 5. Here, the analytical results from DEMATEL and DANP will be integrated with FCA implication rules to form a directional implication relationship map (DIRM), which may thus help DMs to plan for improvements. DMs should plan in a systematic way to avoid merely focusing on the identified performance gap.

19.3 Framework of the Hybrid MRDM Model

The proposed model comprises three parts: 1) generating decision rules for classifying future FP, 2) obtaining implication rules and source criteria for the addressed performance gap, and 3) constructing the DANP model for evaluation and selection. The first two parts can be categorized as MRDM, and the third belongs to MADM. The required steps can be summarized as follows.

Step 1: Define the conditional attributes and the associated decision attributes of the addressed problem, and conduct discretization for all the attributes.

Step 2: Match the values of a stock's conditional attributes in time period $t-1$ with its decision class (DC) in time period t to denote an object (instance).

FCA-Based DANP Model Using the Rough Set Approach ■ 279

The matched data set is devised to predict the FP of a stock in the subsequent period by using its current financial data.

Step 3: Construct the DRSA model and obtain decision rules to identify stocks with plausibly good FP in the next period. These three steps are essential to forming a DRSA model for analyzing imprecise patterns in complex data sets.

The next stage adopts FCA to obtain implication rules. Originating from applied mathematics, FCA was developed based on mathematical order and lattice theory, which has been applied in various fields, such as software engineering, knowledge acquisition, medical classification, and financial investments. FCA can be defined as a set of structure $\Re := (G, M, I)$, in which I denotes the binary relation between the two sets G and M. The elements in set G represent objects, and the elements in M attributes. Thus, a formal context can be established by connecting the objects in G to the attributes in M through the binary relation I (yes or no); that is, $(g, m) \in I$ for $g \subseteq G$ and $m \subseteq M$. If $g = m^I$ and $m = g^I$, then g and m can be called the *extent* and *intent* of a pair of formal concepts. Based on the theorem of concept lattice is (Ganter et al. 1997), while the concept lattice is a complete lattice, it should be made up of closed subsets, which serves to calculate the Duquenne–Guigues base implications with a minimal number of rules. Here, the identified performance gap in a certain criterion can be regarded as m in the attribute set M, and the Duquenne–Guigues implication rules (Ganter et al. 1997; Wille 2005) can be obtained to explore the extents with high supports.

Step 4: Examine a target alternative's performance on the strong decision rules (associated with a positive FP change in the subsequent period), and identify its top-priority gap.

Step 5: Conduct Duquenne–Guigues implication reasoning using FCA to obtain implication rules associated with the source criteria that might lead to the identified performance gap attribute in Step 4. With FCA implication analysis, DMs can receive guidance regarding the source factors (criteria) related to the underperforming criterion (identified by DRSA decision rules) of a company.

Step 6: Conduct DEMATEL analysis using the obtained source criteria from Step 5; this is also the preliminary step for the subsequent DANP model.

Step 7: Use DEMATEL-adjusted dimensional influences to obtain the DANP influential weight for each source criterion (i.e., the FCA-based DANP model). After forming this weighting system, DMs may adopt it to devise and evaluate (or select) improvement plans. Domain experts or DMs can evaluate the potential (available) improvement plans to get the performance scores for each criterion in this hybrid model and synthesize those performance scores with the influential weights from DANP to get the final score for each plausible improvement plan.

The FCA-based DANP model may help a company rank or select improvement plans in a quantitative manner, which further extends the applications of MCDM methods to resolving financial problems.

19.4 Case Study of Semiconductor Companies

This chapter adopts public-listed semiconductor companies in Taiwan as an example. The empirical case study comprised two stages; the first stage analyzed and explored historical patterns to form a DRSA financial model, and a hybrid MCDM model integrating FCA with the DANP method was constructed for improvement planning at the second stage. Furthermore, a sample company was illustrated to identify its critical performance gap, and five suggested improvement plans were evaluated with the ranking result. The research flow is illustrated in Figure 19.2.

To capture the FP change patterns, the major financial indicators of the semiconductor stocks at time t were taken as the conditional attributes, and the FP at time $t + 1$ the decision attribute; the association of those attributes is used to model the imprecise causative relationships between the conditional and decision attributes. All semiconductor stocks in the Taiwan stock market from 2007 to 2012 were included for DRSA modeling. After excluding alternatives with incomplete financial data, 182 instances (from 2007 to 2011) were used for training, and 55 instances (in 2012) were tested. The Taiwan Stock Exchange reports each stock's summary financial result in five aspects with 20 key financial indicators (Table 19.1), and all of the indicators were adopted as the conditional attributes in the DRSA model.

All the financial raw data were retrieved from the Taiwan Economic Journal (TEJ) database and discretized as three values to denote *high* (the top third), *middle*

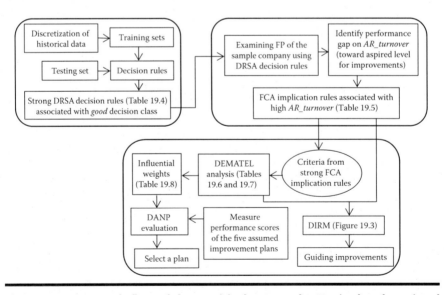

Figure 19.2 Research flow of the empirical case study. (Revised and reprinted with permission from Shen, K. Y., and Tzeng, G.H., *Technological and Economic Development of Economy*, 22(5), 685–714, 2015e.)

FCA-Based DANP Model Using the Rough Set Approach ■ 281

Table 19.1   Definitions of the 20 Conditional Attributes (Financial Indicators)

Dimensions	Financial Ratios	Symbols	Definitions and Brief Explanations
Capital structure	Debt to total assets	Debt	Total debt/total assets
	Long-term capital to total assets	LongCapital	Long-term capital/total assets
Payback capability	Liquidity ratio	Liquidity	Current assets/current liability
	Speed ratio	Speed	(Current assets – inventory)/current liability
	Interest coverage ratio	Interest Coverage	(Net profit before tax + interest expense)/interest expense
Operational efficiency	Accounts receivable ratio	AR_turnover	Net credit sales/average accounts receivable
	Days for collecting AR	AR_days	(Days × AR)/credit sales
	Inventory turnover rate	InvTurnover	Total operational cost/average inventory
	Average days for sales	DAYs	(Average ending inventory/operational cost) × 365 days
	Fixed asset turnover rate	FAssetTurnover	Total revenue/total fixed assets
Profitability	Return on total assets	ROA	Net profit before tax/average total assets
	Return on equity	ROE	Net profit before tax/average total equity
	Operational profit to total capital	OP_capital	Operational profit/total capital
	Net profit before tax to total capital	NP_capital	Net profit before tax/total capital
	Net profit ratio	NetProfit	Net profit/net sales

(Continued)

282 ■ Trends of Hybrid Multiple Criteria Decision Making

Table 19.1 Definitions of the 20 Conditional Attributes (Financial Indicators)

Dimensions	Financial Ratios	Symbols	Definitions and Brief Explanations
	Earnings per share	EPS	(Net income – dividends on preferred stocks) / total outstanding shares
Cash flow	Cash flow ratio	CashFlow	(Operational cash flow – cash dividend for preferred stocks) / weighted average equity
	Cash flow adequacy ratio	CashFlow_adq	Cash flow from operations / annual current maturities
	Cash flow reinvestment ratio	CashFlow_inv	(Increase in fixed assets + increase in working capital) / (net income + noncash expenses – noncash sales – dividends)

Source: Data from Shen, K. Y., and Tzeng, G.H., *Technological and Economic Development of Economy*, 22(5), 685–714, 2015e.

(the middle third), and *low* (the bottom third) as H, M, and L, respectively. The adopted three-level discretization ranked all of the available stocks based on the performance of each attribute in each year, and assigned a corresponding value (i.e., H, M, or L) to each attribute of a stock; therefore, the discretization helped to identify the relative performance of each stock on the 20 financial attributes in each year. The decision classes were also discretized into three states—*Good, Middle,* and *Bad*—according to a stock's corresponding ROA change in the subsequent year. However, since we aim to explore patterns that may lead to improvements or deteriorations in the next period, only the stocks categorized as *Good* and *Bad* were kept for rule induction. Thus, only 145 instances were left as the training set and 35 stocks (in 2012) as the testing set.

The new training set comprised 145 instances, and the classification accuracy (CA) (i.e., the correctly classified percentage) was examined for this data set. In the first step, this training set was examined and compared by three different classifiers: DRSA, support vector machine (SVM), and DT. To measure the stability of each classifier, a fivefold cross-validation was conducted five times for each classifier; the average CA is reported with its standard deviation (SD) in Table 19.2.

The DRSA classifier yielded the highest CA on average, which was regarded as acceptable in this experiment; next, the new testing set (35 instances) was used to

FCA-Based DANP Model Using the Rough Set Approach ■ 283

Table 19.2 Classification Accuracy of the Training Set

	DRSA (%)	SVM (%)	DT (%)
1st	77.93	71.73	73.50
2nd	75.17	69.98	71.71
3rd	76.55	71.96	68.70
4th	78.62	68.67	59.38
5th	77.93	70.67	74.32
Average	77.24 (1)	70.60 (2)	69.52 (3)
SD	1.38	1.35	6.07

Source: Data from Shen, K. Y., and Tzeng, G.H., *Technological and Economic Development of Economy*, 22(5), 685–714, 2015e.

Table 19.3 Decision Rules Associated with the At-Least Good Decision Class

Decisions Rules	Supports
If (*LongCapital* ≥ M and *Speed* ≥ M and *AR_turnover* ≥ M and *AR_days* ≥ H and *ROE* ≥ M and *CashFlow_adq* ≥ M), **then** (DC ≥ *Good*).	17
If (*Speed* ≥ M & *AR_turnover* ≥ H and *EPS* ≥ M and *CashFlow* ≥ H and *CashFlow_adq* ≥ M), **then** (DC ≥ *Good*).	16
If (*Debt* ≥ M and *InterestCoverage* ≥ M and *AR_turnover* ≥ H and *Inventory* ≥ M and *EPS* ≥ H and *CashFlow* ≥ M), **then** (DC ≥ *Good*).	14

Source: Data from Shen, K. Y., and Tzeng, G.H., *Technological and Economic Development of Economy*, 22(5), 685–714, 2015e.

validate the model, which yielded 85.71% CA. This result suggests that the DRSA model has reached certain level of credibility in modeling the FP changes. The trained DRSA model generated 24 decision rules with 118 REDUCTs, and the top-three decision rules, with the highest supports, associated with the *at-least Good* decision class are shown in Table 19.3.

After obtaining the strong decision rules associated with the *at-least Good* decision class, the next step attempted to form a hybrid MCDM model for supporting FP improvements. Here, the FP data of a leading substrate maker in 2012 were adopted as an example; its financial attributes were discretized as: *LongCapital* = M, *Speed* = M, AR_*turnover* = M, AR_days = M, *ROE* = H, *CashFlow* = H, *CashFlow*_adq =

284 ■ *Trends of Hybrid Multiple Criteria Decision Making*

H, *EPS* = H, *Debt* = M, *InterestCoverage* = M, *Inventory* = M, which satisfied all of the conditional requirements of the top-three decision rules, except the attributes *AR_turnover* and *AR_days*. As the attribute *AR_turnover* appeared in two of the top-three decision rules, the subsequent analysis took it as an example to plan for FP improvements.

19.5 FCA-Based DANP Model for Ranking Improvement Plans

To explore the critical attributes that have close relationships with the high *AR_turnover* attribute (*AR_turnover_H* denotes *AR_turnover* with a *high* value in FCA implication rules), all the available instances associated with the *Good* decision class, including the training and the testing sets, were analyzed. The FCA analysis was conducted by using ConExp, which generated 111 implication rules. The top-three highly covered implication rules associated with high *AR_turnover* are shown in Table 19.4.

FCA helps to identify the plausible formal attributes that might cause high *AR_turnover* in the companies in the *Good* decision class, which provides a reasonable and logical foundation to form the subsequent decision model. Here, FCA can be regarded as an MRDM mechanism, and the criteria listed in the top-three decision rules were used to construct an FCA-based DANP model at the next stage; the involved criteria are: *Debt* (C_1), *LongCapital* (C_2), *Speed* (C_3), *InterestCoverage* (C_4), *Inventory* (C_5), *EPS* (C_6), and *CashFlow_adq* (C_7). Referring to Table 19.1, the seven criteria belong to five dimensions; the five dimensions and seven criteria

Table 19.4   Implication Rules Associated with High *AR_turnover*

Implication Rules	Covers
LongCapital_M and *LongCapital_H* and *Speed_M* and *InterestCoverage_M* and *AR_turnover_M* and *Inventory_M* and *EPS_M* and *CashFlow_adq_M* ⇒ **AR_turnover_H**	12
Debt_M and *Debt_H* and *LongCapital_M* and *InterestCoverage_M* and *AR_turnover_M* and *Inventory_M* and *EPS_M* and *EPS_H* and *CashFlow_adq_M* ⇒ **AR_turnover_H**	11
Debt_M and *Debt_H* and *LongCapital_M* and *InterestCoverage_M* and *AR_turnover_M* and *Inventory_M* and *EPS_M* and *EPS_H* and *CashFlow_adq_M* ⇒ **AR_turnover_H**	10

Source: Data from Shen, K. Y., and Tzeng, G.H., *Technological and Economic Development of Economy*, 22(5), 685–714, 2015e.

were adopted to design a questionnaire for retrieving the knowledge/experience of domain experts. In the questionnaire, the questions asked include, Comparing financial attribute *Debt* (C_1) with *Inventory* (C_5), what is the relative influence degree of C_1 to C_5 on attaining a high *AR_turnover* (i.e., a high accounts receivable turnover rate)? After collecting the opinions from eight domain experts, there are two critical steps: using DEMATEL analysis to identify the cause–effect influences among the dimensions or criteria (Step 6), and obtaining DANP influential weights for the seven criteria (Step 7). The initial average matrix A for the seven criteria is shown in Table 19.5, and the required steps to obtain the directional influences among dimensions (observing $r_j^D - c_j^D$) or criteria (observing $r_i^C - c_i^C$) and the DANP influential weights can be found in Chapters 3 and 4.

Table 19.5 Initial Influence Matrix A

Criteria	C_1	C_2	C_3	C_4	C_5	C_6	C_7	Sums
C_1	0.000	3.625	1.250	3.250	1.750	2.625	1.250	13.750
C_2	3.250	0.000	1.250	3.500	1.375	2.750	3.125	15.250
C_3	1.500	1.250	0.000	2.250	3.250	3.000	2.750	14.000
C_4	2.500	2.125	1.375	0.000	1.500	3.500	2.000	13.000
C_5	2.250	2.000	3.500	2.125	0.000	3.000	1.500	14.375
C_6	2.250	2.250	1.750	3.500	2.125	0.000	1.875	13.750
C_7	3.125	2.250	3.000	1.750	2.500	1.375	0.000	14.000
Sums	14.875	13.500	12.125	16.375	12.500	16.250	12.500	

Source: Data from Shen, K. Y., and Tzeng, G.H., *Technological and Economic Development of Economy*, 22(5), 685–714, 2015e.

Table 19.6 Directional Influences of Dimensions and Criteria

Dimensions	r_j^D	c_j^D	$r_j^D - c_j^D$	$r_j^D - c_j^D$	Criteria	r_i^C	c_i^C	$r_i^C - c_i^C$	$r_i^C - c_i^C$
D_1	4.281	4.286	8.567	−0.005	C_1	5.765	6.187	11.952	−0.421
D_2	4.069	4.271	8.340	−0.202	C_2	6.274	5.740	12.014	0.533
D_3	4.208	3.699	7.907	0.510	C_3	5.835	5.014	10.849	0.822
D_4	4.024	4.737	8.761	−0.713	C_4	5.468	6.815	12.283	−1.347
D_5	4.113	3.702	7.815	0.411	C_5	5.981	5.200	11.181	0.780
					C_6	5.734	6.717	12.451	−0.982
					C_7	5.866	5.250	11.115	0.616

Source: Data from Shen, K. Y., and Tzeng, G.H., *Technological and Economic Development of Economy*, 22(5), 685–714, 2015e.

286 ■ *Trends of Hybrid Multiple Criteria Decision Making*

Table 19.7 DANP Influential Weights and the Performance Scores of the Five Assumed Plans

Criteria	Influential Weights	Performance Score* on Each Criterion				
		A	B	C	D	E
Debt (C_1)	0.107	3	5	1	2	4
LongCapital (C_2)	0.100	2	3	4	1	5
Speed (C_3)	0.088	5	3	2	1	4
Interest Coverage (C_4)	0.118	3	2	4	1	5
Inventory(C_5)	0.179	2	3	1	4	5
EPS (C_6)	0.228	3	4	2	1	5
CashFlow_adq (C_7)	0.179	4	2	3	1	5
Summed scores (Ranking)		3.073 (3)	3.142 (4)	2.327 (2)	1.643 (1)	4.800 (5)

Source: Data from Shen, K. Y., and Tzeng, G.H., *Technological and Economic Development of Economy*, 22(5), 685–714, 2015e.

* A five-point Likert scale s is adopted, in ascending order (i.e., 1 is better than 5), as performance scores.

The relative influential degree and cause–effect group indicator for each dimension and criterion are provided in Table 19.6, which will support the drawing of an influential network relationship map (INRM). The DANP influential weights of the seven criteria that will cause the formal concept *AR_turnover_H* are shown in Table 19.7, to illustrate how to select improvement plans in the next section.

19.6 Discussion on Improvement Planning

Since the aim here is to improve the FP of semiconductor companies, five assumed improvement plans (i.e., *A*, *B*, *C*, *D*, and *E* plans) are used to illustrate the selection procedures. The goal for these improvement plans is clear: achieving a high *AR_turnover* rate for this company. From the logical perspective, the implication rules associated with high *AR_turnover* contain the granules of concepts (i.e., criteria) that might cause high *AR_turnover*. Thus, the constructed DANP evaluation model comprises those criteria for attaining high *AR_turnover*, which denotes a weighting system for selecting improvement plans.

In practice, the assumed performance scores of each plan in each criterion should be assessed by the internal management team. Here, a five-point Likert scale was adopted for the five assumed plans. The DANP influential weights and the performance scores of each plan are shown in Table 19.7.

As indicated in Table 19.7, the aggregated score of plan D is 1.643, which is supposed to be the best choice in the proposed model. According to the FCA-based DANP model, the company should select plan D to improve its *AR_turnover* result. Furthermore, referring to the findings in the previous section, the proposed model could further explore the directional influences among the dimensions and criteria that imply high *AR_turnover*. The integration of the implication rules and DEMATEL cause–effect analysis (Table 19.6) may provide more managerial insights regarding FP pattern changes, illustrated as the DIRM in Figure 19.3.

The obtained DIRM not only indicates the directional influences among the five dimensions, but also incorporates implication rules to provide more insights. First, as indicated in Figure 19.3, to achieve high *AR_turnover*, the source dimension should be *Operational efficiency* (D_3), and D_3 will influence the other four dimensions. Therefore, the criterion *Inventory* (C_5) in dimension D_3 should be considered the top priority when planning for an improvements. Second, referring to the DANP weight for each criterion, the top-three influential criteria are C_6 (0.228), C_5 (0.179), and C_7 (0.179); however, the criterion *EPS* (C_6) belongs to the effect group ($r_i^C - c_i^C < 0$), which is influenced by the other dimensions and criteria. Therefore, *Inventory* (C_5) and *CashFlow_adq* (C_7) should be considered the most critical criteria to enhancing performance and may cause relatively superior marginal effects for attaining high *AR_turnover*. Third, dimensions D_1, D_2, and

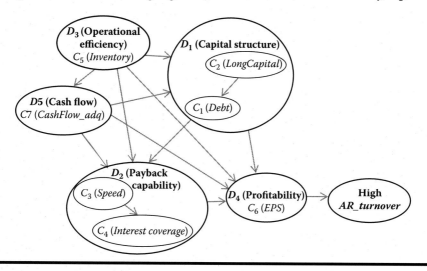

Figure 19.3 DIRM for *AR_turnover_H*. (Reprinted with permission from Shen, K. Y., and Tzeng, G.H., *Technological and Economic Development of Economy*, 22(5), 685–714, 2015e.)

D_5 all have influences on dimension D_4 (*Profitability*); thus, the management team of this company may examine its current performances on the three dimensions and related criteria (i.e., C_1, C_2, C_3, C_4, and C_7) to see if there are any underperforming criterion that should be enhanced. The aforementioned insights cannot be provided by conventional regression models, which is the strength of the combined MRDM and MADM methods.

19.7 Conclusion

To conclude, a combined model integrating computational intelligence (i.e., DRSA and FCA) and MADM methods (i.e., DEMATEL and DANP) for the FP improvement of semiconductor companies is illustrated in this chapter with a numerical example. Compared with previous studies (Hsu et al. 2013; Ou Yang et al. 2013; Shen et al. 2014) that constructed MADM evaluation models using interviews or literature reviews to select the involved criteria/dimensions, the proposed model—conducted using logical reasoning from historical data—is based on the implication rules of FCA. This approach is more objective and goal oriented in constructing a weighting system for the priority gap; FCA can retrieve the other commonly shared attributes based on the fact that those *Good* alternatives have the same formal concept *AR_turnover_H* (i.e., high *AR_turnover*). As a result, the DANP model at the second stage will have a reasonable foundation, which is also critical to transforming the findings of DRSA into a hybrid MCDM model. Furthermore, the obtained figurative result (i.e., the DIRM) may help DMs to gain more understandable insights than with a single approach (e.g., the computational or MCDM approach). The results indicate that the integrated model may resolve the FP diagnosis problem and evaluate future improvement plans in a reasonable manner.

The case illustrated here has shown how to construct an integrated approach to resolving complex financial problems in real business environments. Nevertheless, several limitations remain. First, the MRDM approach, using either DRSA or FCA, is based on the assumption that historical financial patterns will reoccur in the near future, and the model requires recent data to form patterns and decision rules. Second, the adopted three-level discretization method in DRSA might generate different results by using the other discretization methods. More discretization methods are suggested to be examined to explore their effects on DRSA models in the future. Third, only the DRSA decision rules associated with the at-least *Good* decision class were used to highlight the FP patterns for improvement; future studies might examine the rules associated with the *at-most Bad* decision class to explore FP patterns that lead to performance deteriorations, which could serve as warning mechanisms.

Chapter 20

Hybrid Bipolar MRDM Model For Business Analytics

In a fast-changing and highly competitive environment, nearly all companies attempt to retrieve valuable information from business data to increase their operational efficiency or value, and the rising interest of research in this field could be termed *business analytics* (BA). There are various applications for BA (ranging from marketing to manufacturing) that require forecasting or retrieving patterns or logic from historical data. Nevertheless, most of the problems or applications confronted by businesses involve a group of complex and interrelated factors, and conventional statistics models (e.g., regression) have constraints to resolving them. To tackle or model the complexity of business data, certain computational intelligence or artificial intelligence (AI) techniques have recently been adopted. However, previous studies based on those techniques mainly focus on increasing the accuracy of classification for prediction (Sharma et al. 2010); how to retrieve implicit and complicated or patterns from data to support performance improvement is underexplored.

To bridge the gap, a multiple rules–based decision making (MRDM) model is introduced in this chapter; a financial application revised from a previous work by the authors (Shen and Tzeng 2016) is used as an example to illustrate this hybrid bipolar model. The essential ideas of this bipolar approach are based on similarity to the positive contexts (rules) and dissimilarity to the negative ones. The details of this model will be introduced in Section 20.2, and a numerical example of improving the financial performance (FP) of semiconductor companies is provided in the second half of this chapter.

20.1 Background of Business Analytics

Using an analytical approach to obtain superior performance outcomes is not a new idea; however, the two influential works of Davenport (2006) and Davenport and Harris (2013) have attracted more attention from researchers and business professionals to the term BA. In the so-called Big Data era, the importance of identifying, assimilating, and utilizing crucial information to increase the performance or value of businesses has gained interest (Delen and Demirkan 2013). On the one hand, investments in information technology infrastructure for collecting data are inevitable; on the other hand, how to recognize critical information and transform BA to guide strategy or planning might be more challenging for businesses.

Based on the proposed taxonomy of BA (Delen and Demirkan 2013), it could be categorized into three types of analytics: descriptive, predictive, and prescriptive. Descriptive analytics often takes the form of business reporting or data warehousing, commonly presented and analyzed as statistics. This type of BA is relatively simple and straightforward but has difficulty generating predictions or implications. Compared with the descriptive type of BA, the predictive and the prescriptive types may provide much more knowledge for businesses and are the main focus of academic studies. Not surprisingly, a recent survey from business professionals also indicated that the top obstacle of organizations to BA is a lack of understanding of how to use analytics to improve their business (LaValle et al. 2010; LaValle et al. 2011). To resolve this problem, the fusion (or combination) of data-mining techniques and MADM methods (Doumpos and Zopounidis 2011) might be a plausible approach; in this context, a reasonable framework that can leverage the strengths of these two methods is required.

Although BA mainly involves the analysis of numerical data from operations, the selected variables and the expected outputs vary in different circumstances. Among the various analytics, financial management (Delen and Demirkan 2013) may be the most prevalent. Statistics is the most common approach for businesses to analyze financial data; it includes descriptive statistics (e.g., averages, modes, standard deviations), charts (bar charts, histograms), correlations, regression models, and time series analyses, to observe trends of changes or to capture the relationships among variables. On the academic side, much more emphasis is put on modeling the patterns or making predictions using statistical inferences. Nevertheless, the regression model, for instance, is often constrained by certain unrealistic assumptions, such as the independence of the variables involved, the linear relationship between the criteria under consideration (independent variables) and the target output (dependent variable), and so on.

As the statistics approach has certain constraints/limitations, researchers have attempted to adopt data-mining techniques for BA. In the field of financial applications, the commonly adopted machine learning approach, including various types of artificial neural networks (ANNs), has strengths in dealing with nonlinear data to minimize modeling errors (Lam 2004; Shen 2013). However, the black-box characteristics of its learning outcomes are criticized for lacking implications for decision makers (DMs). Therefore, a rising trend of taking the rules-based approach to obtain easy-to-understand and applicable guidance in practice has emerged. Certain computational intelligence techniques (e.g., decision tree, formal concept analysis, the rough set approach) have shown positive results in previous studies (Tay and Shen 2002; Ganter et al. 2005; Shyng et al. 2010a,b). However, these rule-based computing techniques mainly provide classification outcomes, such as, "If CONDITION$_1$ was satisfied and CONDITION$_2$ was satisfied and CONDITION$_3$ was satisfied, then the alternative can be classified into the 1st group." For alternatives that might fall within the same category or into an undefined category, the aforementioned rule-based techniques still have difficulty ranking precisely. Moreover, how to transform the analytics into strategy formation or improvement planning—the major concern of business—is still unanswered.

To facilitate our understanding of how to integrate the two approaches—computational intelligence and MADM—to resolving BA problems, a hybrid bipolar decision model is introduced to tackle the FP prediction problem and obtain guidance on improving business performance. A famous finding (Tversky and Kahneman 1992) from research into human behavior and psychology suggested that human minds process information by distinguishing between positive and negative sides, and reasoning or making decisions based on these two sides is termed the *propensity of bipolarity* (Dubois and Prade 2012). In this context, the term *bipolar* denotes the idea that decision rules are divided into two groups, positive and negative, and alternatives can be ranked based on their similarity with the positive group and dissimilarity with the negative one. The requirements (i.e., conditions or antecedents or premises) in each decision rule may be regarded as a specific context associated with the positive or the negative side; the two groups of decision rules will form a weighting system according to the supported instances. In addition, the modified VIKOR method is integrated with the bipolar weighting system to support decisions such as ranking, selection, and improvement planning. Although other MADM techniques, such as DEMATEL and DEMATEL-based ANP (DANP) (Chapters 3–5), have been applied to explore interrelationships among criteria, the proposed bipolar model offers a different approach, evaluating plausible improvement results by considering the interrelated influences among criteria. The detailed procedures of ranking and improvement planning will be illustrated by an empirical example from the semiconductor industry in Sections 20.3 through 20.5. The aforementioned bipolar approach can be categorized into the emerging field of MRDM, which attempts to provide insightful and transparent analyses for businesses to plan for performance improvements.

20.2 Hybrid MRDM Model Using the Bipolar Approach

The framework of the proposed model comprises six main parts: (1) DRSA decision rules, (2) a threshold for selecting positive and negative rules and forming new contexts, (3) calculating support weights for new contexts, (4) transforming DRSA decision rules (contexts) into a bipolar model, (5) fuzzy evaluations of alternatives,

292 ■ *Trends of Hybrid Multiple Criteria Decision Making*

(6) fusing the bipolar model, using the modified VIKOR method, with fuzzy evaluations.

The proposed model begins by obtaining decision rules from DRSA, which extends the non-ordered attributes of classical RSA to consider the preferential characteristics of attributes, suitable for resolving MADM problems. DRSA begins by organizing alternatives in an information system (*IS*) in the form of an information table, where all attributes are located in rows. The table of DRSA is a 4-tuple *IS*; that is,

$$IS = (U, Q, V, f),$$

where:

U	is a finite state of the universe
$Q = \{q_1, q_2, ..., q_p\}$	is a finite set of attributes, comprised of conditional attributes (a set C) and a decision attribute (D)
V_q	is the value domain of attribute q (V is the union of all value domains of q_i, for $i = 1$ to p)
f	is a total function, such that $f : U \times Q \rightarrow V$, where $f(x,q) \in V_q$, for each $x \in U$ and $q \in Q$

In typical MADM applications of DRSA, there is only a single decision attribute in D. Decision classes (DCs) that belong to a decision attribute can be denoted as $Cl = \{Cl_t, t = 1, ..., n\}$ in a general case. The required steps for forming a hybrid bipolar model are as follows:

Step 1: Discretize all of the conditional attributes (C). This step will influence the following granular computing. The more classes that are defined for an attribute, the smaller the size of the pertaining granules, and the larger the total number of cells (partitioned by multiple dimensions).

Next, $\succeq q$ is defined as a weak preference relation on U, considering a criterion q ($q \in Q$). Objects $x, y \in U$, if $x \succeq_q y$, which denotes that "x is at least as good as y considering attribute q," and the weak preference relation means that x and y are always comparable on attribute q. Assume that DCs are all ordered by preference (i.e., for all $r, s = 1, ..., n$, if $r \succ s$, then Cl_r is preferred to Cl_s). Then, given a set of Cl_s (the upward union is explained here), the upward and downward union of positive DCs can be defined as follows:

$$Cl_s^{\geq} = \bigcup_{r \geq s} Cl_r, \tag{20.1}$$

$$Cl_s^{\leq} = \bigcup_{r \leq s} Cl_r. \tag{20.2}$$

The upward and downward unions of DCs help to define the dominance relation for $G \subseteq C$, where D_G represents the dominance relation regarding G, and $x D_G y$ denotes that x G-dominates y with regard to any subset criteria in G. Accordingly, the G-dominating set and G-dominated set can be denoted as follows:

$$D_G^+(x) = \{y \in U : y D_G x\}, \tag{20.3}$$

$$D_G^-(x) = \{y \in U : x D_G y\}. \tag{20.4}$$

Then, sets of collected unions of upward DCs can be used to define G-lower and G-upper approximations, as in Equations 20.5 and 20.6. For brevity, only the upward unions of DCs are discussed hereafter, and the downward unions of DCs can be defined by analogy.

$$\underline{G}(Cl_r^{\geq}) = \{x \in U : D_G^+(x) \subseteq Cl_r^{\geq}\}, \tag{20.5}$$

$$\overline{G}(Cl_r^{\geq}) = \{x \in U : D_G^-(x) \cap Cl_r^{\geq} \neq \varnothing\}. \tag{20.6}$$

The G-lower approximation represents the certain classification to categorize an alternative in $Cl_r^{\geq}$, the G-upper approximation the uncertain one. Therefore, $\underline{G}(Cl_r^{\geq})$ is a subset of $\overline{G}(Cl_r^{\geq})$; that is, $\underline{G}(Cl_r^{\geq}) \subseteq \overline{G}(Cl_r^{\geq})$. The doubtful region can be defined as follows, which is called the *G-boundary* regarding the criteria set G $(G \subseteq C)$:

$$Bn_G(Cl_r^{\geq}) = \overline{G}(Cl_r^{\geq}) - \underline{G}(Cl_r^{\geq}). \tag{20.7}$$

In DRSA, the quality of approximation $\gamma_G(Cl)$ for every $G \subseteq C$ for ordinal DCs with respect to a set of attributes G can be defined as follows:

$$\gamma_G(Cl) = \left| U - \left(\bigcup_{r \in \{2, \dots, n\}} Bn_G\left(Cl_r^{\geq}\right) \right) \right| \Big/ |U|. \tag{20.8}$$

The quality of approximation defines the ratio of the objects G-consistent with the dominance relationship, divided by the total number of objects in U; $|\cdot|$ denotes the cardinality of a set in Equation 20.8. Here, a bipolar approach is adopted to position an alternative, which may be regarded as a special case of DRSA. The bipolar approach mentioned here refers to the defined DCs on two opposite sides (e.g., positive and negative). Decision rules associated with the positive side will give

294 ■ *Trends of Hybrid Multiple Criteria Decision Making*

credits to an alternative, and rules associated with the negative side will decrease the credits of an alternative.

Step 2: Define the decision attribute (D) using the bipolar approach. The DCs are ordered into three types: positive, neutral, and negative; positive DCs should be preferred by DMs. Therefore, a set of DCs can be defined as $Cl = \{Cl_{POS}, Cl_{Neutral}, Cl_{NEG}\}$, where Cl_{POS} denotes the positive DC, Cl_{NEG} the negative one, and $Cl_{Neutral}$ is neither preferred nor discarded (i.e., $Cl_{POS} \succ Cl_{Neutral} \succ Cl_{NEG}$). For each object (alternative) $x \in U$, it only belongs to one DC in Cl.

The dominance-based rough approximation of DCs may yield a set of decision rules in the form of "If *antecedents*, then *consequence*" (i.e., the alternative belongs to a specific DC). To define the concept of *support* used in the subsequent discussion, the general form of a decision rule can be indicated as follows:

$Rule_\phi \equiv$ "If $f_{c_1}(x) \geq v_{c_1}$ and $f_{c_2}(x) \geq v_{c_2}$ and ... and $f_{c_g}(x) \geq v_{c_g}$, then $x \in Cl_r^{\geq}$,;

where $f_{c_1}(x)$ denotes the value of alternative x on criterion c_1, $v_{c_1} \subseteq V_{c_1}$, $v_{c_2} \subseteq V_{c_2}$

,..., $v_{c_g} \subseteq V_{c_g}$; $\{c_1, c_2, ..., c_p\} \subseteq C$ and $r = 3$ (i.e., three DCs: *POS*, *Neutral*, and

NEG)."

For any $z \in U$, if $f_{c_1}(z) \geq v_{c_1}$ and $f_{c_2}(z) \geq v_{c_2}$ and ... and $f_{c_g}(z) \geq v_{c_g}$ (i.e., all of the requirements in the attributes of $rule_\phi$ are satisfied), then object z is called a *support* of $rule_\phi$. In this case, the total number of supports for $rule_\phi$ in U is used to calculate the support weight of $rule_\phi$ (in Step 5). Detailed discussions on DRSA decision rules can be found in previous research (Greco et al. 2001; Greco et al. 2002b), and the rules obtained at this stage are the main inputs for constructing a bipolar model at the next stage.

Step 3: Apply DRSA algorithms to obtain positive and negative decision rules; also, count the number of supports for each rule to obtain the support weight for each new context.

Once the two types (i.e., $Rule_{POS_i}$ and $Rule_{NEG_j}$) of decision rules are obtained from DRSA, the rules associated with positive or negative DCs should be further selected. To select the rules with a high number of supports (i.e., that provide high confidence according to available data), it is necessary to set a threshold to select the covered objects in the two groups of rules; the required processes are explained in Steps 4 and 5.

The meaning of *support weights* needs some further explanation here. Each rule, whether in the positive or negative group, denotes a specific context that associates the premises with the outcome (i.e., a positive or negative DC). The support weight of each rule indicates the relative occurrence frequency of a specific context of two types of DCs (positive and negative); therefore, the higher the support weight of a rule, the higher the confidence that DMs will have in the rule (based on the observed data), which may be regarded as learned experience from the past.

Step 4: Set a threshold (Γ) to select decision rules for the bipolar model.

The decision rules from DRSA are induced from available data, which may be regarded as known facts/experience. Also, as DCs are divided into two types ($Cl_{Neutral}$ will not be included in the bipolar model), rules associated with positive or negative DCs can be ranked by the number of its total supports, from high to low. Then, the required objects are calculated to satisfy a threshold Γ for $Rule_{POS}$ and $Rule_{NEG}$, respectively.

$$|O_{POS}|/h \geq \Gamma \text{ and } |O_{NEG}|/l \geq \Gamma \tag{20.9}$$

In Equation 20.9, h and l denote the total number of objects categorized as Cl_{POS} and Cl_{NEG} in the available data, $|O_{POS}|$ and $|O_{NEG}|$ represent the required numbers of objects covered in $Rule_{POS}$ and $Rule_{NEG}$, respectively.

With the numbers of required O_{POS} and O_{NEG} in two types of rules, a bipolar model needs to accumulate the covered objects in each type of decision rule, from high supports to low supports. And the minimal decision rules in $Rule_{POS}$ and $Rule_{NEG}$ that satisfy Equation 20.9 can be transformed into new contexts in a bipolar model.

Step 5: Calculate the support weight of each new context in the bipolar model.

After transforming decision rules from DRSA into new contexts (i.e., the new positive and negative contexts, C_{POS} and C_{NEG}) in the bipolar model in Step 4, the support weight for C_{POS} and C_{NEG} can be defined as follows:

$$Supp.weight\left(C_{POS_s}\right) = Supp\left(C_{POS_s}\right)/h = w_{POS_s}, \tag{20.10}$$

$$Supp.weight\left(C_{NEG_t}\right) = Supp\left(C_{NEG_t}\right)/l = w_{NEG_t} \tag{20.11}$$

In Equations 20.10 and 20.11, $s = 1, 2, \ldots, S$, and S is the number of total $Rule_{POS}$ in the bipolar model; $t = S + 1, S + 2, \ldots, T$, and T is the number of total rules in the bipolar model; $Supp.weight(\bullet)$ denotes the number of supports for a rule; h and l are the same as defined in Equation 20.9.

The obtained support weights need to be normalized to sum up to 1 to be synthesized with the following modified VIKOR method.

296 ■ *Trends of Hybrid Multiple Criteria Decision Making*

To rank alternatives using the proposed model, the modified VIKOR is adopted as an aggregator to synthesize the support weights from the bipolar model. The essential concepts of VIKOR have been introduced in Chapter 5, and it begins with an L_p-*metric*, as in Equation 20.12, and o alternatives are assumed to be $A_1, A_2, \ldots, A_o$.

$$L_k^p = \{\sum_{a=1}^{T} [w_a \left(\left|f_a^{aspire} - f_{ka}\right|\right) \big/ \left(\left|f_a^{aspire} - f_a^{worst}\right|\right)]^p \}^{1/p}, \ 1 \le p \le \infty, \ k = 1, \ldots, o, \ (20.12)$$

In Eq. (20.12), w_a represents the normalized support weight of the new context C_a (i.e., combines C_{POS} with C_{NEG}, denoted as C_a, for $a = 1, \ldots, S, \ldots, T$), f_a^{aspire} represents the best performance score (i.e., the aspiration level) among the o alternatives on C_a, f_a^{worst} is the worst performance score of the o alternatives on C_a, and f_{ka} denotes the performance score of alternative k on C_a.

While $p = 1$ and $p = \infty$, indices S_k and R_k can be defined as in Equations 20.13 and 20.14, which represent the total performance gap and the individual regret in a certain context, respectively. For example, assume that the performance scores are from 0 to 1 in data sets (i.e., from very bad to very good). We may set $f_a^{aspire} = 1$ as the best performance score (i.e., the aspiration level) and $f_a^{worst} = 0$ as the worst performance score, also called the modified VIKOR method.

$$S_k = L_k^{p=1} = \sum_{a=1}^{T} w_a \left(\left|f_a^{aspire} - f_{ka}\right|\right) \big/ \left(\left|f_a^{aspire} - f_a^{worst}\right|\right), \ for \ a = 1, 2, \ldots, T. \quad (20.13)$$

$$R_k = \max_a \left\{\left(\left|f_a^{aspire} - f_{ka}\right|\right) \big/ \left(\left|f_a^{aspire} - f_a^{worst}\right|\right) \middle| a = 1, 2, \ldots, T\right\}. \quad (20.14)$$

where:

$r_{ka} = \left(\left|f_a^{aspire} - f_{ka}\right|\right) \big/ \left(\left|f_a^{aspire} - f_a^{worst}\right|\right)$ is the normalized gap of each criterion on alternative k

S_k is the average ratio of the performance gap in alternative k

R_k is the maximal ratio of performance gap in alternation k

T is the total number of new contexts in a bipolar model

The performance score f_{ka} of the kth alternative on each context ($a = 1, \ldots, T$) is defined as the percentage of satisfied requirements. For example, suppose C_1 is "If attribute$_1 \ge 2$ and attribute$_2 \ge 3$ and attribute$_3 \ge 4$, then DC = *Good*." If two of the three requirements were satisfied (e.g., attribute$_1$ = attribute$_2$ = attribute$_3$ = 3), the performance score of alternative k on this criterion would be $2/3 = 0.67$ (CONDITION$_3$ was not satisfied). This is based on the equal-weighted assumption of each requirement in a rule. Setting the aspiration

Hybrid Bipolar MRDM Model For Business Analytics ■ 297

level to its maximal level, all of the calculations in the following examples have the same settings: the best outcome ratio in S_k or R_k is a zero gap (i.e., $S^{aspire} = R^{aspire} = 0$) and the worst outcome ratio in S_k or R_k is a full gap (i.e., $S^{worst} = R^{worst} = 1$). A new ranking index Q_k can be defined accordingly, as in Equation 20.15:

$$Q_k = v \times S_k + (1-v) \times R_k, \tag{20.15}$$

where:

v $(0 \le v \le 1)$ denotes the relative importance that DMs put on group utility,
$(1 - v)$ denotes the relative importance of individual regret.

Step 6: Using the same approach as in Step 1, discretize the raw figures from new alternatives for each attribute (applying the fused bipolar model for rating new data) to get their performance scores in each new context.

While applying the proposed model, a performance evaluation of each alternative is needed, and fuzzy evaluation is suitable for DMs to give opinions in the form of natural language. Unlike performance evaluations in conventional MADM models, the evaluations of each alternative on each required condition in different rules have to be collected separately from attributes with different requirements. For example, if two requirements (e.g., $A_1 \ge$ high and $A_1 \le$ low) have to be satisfied in Rule 1 and Rule 2, respectively, then DMs will have to give opinions regarding the degree of satisfaction with an alternative on the two requirements separately. The widely used triangular fuzzy membership function is adopted with three levels (unsatisfied [U], neutral [N], and satisfied [S]) in this model, and the performance score of the ath alternative on the first requirement of criterion C_1 by m DMs is calculated as follows:

$$f_{C_1}^a = (f_{1,C_1}^a \oplus \ldots \oplus f_{i,C_1}^a \oplus \ldots \oplus f_{m,C_1}^a) / m. \tag{20.16}$$

In addition, the importance of each requirement in a criterion (rule) is assumed to be equal; therefore, if there are n requirements in criterion C_1, the performance score of the ath alternative on C_1 would be calculated as follows:

$$f_{C_1}^a = \sum_{j=1}^{n} f_{C_1^j} / n, \text{ for } j = 1, \ldots, n. \tag{20.17}$$

Finally, the performance score in each criterion can be synthesized by modified VIKOR with support weights from the bipolar model for performance evaluation.

Step 7: Synthesize the final performance gap of each alternative by modified VIKOR and determine the final ranking; also, use the priority gap obtained in each context (termed the *priority contexts*) to plan for improvements. The processes will be illustrated and discussed in detail in Sections 20.3 and 20.4.

20.3 A Case from the Semiconductor Industry

A group of semiconductor companies listed on the Taiwan stock market were analyzed, and the real financial results of three semiconductor stocks in 2011 were examined to generate the model's outputs for comparison with their actual FP in 2012. In addition, to illustrate the improvement planning procedures, three example companies were given in-depth analyses. The three companies were explored to identify their performance gaps in those contexts. The research flow of this case study is shown in Figure 20.1.

20.3.1 Data

The data came from the Taiwan Economics Journal (TEJ) database, which reports the historical data of public-listed stocks on the Taiwan stock market to date. To capture and predict the FP changes of the semiconductor companies, a one-period lagged model was adopted in this case. Each company's financial results in time period t were used as conditional attributes to match its own FP change (as a decision attribute) in period $t + 1$. After excluding incomplete

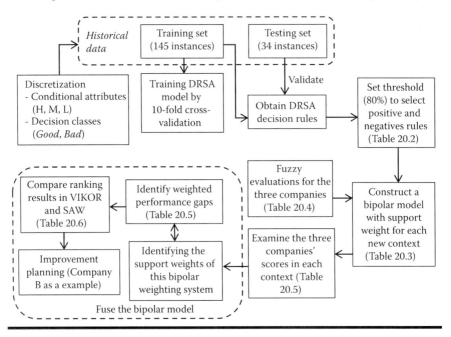

Figure 20.1 Research flows of the case study.

data, there were a total of 179 instances used; 145 instances (with conditional attributes in 2007 to 2010) were trained and 34 instances (with conditional attributes in 2011) were validated as the testing set. Fuzzy evaluations on alternatives were conducted by collecting opinions from five domain experts, all of them with more than 15 years' working experience in IT or the financial industry.

The indicators that are included in the quarterly financial summary requested by the Taiwan Stock Exchange were used for DRSA modeling. There were 20 financial indicators used as conditional attributes, and the change of return on assets (ΔROA) was used as the decision attribute. A summary and brief definitions of the attributes are shown in Table 20.1.

To conduct DRSA, the raw financial indicators were transformed (i.e., discretized) into three ordered states—to denote *high* (H), *middle* (M), and *low* (L)—for the 20 conditional attributes. In each year, all of the sample companies were ranked from high to low on each attribute, and the top third, middle third, and bottom third were classified as H, M, and L, respectively. The decision attribute, following a similar approach of observing the ΔROA in each year, was also ranked and categorized into three groups; nevertheless, only the top- and bottom-third stocks were defined as *Good* and *Bad* DCs, respectively. In addition, instances that ranked in the bottom third with a positive ΔROA were excluded from the *Bad* DC.

20.3.2 Bipolar Weighting System

The rough set classification engine jMAF (developed by the Laboratory of Intelligent Decision Support Systems at the Institute of Computing Science, Poznan University of Technology, Poland) was used to obtain DRSA decision rules. The training set was examined by conducting a 10-fold cross-validation five times. The 10-fold cross-validation experiment indicated that the DRSA model generated an acceptable CA on average (79.31%). The whole training set was then analyzed to obtain the DRSA decision rules, and the untouched testing set was validated. The CA of the testing set was 88.24%, which validates the training result. The obtained decision rules from the training set comprised 24 decision rules; among the 24 rules, 13 rules associated with the *Good* DC, 11 rules the *Bad* DC.

To select decision rules with a certain level of coverage, the threshold (Γ) was set as 80% to select the rules associated with the *Good* and *Bad* DCs. Since there are 52 *Good* and 93 *Bad* instances in the training set (145 in total), to reach the 80% threshold for each DC, the selected decision rules associated with the *Good* and *Bad* DCs should cover at least 42 and 75 instances in each DC, respectively. The selection process was done by accumulating the instances covered in each decision rule (from high-support to low-support decision rules) for each DC. Once the total number of the covered instances reached 42 for

300 ■ *Trends of Hybrid Multiple Criteria Decision Making*

Table 20.1   Definitions and Brief Explanations of the Attributes

	Decision Attribute	*Symbols*	*Definition*
	Change of ROA	ΔROA	$[(ROA_{t+1} - ROA_t) / ROA_t] \times 100\%$
Dimensions	**Conditional Attributes**	**Symbols**	**Definitions and Brief Explanations**
Capital structure	Debt to total assets	*Debt*	Total debt/total assets
	Long-term capital to total assets	*LongCapital*	Long-term capital/total assets
Payback capability	Liquidity ratio	*Liquidity*	Current assets/current liability
	Speed (quick) ratio	*Speed*	(Current assets/inventory)/ current liability
	Interest coverage ratio	*Interest Coverage*	(Net profit before tax + interest expenses)/interest expenses
Operational efficiency	Accounts receivable	*AR_ turnover*	Net credit sales/average accounts receivable
	Days for collecting AR	*AR_days*	(Days × AR)/credit sales
	Inventory turnover rate	*InvTurnover*	Total operational costs/average inventory
	Average days for sales	*DAYs*	(Average ending inventory/ operational costs) × 365
	Fixed asset turnover	*FAsset Turnover*	Total revenue/Total fixed assets
Profitability	Return on total assets	*ROA*	Net profit before tax/average total assets

(Continued)

Table 20.1 (Continued)   Definitions and Brief Explanations of the Attributes

	Decision Attribute	Symbols	Definition
	Change of ROA	ΔROA	$[(ROA_{t+1} - ROA_t) / ROA_t] \times 100\%$
	Return on equity	ROE	Net profit before tax/average total equity
	Operational profit to total capital	OProfit_capital	Operational profit/total capital
	Net profit before tax to total capital	NProfit_capital	Net profit before tax/total capital
	Net profit ratio	NetProfit	Net profit/net sales
	Earnings per share	EPS	(Net income – dividends on preferred stocks)/total outstanding shares
Cash flow	Cash flow ratio	CashFlow	(Operational cash flow – cash dividend for preferred stocks)/ weighted average equity
	Cash flow adequacy ratio	CashFlow_adq	Cash flow from operations/ annual current maturities
	Cash flow reinvestment ratio	CashFlow_reinv	(Increase in fixed assets + increase in working capital)/net income + noncash expenses – noncash sales – dividends)

Source: Data from Shen, K.Y., and Tzeng, G.H. *International Journal of Fuzzy Systems*, 18(6), 940–955, 2016.

the *Good* DC and 75 for the *Bad* DC, the rule selection process would stop. The selected rules for each DC are shown in Table 20.2 with the number of supports.

The selected rules associated with *Good* and *Bad* DCs were renamed as new contexts (i.e., C_1, C_2,..., C_{10}) and placed into two distinct groups; the number

302 ■ *Trends of Hybrid Multiple Criteria Decision Making*

Table 20.2 Selected Decision Rules While Γ = 80%

	Decision Rules Associated with Good DC	Supports
C_1	(AR_turnover ≥ H) and (Inventory ≥ H) and (ROA ≥ M) and (EPS ≥ M)	12
C_2	(Speed ≥ M) and (AR_turnover ≥ H) and (EPS ≥ M) and (CashFlow ≥ H) and (CashFlow_adq ≥ M)	16
C_3	(Debt ≥ M) and (InterestCoverage ≥ M) and (AR_turnover ≥ H) and (Inventory ≥ M) and (EPS ≥ H) and (CashFlow ≥ M)	14
C_4	(LongCapital ≥ M) and (AR_days ≥ H) and (ROA ≥ M) and (CashFlow ≥ H)	13
C_5	(LongCapital ≥ M) and (Speed ≥ M) and (AR_turnover ≥ M) and (AR_days ≥ H) and (ROE ≥ M) and (CashFlow_adq ≥ M)	17
C_6	(FixAssetTurnover ≥ M) and (EPS ≥ M) and (CashFlow ≥ H) and (CashFlow_adq ≥ H)	9
	Decision Rules Associated with Bad DC	
C_7	(EPS ≤ L)	34
C_8	(LongCapital ≤ L) and (Liquidity ≤ M)	30
C_9	(Speed ≤ L) and (ROA ≤ M)	28
C_{10}	(AR_turnover ≤ M) and (AR_days ≤ L)	29

Source: Data from Shen, K.Y., and Tzeng, G.H. *International Journal of Fuzzy Systems*, 18(6), 940–955, 2016.

of attributes involved went down from 20 to 14. In addition, the number of supports divided by the total number in each DC was regarded as the support weight for each context. Take C_1, for example: its support weight was 23.08% (12 / 52 = 23.08%). Adopting this transformation, the 10 new contexts form a bipolar weighting system comprising 10 rules. The support weight of each context is shown in Table 20.3.

20.3.3 Aggregate Fuzzy Performance Evaluations Using Modified VIKOR

With the rules obtained in Table 20.3, three semiconductor companies, MOSPEC Semiconductor (*A*), SITRONIX (*B*), and KINSUS (*C*), were used as

Table 20.3   New Contexts and Involved Attributes of the Bipolar Model

New Contexts	Contexts Associated with Good						Contexts Associated with Bad			
	C_1	C_2	C_3	C_4	C_5	C_6	C_7	C_8	C_9	C_{10}
Support Weights (%)	23.08	30.77	26.92	25.00	32.69	17.31	36.56	32.26	30.11	31.18
	(12/52)	(16/52)	(14/52)	(13/52)	(17/52)	(9/52)	(34/93)	(30/93)	(28/93)	(29/93)
Attributes										
Debt (A_1)			$\geq M$							
LongCapital (A_2)				$\geq M$	$\geq M$				$\leq L$	
Liquidity (A_3)								$\leq M$		
Speed (A_4)		$\geq M$			$\geq M$				$\leq L$	
InterestCoverage (A_5)			$\geq M$							
AR_turnover (A_6)	$\geq H$	$\geq H$	$\geq H$		$\geq M$					$\leq M$
AR_days (A_7)				$\geq H$	$\geq H$					$\leq L$
Inventory (A_8)	$\geq H$		$\geq M$							
FixAssetTurnover (A_9)						$\geq M$				
ROA (A_{10})	$\geq M$			$\geq M$					$\leq M$	
ROE (A_{11})					$\geq M$					
EPS (A_{12})	$\geq M$	$\geq M$	$\geq H$			$\geq M$	$\leq L$			
CashFlow (A_{13})		$\geq H$	$\geq M$	$\geq H$		$\geq H$				
CashFlow_adq (A_{14})		$\geq M$			$\geq M$	$\geq H$				

Source: Data from Shen, K.Y., and Tzeng, G.H. *International Journal of Fuzzy Systems*, 18(6), 940–955, 2016.

Note: The shaded region denotes the negative group.

304 ■ *Trends of Hybrid Multiple Criteria Decision Making*

Table 20.4 Defuzzified Financial Evaluations of the Three Companies in 2011

		Company		
Conditions of attributes		*A*	*B*	*C*
Debt(A_1)	≥ M	0.34	0.34	0.57
LongCapital (A_2)	≥ M	0.10	0.79	0.26
LongCapital (A_2)	≤ L	0.86	0.10	0.10
Liquidity (A_3)	≤ M	0.79	0.10	0.19
Speed (A_4)	≥ M	0.78	0.78	0.86
Speed (A_4)	≤ L	0.10	0.10	0.10
InterestCoverage (A_5)	≥ M	0.35	0.26	0.86
AR_turnover (A_6)	≥ H	0.10	0.18	0.79
AR_turnover (A_6)	≥ M	0.10	0.57	0.86
AR_turnover (A_6)	≤ M	0.78	0.57	0.18
AR_days (A_7)	≥ H	0.10	0.18	0.86
AR_days(A_7)	≤ L	0.86	0.10	0.10
Inventory (A_8)	≥ H	0.10	0.10	0.10
Inventory (A_8)	≥ M	0.10	0.16	0.80
FixAssetTurnover (A_9)	≥ M	0.79	0.10	0.79
ROA (A_{10})	≥ M	0.10	0.34	0.79
ROA (A_{10})	≤ M	0.86	0.71	0.19
ROE (A_{11})	≥ M	0.73	0.86	0.86
EPS (A_{12})	≥ H	0.10	0.71	0.71
EPS (A_{12})	≥ M	0.50	0.86	0.86
EPS (A_{12})	≤ L	0.26	0.10	0.10
CashFlow (A_{13})	≥ H	0.10	0.10	0.10
CashFlow (A_{13})	≥ M	0.10	0.86	0.71
CashFlow_adq (A_{14})	≥ H	0.57	0.10	0.10
CashFlow_adq (A_{14})	≥ M	0.79	0.26	0.80

Source: Data from Shen, K.Y., and Tzeng, G.H. *International Journal of Fuzzy Systems*, 18(6), 940–955, 2016.

Hybrid Bipolar MRDM Model For Business Analytics ▪ 305

Table 20.5 Performance Scores (f_{ka}) and Gaps

Criteria (Rules)	Normalized Supp.weights	Performance Scores			Performance Gaps		
		A	B	C	A	B	C
C_1	8.07%	0.20*	0.37	0.64	0.66*	0.49	0.23
C_2	10.76%	0.45	0.44	0.68	0.41	0.42	0.18
C_3	9.42%	0.18	0.42	0.74	0.68	0.44	0.12
C_4	8.74%	0.10	0.35	0.50	0.76	0.51	0.36
C_5	11.43%	0.43	0.57	0.75	0.43	0.29	0.11
C_6	6.05%	0.49	0.29	0.46	0.37	0.57	0.40
C_7	12.79%	0.26	0.10	0.10	0.26	0.10	0.10
C_8	11.28%	0.83	0.10	0.15	0.83	0.10	0.15
C_9	10.53%	0.48	0.41	0.15	0.48	0.41	0.15
C_{10}	10.91%	0.82	0.34	0.14	0.82	0.34	0.14

Source: Data from Shen, K.Y., and Tzeng, G.H. *International Journal of Fuzzy Systems,* 18(6), 940–955, 2016.

* The highest defuzzified performance scores from the experts is 0.86; therefore, the performance gap of company *A* is 0.86 – 0.20 = 0.66.

empirical case studies to illustrate this hybrid bipolar approach. The defuzzified financial results of the three companies on the 14 conditional attributes in 2011 are summarized in Table 20.4, and the three companies' performance scores and gaps on the 10 new contexts are shown in Table 20.5. One thing needs to be noted here: the performance scores in the negative contexts (rules) were regarded as the performance gaps for forming the final VIKOR aggregations.

The performance gap of each company was synthesized with the normalized support weight of each context by modified VIKOR; the S_k and R_k for each company can be obtained using Equations 20.13 and 20.14. In this case, the modified VIKOR method uses the score 0.86 (the highest defuzzified value from the five domain experts) as the aspiration level in each context.

20.4 Results of the Evaluations

The three companies' performance scores and transformed gap values in the criteria are shown in Table 20.5. To compare the ranking results, the modified

306 ■ *Trends of Hybrid Multiple Criteria Decision Making*

VIKOR method ($v = 0.5$) with different values in v (i.e., $v = 0.5$ and $v = 0.8$) and the simple additive weight (SAW) method were used; the results are summarized in Table 20.6 along with the three companies' actual ROA results in 2012.

The modified VIKOR and SAW methods both indicated the same top-ranking choice: company C. In addition, the actual ROA of company C in 2012 was 27.96%, the best outcome among the three companies; therefore, the fused bipolar model has shown its capability in making selections. The actual ranking sequence of the three companies' FPs in 2012 was $C > B > A$, which is consistent with the ranking from the modified VIKOR and SAW aggregation methods. Therefore, the results imply the effectiveness of the fused bipolar model. In addition, in the trained DRSA model, companies B and C were not distinguishable (they were all categorized into as the *Good* DC); however, they could still be ranked using the bipolar approach, which is another advantage of the proposed model.

20.5 Discussion on Contextual Improvement Planning

In addition, the weighted gaps in each context (i.e., the new criteria, from C_1 to C_{10}) are summarized in Table 20.6 to support improvement planning. Using the data from company B in Table 20.5 as an example, the weighted performance gap in each context for company B can be obtained. The top performance gap of company B is 0.05, which may be regarded as its *priority context*. If company B plans to improve its FP, its top-priority gap will be C_2, and the associated attributes are A_4, A_6, A_{12}, A_{13}, and A_{14} (Table 20.3) in the context of C_2. In addition, from Table 20.4, we may learn that the lowest defuzzified performance score for the five requirements (i.e., $A_4 \geq M$, $A_6 \geq H$, $A_{12} \geq M$, $A_{13} \geq H$, and $A_{14} \geq M$) is 0.1 (i.e., the defuzzified evaluation of B on "$A_{13} \geq H$" is 0.10); therefore, the management team should devise plans to improve the attribute *CashFlow* (A_{13}).

Each context in the fused bipolar model denotes a scenario for a company to improve (i.e., make it more similar to the positive contexts or more dissimilar to the negative ones), and the priority contexts obtained may guide a company to select targets (i.e., contexts/rules) for improvement while considering limited resources. In other words, by fusing the bipolar decision model with the modified VIKOR method, the priority contexts can be identified, and a company may devise plans to minimize its average performance gap. This concept highlights the importance of context in improvement planning, which is underaddressed in previous research.

Table 20.6 Weighted Gaps and the Priority Contexts of the Example Companies

	C_1	C_2	C_3	C_4	C_5	C_6	C_7	C_8	C_9	C_{10}	S_i (Rank)	R_i (Rank)	Q_i $v = 0.5$	Priority Contexts	Associated Attributes
A	0.05	0.04	0.06	0.07	0.05	0.02	0.03	0.09	0.05	0.09	0.56 (3)	0.09	0.33 (3)	C_8, C_{10}	A_2, A_3 or A_7, A_8
B	0.04	0.05	0.04	0.04	0.03	0.03	0.01	0.01	0.04	0.04	0.34 (2)	0.05	0.19 (2)	C_2	$A_4, A_6, A_{12}, A_{13}, A_{14}$
C	0.02	0.02	0.01	0.03	0.01	0.02	0.01	0.02	0.02	0.02	0.18 (1)	0.03	0.10 (1)	C_4	A_2, A_7, A_{10}, A_{13}

Note: The ranking result of SAW is the same as that of S_i; all of the rankings (i.e., SAW and modified VIKOR) are consistent with the ROA results ($A = 1.08\%$, $B = 8.95\%$, $C = 27.96\%$) of the three companies in 2012 (i.e., $C \succ B \succ A$).

20.6 Conclusion

To conclude, a hybrid bipolar decision model was introduced in this chapter with an empirical case study, extended from the obtained DRSA decision rules. This bipolar decision model has the advantages of MRDM in retrieving understandable knowledge (logic or rules) to support decision making; in addition, it highlights the importance of *contexts* to performance evaluation and improvement planning. The hybrid model comprises multiple soft computing techniques: DRSA, fuzzy evaluation, and the threshold-based bipolar model. These techniques bridge the imprecise judgments of human beings with induce rough knowledge from historical data, which is crucial to modeling practical business dynamics.

Though the extended work on DRSA in previous research may provide ranking or selection, alternatives in the same DC will encounter difficulties being ranked. Compared with previous works, the proposed hybrid bipolar approach has at least four advantages: (1) alternatives within the same or an undefined DC can still be ranked precisely; (2) transparent knowledge in the form of contexts/rules for decision aids can be obtained; (3) intercontext influences can be observed; and (4) the approach covers a user-defined level of confidence (i.e., the threshold Γ) from the learned instances—for ranking or selection—from historical data. The idea of bipolarity can be intuitively comprehended: an alternative should be ranked higher if it is more similar to the positive rules and dissimilar to the negative ones, which is close to the judgments made in a real business environment. By setting a threshold, DMs can comprehend the plausible credibility of the ranking result to make decisions. The aforementioned advantages are the major novelties and contributions of this bipolar approach.

Aside from ranking or selection, this chapter extended the focus to guiding improvement planning. How to rank correctly is important; nevertheless, when applying BA logic to a real business environment, how to use the obtained results to guide improvements may be more critical. Three example companies were evaluated using the hybrid bipolar model. The modified VIKOR method can help a company to identify its priority contexts; the assumed improvements in performance gaps may help management teams estimate plausible improvements for the overall FP. The modified VIKOR method adopts the idea of aspiration levels in each context, which may help a company prioritize its performance gaps and pursue excellence. This work has taken a step in the direction of transforming analytical results into managerial insights and shows how new MCDM methods may help to resolve practical business problems.

References

Aaker, D.A. (1991). *Managing Brand Equity*. New York: The Free Press.

Ahn, B.S., Cho, S.S., and Kim, C.Y. (2000). The integrated methodology of rough set theory and artificial neural network for business failure prediction. *Expert Systems with Applications*, 18(2): 65–74.

Al-Qirim, N. (2006). The role of the government and e-commerce adoption in small businesses in New Zealand. *International Journal of Internet and Enterprise Management*, 4(4): 293–313.

Amindoust, A., Ahmed, S., Saghafinia, A., and Bahreininejad, A. (2012). Sustainable supplier selection: A ranking model based on fuzzy inference system. *Applied Soft Computing*, 12(6): 1668–1677.

Andy, L., Lee, R.Y., and Tzeng, G.H. (2014). Characteristics of professional scuba dive guides. *Tourism in Marine Environments*, 10(1/2): 85–100.

Ao, S.I. (2011). A hybrid neural network cybernetic system for quantifying cross-market dynamics and business forecasting. *Soft Computing*, 15(6): 1041–1053.

Asllani, A., and Lari, A. (2007). Using genetic algorithm for dynamic and multiple criteria web-site optimizations. *European Journal of Operational Research*, 176: 1767–1777.

Awasthi, A., Chauhan, S.S., and Goyal, S.K. (2010). A fuzzy multicriteria approach for evaluating environmental performance of suppliers. *International Journal of Production Economics*, 126(2): 370–378.

Aydogan, E.K. (2011). Performance measurement model for Turkish aviation firms using the rough-AHP and TOPSIS methods under fuzzy environment. *Expert Systems with Applications*, 38(4): 3992–3998.

Ažman, S., and Gomišček, B. (2012). Asymmetric and nonlinear impact of attribute-level performance on overall customer satisfaction in the context of car servicing of four European automotive brands in Slovenia. *Organizacija*, 45: 75–86.

Babic, Z., and Pavic, I. (1996). Multicriterial production planning by De Novo programming approach. *International Journal of Production Economics*, 43(1): 59–66.

Barney, J., Wright, M., and Ketchen, D.J. (2001). The resource-based view of the firm: Ten years after 1991. *Journal of Management*, 27(6): 625–641.

Bearden, W.O., and Teel, J.E. (1983). Selected determinants of consumer satisfaction and complaint reports. *Journal of Marketing Research*, 20: 21–28.

Bellman, R.E., and Zadeh, L.A. (1970). Decision-making in a fuzzy environment. *Management Science*, 17(4): 141–164.

Belton, V., and Stewart, T.J. (2002). *Multiple Criteria Decision Analysis: An Integrated Approach*. Boston, MA: Kluwer Academic.

310 ■ References

Benayoun, R., Roy, B., and Sussman, N. (1966). Manual de reference du programme ELECTRE. Note de synthese et formation, no. 25. Paris, France.

Berk, R.A., and Freedman, D.A. (2003). Statistical assumptions as empirical commitments. In: Blomberg, B.G., and Cohen, S. (eds), *Law, Punishment, and Social Control: Essays in Honor of Sheldon Messinger* (pp. 235–234), 2nd edn. Aldine de Gruyter.

Bernoulli, D. (1738). Specimen theoriae novae de mensura sortis. *Commentarri academiae scientiarum imperialis Petropolitanae*, 5: 175–192.

Bertolini, M., Braglia, M., and Carmignani, G. (2006). Application of the AHP methodology in making a proposal for a public work contract. *International Journal of Project Management*, 24(5): 422–430.

Biery, K. (2006). Aligning an information risk management approach to BS 7799-3: 2005. SANS Institute InfoSec Reading Room.

Błaszczyński, J., Greco, S., Matarazzo, B., Słowiński, R., and Szeląg, M. (2013). jMAF dominance-based rough set data analysis framework. In: Skowron, A., and Suraj, Z. (eds), *Rough Sets and Intelligent Systems: Professor Zdzisław Pawlak in Memoriam* (pp. 185–209). Berlin Heidelberg: Springer.

Błaszczyński, J., Słowiński, R., and Szeląg, M. (2011). Sequential covering rule induction algorithm for variable consistency rough set approaches. *Information Sciences*, 181(5): 987–1002.

Bouyssou, D., and Pirlot, M. (2005). Conjoint measurement tools for MCDM. In: Figueira, J., Greco, S., and Ehrgott, M. (eds), *Multiple Criteria Decision Analysis: State of the Art Surveys* (pp. 73–130). Springer Science and Business.

Bouyssou, D., and Vansnick, J.C. (1986). Noncompensatory and generalized noncompensatory preference structures. *Theory and Decision*, 21(3): 251–266.

Boyacioglu, M.A., and Avci, D. (2010). An adaptive network-based fuzzy inference system (ANFIS) for the prediction of stock market return: The case of the Istanbul stock exchange. *Expert Systems with Applications*, 37(12): 7908–7912.

Brans, J.P., Mareschal, B., and Vincke, P. (1984). PROMETHEE: A new family of outranking methods in multicriteria analysis. In: Brans, J.P. (ed), *Operational Research* (pp. 408–421). Amsterdam, the Netherlands: North-Holland.

Brav, A., and Heaton, J.B. (2002). Competing theories of financial anomalies. *Review of Financial Studies*, 15(2): 575–606.

Buil, I., De Chernatony, L., and Martínez, E. (2013). Examining the role of advertising and sales promotions in brand equity creation. *Journal of Business Research*, 66(1): 115–122.

Büyüközkan, G., and Çifçi, G. (2011). A novel fuzzy multi-criteria decision framework for sustainable supplier selection with incomplete information. *Computers in Industry*, 62(2): 164–174.

Büyüközkan, G., and Çifçi, G. (2012). An integrated fuzzy multi-criteria group decision-making approach for green supplier evaluation. *International Journal of Production Research*, 50(11): 2892–2909.

Campanella, G., and Ribeiro, R.A. (2011). A framework for dynamic multiple-criteria decision making. *Decision Support Systems*, 52(1): 52–60.

Carini, A., Sehgal, V., Evans, P.F., and Roberge, D. (2011). European online retail forecast, 2010 to 2015. Reported by InternetRetailing.net.

Chan, C.C., and Tzeng, G.H. (2008). Computing approximations of dominance-based rough sets by bit-vector encodings. In: *International Conference on Rough Sets and Current Trends in Computing* (pp. 131–141). Berlin Heidelberg: Springer.

Chan, C.C., and Tzeng, G.H. (2009). Dominance-based rough sets using indexed blocks as granules. *Fundamenta Informaticae*, 94(2): 133–146.

Chan, C.C., and Tzeng, G.H. (2011). Bit-vector representation of dominance-based approximation space. In: *Transactions on Rough Sets XIII* (pp. 1–16). Berlin Heidelberg: Springer.

Chan, F.T., and Kumar, N. (2007). Global supplier development considering risk factors using fuzzy extended AHP-based approach. *Omega*, 35(4): 417–431.

Chan, K.C., and Lakonishok, J. (2004). Value and growth Investing: Review and update. *Financial Analyst Journal*, 60(1): 71–86.

Chan, L.K., Lakonishok, J., and Sougiannis, T. (2001). The stock market valuation of research and development expenditures. *The Journal of Finance*, 56(6): 2431–2456.

Chang, B., Kuo, C., and Tzeng, G.H. (2015a). Using fuzzy analytic network process to assess the risks in enterprise resource planning system implementation. *Applied Soft Computing*, 28: 196–207.

Chang, D.S., Chen, S.H., Hsu, C.W., Hu, A.H., and Tzeng, G.H. (2015b). Evaluation framework for alternative fuel vehicles: Sustainable development perspective. *Sustainability*, 7(9): 11570–11594.

Chang, H.F., and Tzeng, G.H. (2010). A causal decision making model for knowledge management capabilities to innovation performance in Taiwan's high-tech industry. *Journal of Technology Management and Innovation*, 5(4): 137–146.

Chang, H.H., and Chen, S.W. (2009). Consumer perception of interface quality, security, and loyalty in electronic commerce. *Information and Management*, 46: 411–417.

Chang, M.C., Hu, J.L., and Tzeng, G.H. (2009). Decision making on strategic environmental technology licensing: Fixed-fee versus royalty licensing methods. *International Journal of Information Technology and Decision Making*, 8(3): 609–624.

Charnes, A., and Cooper, W.W. (1955). Generalizations of the warehousing model. *OR*, 6(4): 131–172.

Charnes, A., and Cooper, W.W. (1961). *Management Models and Industrial Applications of Linear Programming*. New York: John Wiley.

Charnes, A., Cooper, W.W., and Rhodes, E. (1978). Measuring the efficiency of decision making units. *European Journal of Operational Research*, 2(6): 429–446.

Chau, P., and Tam, K. (1997). Factors affecting the adoption of open systems: An exploratory study. *MIS Quarterly*, 21(1): 1–24.

Chen, C.H., and Tzeng, G.H. (2011). Creating the aspired intelligent assessment systems for teaching materials. *Expert Systems with Applications*, 38(10): 12168–12179.

Chen, F.H., Hsu, T.S., and Tzeng, G.H. (2011). A balanced scorecard approach to establish a performance evaluation and relationship model for hot spring hotels based on a hybrid MCDM model combining DEMATEL and ANP. *International Journal of Hospitality Management*, 30(4): 908–932.

Chen, F.H., and Tzeng, G.H. (2015). Probing organization performance using a new hybrid dynamic MCDM method based on the balanced scorecard approach. *Journal of Testing and Evaluation*, 43(4): 1–14.

Chen, F.H., Tzeng, G.H., and Chang, C.C. (2015). Evaluating the enhancement of corporate social responsibility website quality based on a new hybrid MADM model. *International Journal of Information Technology and Decision Making*, 14(3): 697–724.

Chen, H.C., Hu, Y.C., Shyu, J.Z., and Tzeng, G.H. (2005a). Comparing possibility grey forecasting with neural network–based fuzzy regression by an empirical study. *Journal of Grey System*, 18(2): 93–106.

312 ■ *References*

Chen, H.C., Hu, Y.C., Shyu, J.Z., and Tzeng, G.H. (2005b). Comparison analysis of possibility grey forecasting and fuzzy regression to the stock-market price in Taiwan. *Journal of Information Management*, 12(1): 195–214.

Chen, K.C., and Tzeng, G.H. (2009). Perspective strategic alliances and resource allocation in supply chain systems through the De Novo programming approach. *International Journal of Sustainable Strategic Management*, 1(3): 320–339.

Chen, M.F., and Tzeng, G.H. (2004). Combining grey relation and TOPSIS concepts for selecting an expatriate host country. *Mathematical and Computer Modelling*, 40(13): 1473–1490.

Chen, M.F., Tzeng, G.H., and Ding, C.G. (2008). Combing fuzzy AHP with MDS in identifying the preference similarity of alternatives. *Applied Soft Computing*, 8(1): 110–117.

Chen, M.F., Tzeng, G.H., and Tang, T.I. (2005). Fuzzy MCDM approach for evaluation of expatriate assignments. *International Journal of Information Technology and Decision Making*, 4(2): 277–296.

Chen, P.T., Lee, Z.Y., Yu, H.C., and Tzeng, G.H. (2003). Analysis of the theoretical basis and practical applicability of a college teacher's achievement evaluation model: The case study of a national university in Hsinchu. *Bulletin of Educational Research*, 49(4): 191–218.

Chen, S.C., and Dhillon, G.S. (2003). Interpreting dimensions of consumer trust in e-commerce. *Information Technology and Management*, 4(2): 303–318.

Chen, T.Y., Chang, C.C., and Tzeng, G.H. (2001a). Applying fuzzy measures to establish priority-setting procedures for the pavement management system. *Pan-Pacific Management Review*, 4(1): 23–33.

Chen, T.Y., Chang, H.L., and Tzeng, G.H. (2001b). Using a weight-assessing model to identify route choice criteria and information effects. *Transportation Research, Part A*, 35(3): 197–224.

Chen, T.Y., Chang, H.L., and Tzeng, G.H. (2002). Using fuzzy measure and habitual domains to analyze the public attitude and apply to the gas taxi policy. *European Journal of Operational Research*, 137(2): 145–161.

Chen, T.Y., and Tzeng, G.H. (2003). Exploring the public attitude using fuzzy integrals. *Pan-Pacific Management Review*, 6(1): 45–58.

Chen, T.Y., Wang, J.C., and Tzeng, G.H. (2000). Identification of general fuzzy measures by genetic algorithms based on partial information. *IEEE Transactions on Systems, Man, and Cybernetics (Part B): Cybernetics*, 30B(4): 517–528.

Chen, Y.C., Lien, H.P., Liu, C.H., Liou, J.J.H., Tzeng, G.H., and Yang, L.S. (2011a). Fuzzy MCDM approach for selecting the best environment-watershed plan. *Applied Soft Computing*, 11(1): 265–275.

Chen, Y.C., Lien, H.P., and Tzeng, G.H. (2011b). Measures and evaluation for environment watershed plans using a novel hybrid MCDM model. *Expert Systems with Applications*, 37(2): 926–938.

Chen, Y.W., and Tzeng, G.H. (1999). A fuzzy multi-objective model for reconstructing post-earthquake road-network by genetic algorithm. *International Journal of Fuzzy Systems*, 1(2): 85–95.

Chen, Y.W., and Tzeng, G.H. (2000). Fuzzy multi-objective approach to the supply chain model. *International Journal of Fuzzy Systems*, 1(3): 220–227.

Chen, Y.W., and Tzeng, G.H. (2001). Using fuzzy integral for evaluating subjectively perceived travel costs in a traffic assignment model. *European Journal of Operational Research*, 130(3): 653–664.

Chen, Z., and Dubinsky, A.J. (2003). A conceptual model of perceived customer value in e-commerce: A preliminary investigation. *Psychology and Marketing*, 20: 323–347.

Cheng, C.H., Chen, T.L., and Wei, L.Y. (2010). A hybrid model based on rough sets theory and genetic algorithms for stock price forecasting. *Information Sciences*, 180(9): 1610–1629.

Cheng, J.W., Chiu, W.L., and Tzeng, G.H. (2013). Do impression management tactics and/or supervisor–subordinate *guanxi* matter? *Knowledge-Based Systems*, 40: 123–133.

Cheng, J.Z., Yu, Y.W., Tsai, M.J., and Tzeng, G.H. (2004). Setting a business strategy to weather the telecommunications industry downturn by using fuzzy MCDM. *International Journal of Services Technology and Management*, 5(4): 346–361.

Chiang, C.I., Li, J.M., Tzeng, G.H., and Yu, P.L. (2000). A revised minimum spanning table method for optimal expansion of competence sets. In: *Research and Practice in Multiple Criteria Decision Making: Proceedings of the XIVth International Conference on Multiple Criteria Decision Making (MCDM)* (pp. 238–247). Berlin Heidelberg: Springer.

Chiang, C.I., and Tzeng, G.H. (2000a). A multiple objective programming approach to data envelopment analysis. In: *New Frontiers of Decision Making for the Information Technology Era* (pp. 270–285). Singapore: World Science.

Chiang, C.I., and Tzeng, G.H. (2000b). A new efficiency measure for DEA: Efficiency achievement measure established on fuzzy multiple objectives programming. *Journal of Management*, 17(2): 369–388 (in Chinese).

ChiangLin, C.Y., Lai, T.C., and Yu, P.L. (2007). Linear programming models with changeable parameters: Theoretical analysis on "taking loss at the ordering time and making profit at the delivery time." *International Journal of Information Technology and Decision Making*, 6(4): 577–589.

Chin, Y.C., Chang, C.C., Lin, C.S., and Tzeng, G.H. (2010). The impact of recommendation sources on the adoption intention of microblogging based on dominance-based rough sets approach. In Szczuka, M., et al. (eds), *RSCTC 2010* (pp. 514–523), LNAI 6086. Berlin Heidelberg: Springer.

Chiou, H.K., and Tzeng, G.H. (2002). Fuzzy multiple-criteria decision-making approach for industrial green engineering. *Environmental Management*, 30(6): 816–830.

Chiou, H.K., and Tzeng, G.H. (2003). An extended approach of multicriteria optimization for MODM problems. In Tanino, T., Tanaka, T., and Inuiguchi, M. (eds), *Multiobjective Programming and Goal-Programming: Theory and Applications* (pp. 111–116). New York: Springer-Verlag.

Chiou, H.K., Tzeng, G.H., and Cheng, D.C. (2005). Evaluating sustainable fishing development strategies using a fuzzy MCDM approach. *OMEGA-International Journal of Management Science*, 33(3): 223–234.

Chiu, W.Y., Tzeng, G.H., and Li, H.L. (2013). A new hybrid MCDM model combining DANP with VIKOR to improve e-store business. *Knowledge-Based Systems*, 37: 48–61.

Chiu, W.Y., Tzeng, G.H., and Li, H.L. (2014). Developing e-store marketing strategies to satisfy customers' needs using a new hybrid grey relational model. *International Journal of Information Technology and Decision Making*, 13(2): 231–261.

Chiu, Y.C., Chen, B., Shyu, J.Z., and Tzeng, G.H. (2006). An evaluation model of new product launch strategy. *Technovation*, 26(11): 1244–1252.

314 ■ *References*

Chiu, Y.C., Shyu, J.Z., and Tzeng, G.H. (2004). Fuzzy MCDM for evaluating the e-commerce strategy. *International Journal of Computer Applications in Technology*, 19(1): 12–22.

Chiu, Y.C., and Tzeng, G.H. (1999). The market acceptance of electric motorcycles in Taiwan: Experience through a stated preference analysis. *Transportation Research (Part D)*, 4(2): 127–146.

Chiu, Y.J., Chen, H.C., Tzeng, G.H., and Shyu, J.Z. (2006). Marketing strategy based on customer behavior for the LCD-TV. *International Journal of Management and Decision Making*, 7(2–3): 143–165.

Chloe, R. (2012). European online shopping to grow by 12% a year: Predictions. Reported by InternetRetailing.net, http://internetretailing.net/2012/02/european-online-shopping-to-grow-by-12-a-year-predictions/.

Chon, K.S. (1990). The role of destination image in tourism: A review and discussion. *Tourism Review*, 45: 2–9.

Choquet, G. (1953). Theory of capacities. In: *Annales de l'Institut Fourier*, vol. 5 (pp. 131–295). DOI: 10.5802/aif.53, http://www.numdam.org/item?id=AIF_1954__5__131_0.

Chou, S.Y., Yu, C.C., and Tzeng, G.H. (2016). A novel hybrid MCDM procedure for achieving aspired earned value management applications. *Journal of Mathematical Problems in Engineering*, Article ID 9721726, 16 pages. DOI: 10.1155/2016/9721726, https://www.hindawi.com/journals/mpe/2016/9721726/abs/.

Chou, T.Y., Chou, S.T., and Tzeng, G.H. (2006). Evaluating IT/IS investments: A fuzzy multi-criteria decision model approach. *European Journal of Operational Research*, 173(3): 1026–1046.

Chou, W.C., and Cheng, Y.P. (2012). A hybrid fuzzy MCDM approach for evaluating website quality of professional accounting firms. *Expert Systems with Applications*, 39(3): 2783–2793.

Chu, M.T., Shyu, J., Tzeng, G.H., and Khosla, R. (2007a). Comparison among three analytical methods for knowledge communities group-decision analysis. *Expert Systems with Applications*, 33(4): 1011–1024.

Chu, M.T., Shyu, J., Tzeng, G.H., and Khosla, R. (2007b). Using nonadditive fuzzy integral to assess performances of organizational transformation via communities of practice. *IEEE Transactions on Engineering Management*, 54(2): 327–339.

Chu, P.Y., Tzeng, G.H., and Teng, M.J. (2004). A multivariate analysis of the relationship among market share, growth and profitability: The case of the science-based industrial park. *Yat-Sen Management Review*, 12(3): 507–534.

Coase, R.H. (1937). The nature of the firm. *Economica*, 4(16): 386–405.

Coello, C.A.C., Pulido, G.T., and Lechuga, M.S. (2004). Handling multiple objectives with particle swarm optimization. *IEEE Transactions on Evolutionary Computation*, 8(3): 256–279.

Contractor, F.J., and Lorange, P. (1988). Why should firms cooperate? The strategy and economics basis for cooperative ventures. *Cooperative Strategies in International Business*, 27(1): 77–93.

Cook, W.D., Liang, L., and Zhu, J. (2010). Measuring performance of two-stage network structures by DEA: A review and future perspective. *Omega*, 38(6): 423–430.

Dantzig, G.B. (1951). Application of the simplex method to a transportation problem. *Activity Analysis of Production and Allocation*, 13: 359–373.

Davenport, T.H. (2006). Competing on analytics. *Harvard Business Review*, 84: 98–107.

Davenport, T.H., and Harris, J.G. (2013). *Competing on Analytics: The New Science of Winning*. Cambridge, MA: Harvard Business Press.

Delen, D., and Demirkan, H. (2013). Data, information and analytics as services. *Decision Support Systems*, 55: 359–363.

Deng, J.L. (1982). Control problems of grey systems. *Systems and Control Letters*, 1(5): 288–294.

Deng, Y., and Chan, F.T. (2011). A new fuzzy Dempster MCDM method and its application in supplier selection. *Expert Systems with Applications*, 38(8): 9854–9861.

Diakoulaki, D., Mavrotas, G., and Papayannakis, L. (1992). A multicriteria approach for evaluating the performance of industrial firms. *Omega*, 20(4): 467–474.

Dimitras, A.I., Slowinski, R., Susmaga, R., and Zopounidis, C. (1999). Business failure prediction using rough sets. *European Journal of Operational Research*, 114(2): 263–280.

Doumpos, M., and Grigoroudis, E. (eds) (2013). *Multicriteria Decision Aid and Artificial Intelligence: Links, Theory and Applications*. New York: John Wiley.

Doumpos, M., and Zopounidis, C. (2011). Preference disaggregation and statistical learning for multicriteria decision support: A review. *European Journal of Operational Research*, 209: 203–214.

Dubois, D., and Prade, H. (1980). *Fuzzy Sets and Systems*. New York: Academic Press.

Dubois, D., and Prade, H. (2012). Gradualness, uncertainty and bipolarity: Making sense of fuzzy sets. *Fuzzy Sets and Systems*, 192: 3–24.

Eiben, A.E., and Smith, J.E. (2003). Evolutionary programming. In: *Introduction to Evolutionary Computing* (pp. 89–99). Berlin Heidelberg: Springer-Verlag.

Ellinger, A., Shin, H., Northington, W.M., Adams, F.G., Hofman, D., and O'Marah, K. (2012). The influence of supply chain management competency on customer satisfaction and shareholder value. *Supply Chain Management: An International Journal*, 17: 249–262.

Ellinger, A.E., Lynch, D.F., and Hansen, J.D. (2003). Firm size, web site content, and financial performance in the transportation industry. *Industrial Marketing Management*, 32: 177–185.

Ergul, N., and Oktem, R. (2011). Searching of usability of TOPSIS and ELECTRE methods in measurement and evaluation of financial performance of construction and public works companies. *Applied Finance*, 2(9): 1086–1100.

Falagario, M., Sciancalepore, F., Costantino, N., and Pietroforte, R. (2012). Using a DEA-cross efficiency approach in public procurement tenders. *European Journal of Operational Research*, 218(2): 523–529.

Fama, E. F. (1970). Efficient capital markets: A review of theory and empirical work. *The Journal of Finance*, 25(2): 383–417.

Fama, E.F., and French, K.R. (2006). The value premium and the CAPM. *The Journal of Finance*, 61(5): 2163–2185.

Fan, T.F., Liu, D.R., and Tzeng, G.H. (2006). Arrow decision logic for relational information systems. In: Peters, J.F., and Skowron, A. (eds), *Transactions on Rough Sets V, LNCS* 4100: 240–262. Berlin Heidelberg: Springer-Verlag.

Fan, T.F., Liu, D.R., and Tzeng, G.H. (2007). Rough set-based logics for multicriteria decision analysis. *European Journal of Operational Research*, 182(1): 340–355.

Fang, S.K., Shyng, J.Y., Lee, W.S., and Tzeng, G.H. (2012). Exploring the preference of customers between financial companies and agents based on TCA. *Knowledge-Based Systems*, 27(2): 137–151.

316 ■ *References*

Färe, R., and Grosskopf, S. (1997). Intertemporal production frontiers: With dynamic DEA. *Journal of the Operational Research Society*, 48(6): 656–656.

Färe, R., and Grosskopf, S. (2012). *Intertemporal Production Frontiers with Dynamic DEA*. Norwell, MA: Kluwer Academic Publishers.

Fethi, M.D., and Pasiouras, F. (2010). Assessing bank efficiency and performance with operational research and artificial intelligence techniques: A survey. *European Journal of Operational Research*, 204(2): 189–198.

Figueira, J., Greco, S., and Ehrgott, M. (eds) (2005). *Multiple Criteria Decision Analysis: State of the Art Surveys*. New York: Springer Science and Business.

Flavell, R.B. (1976). A new goal programming formulation. *Omega*, 4(6): 731–732.

Fontela, E., and Gabus, A. (1974). DEMATEL, innovative methods. Technical Report no. 2, Structural analysis of the world problematique. Battelle Geneva Research Institute.

Fontela, E., and Gabus, A. (eds) (1976). *The DEMATEL observe*. Battelle Institute, Geneva Research Center.

Gan, C., Clemes, M., Limsombunchai, V., and Weng, A. (2006). A logit analysis of electronic banking in New Zealand. *International Journal of Bank Marketing*, 24: 360–383.

Ganter, B., Wille, R., and Franzke, C. (1997). *Formal Concept Analysis: Mathematical Foundations*. New York: Springer.

Gill, M., Evans, P.F., Sehgal, V., and Costa, M.D. (2012). European online retail forecast: 2011 to 2016. Reported by InternetRetailing.net, http://www.forrester.com/Europea n+Online+Retail+Forecast+2011+To+2016/fulltext/-/E-RES60745?docid=60745&src =RSS_2&cm_mmc=Forrester-_-RSS-_-Document-_-23.

Gölcük, İ., and Baykasoğlu, A. (2016). An analysis of DEMATEL approaches for criteria interaction handling within ANP. *Expert Systems with Applications*, 46: 346–366.

Goldberg, D.E. (1989). *GA in Search, Optimization, and Machine Learning*. Boston, MA: Addison-Wesley Publishing Company.

Grabisch, M. (1995). Fuzzy integral in multicriteria decision making. *Fuzzy Sets and Systems*, 69(3): 279–298.

Gradojevic, N., and Gençay, R. (2013). Fuzzy logic, trading uncertainty and technical trading. *Journal of Banking and Finance*, 37(2): 578–586.

Grant, R.M. (1991). The resource-based theory of competitive advantage: Implications for strategy formulation. *California Management Review*, 33(3): 114–135.

Greco, S., Ehrgott, M., and Figueira, J. (eds) (2016). *Multiple Criteria Decision Analysis: State of the Art Surveys*, 2nd edn. New York: Springer Science & Business Media.

Greco, S., Matarazzo, B., and Słowiński, R. (1997). Rough set approach to multi-attribute choice and ranking problems. In: Fandel, G., and Gal, T. (eds), *Multiple Criteria Decision Making* (pp. 318–329). Berlin Heidelberg: Springer.

Greco, S., Matarazzo, B., and Słowiński, R. (1999). Rough approximation of a preference relation by dominance relations. *European Journal of Operational Research*, 117(1): 63–83.

Greco, S., Matarazzo, B., and Słowiński, R. (2001). Rough sets theory for multicriteria decision analysis. *European Journal of Operational Research*, 129(1): 1–47.

Greco, S., Matarazzo, B., and Słowiński, R. (2002a). Rough approximation by dominance relations. *International Journal of Intelligent Systems*, 17(2): 153–171.

Greco, S., Matarazzo, B., and Słowiński, R. (2002b). Rough sets methodology for sorting problems in presence of multiple attributes and criteria. *European Journal of Operational Research*, 138(2): 247–259.

References ■ 317

Greco, S., Matarazzo, B., and Słowiński, R. (2005). Decision rule approach. In: Figueira, J., Greco, S., and Ehrgott, M. (eds), *Multiple Criteria Decision Analysis: State of the Art Surveys* (pp. 507–557). New York: Springer Science & Business.

Greco, S., Matarazzo, B., and Słowiński, R. (2008). Dominance-based rough set approach to interactive multiobjective optimization. In: Branke, J., Deb, K., Miettinen, K., and Slowiński, R. (eds), *Multiobjective Optimization* (pp. 121–155). Berlin Heidelberg: Springer.

Greco, S., Matarazzo, B., Słowiński, R., and Stefanowski, J. (2000). Variable consistency model of dominance-based rough sets approach. In: *International Conference on Rough Sets and Current Trends in Computing* (pp. 170–181). Berlin Heidelberg: Springer.

Greco, S., Matarazzo, B., Słowiński, R., and Tsoukiàs, A. (1998). Exploitation of a rough approximation of the outranking relation in multicriteria choice and ranking. In: *Trends in Multicriteria Decision Making* (pp. 45–60). Berlin Heidelberg: Springer.

Hillier, F.S. (2001). *Evaluation and Decision Models: A Critical Perspective*. Boston, MA: Kluwer Academic.

Ho, W., Xu, X., and Dey, P.K. (2010). Multi-criteria decision making approaches for supplier evaluation and selection: A literature review. *European Journal of Operational Research*, 202(1): 16–24.

Hobday, M. (1995). East Asian latecomer firms: Learning the technology of electronics. *World Development*, 23(7): 1171–1193.

Holland, J.M. (1975). *Adaptation in Natural and Artificial Systems*. Ann Arbor, MI: University of Michigan Press.

Hotho, J.J., Lyles, M.A., and Easterby-Smith, M. (2015). The mutual impact of global strategy and organizational learning: Current themes and future directions. *Global Strategy Journal*, 5(2): 85–112.

Hsieh, T.Y., and Liu, H.L. (2004). Genetic algorithm for optimization of infrastructure investment under time-resource constraints. *Computer-Aided Civil and Infrastructure Engineering*, 19(3): 203–212.

Hsieh, T.Y., Lu, S.T., and Tzeng, G.H. (2004). Fuzzy MCDM approach for planning and design tenders selection in public office buildings. *International Journal of Project Management*, 22(7): 573–584.

Hsieh, Y.L., Tzeng, G.H., Lin, T.R., and Yu, H.C. (2010). Wafer sort bitmap data analysis using the PCA-based approach for yield analysis and optimization. *IEEE Transactions on Semiconductor Manufacturing*, 24(4): 493–502.

Hsu, C.C., Liou, J.J.H., and Chuang, Y.C. (2013). Integrating DANP and modified grey relation theory for the selection of an outsourcing provider. *Expert Systems with Applications*, 40(6): 2297–2304.

Hsu, C.H., Wang, F.K., and Tzeng, G.H. (2012). The best vendor selection for conducting the recycled material based on a hybrid MCDM model combining DANP with VIKOR. *Resources, Conservation and Recycling*, 66: 95–111.

Hsu, C.S., Lee, Z.Y., Hung, C.Y., Shih, C., Yu, H.C., and Tzeng, G.H. (2004). Key factors in performance appraisal for R&D organizations: The case of the industrial technology research institute in Taiwan. *Journal of Biomedical Fuzzy Systems*, 10(1/2): 19–29.

Hsu, C.W., and Hu, A.H. (2009). Applying hazardous substance management to supplier selection using analytic network process. *Journal of Cleaner Production*, 17(2): 255–264.

318 ■ *References*

Hsu, C.Y., Chen, K.T., and Tzeng, G.H. (2007). FMCDM with fuzzy DEMATEL approach for customers' choice behavior model. *International Journal of Fuzzy Systems*, 9(4): 236–246.

Hsu, W.C.J., Tsai, M.H., and Tzeng, G.H. (2016). Exploring the best strategy plan for improving the digital convergence by using a hybrid MADM model. *Technological and Economic Development of Economy* (in press).

Hsu, Y.G., Shyu, J.Z., and Tzeng, G.H. (2005). Policy tools on the formation of new biotechnology firms in Taiwan. *Technovation*, 25(3): 281–292.

Hsu, Y.G., Tzeng, G.H., and Shyu, J.Z. (2003). Fuzzy multiple criteria selection of government-sponsored frontier technology R&D projects. *R&D Management*, 33(5): 539–551.

Hu, K.H., Chen, F.H., and Tzeng, G.H. (2016). Evaluating the improvement of sustainability of sports industry policy based on MADM. *Sustainability*, 8(7): 606–632.

Hu, K.H., Chen, F.H., Tzeng, G.H., and Lee, J.D. (2015). Improving corporate governance effects on an enterprise crisis based on a new hybrid DEMATEL with the MADM model. *Journal of Testing and Evaluation*, 43(6): 1395–1412.

Hu, K.H., Wei, J., and Tzeng, G.H. (2016). Improving China's regional financial center modernization development using new hybrid MADM model. *Technological and Economic Development of Economy* (in press).

Hu, S.K., Chuang, Y.C., Yeh, Y.F., and Tzeng, G.H. (2012). Hybrid MADM with fuzzy integral for exploring the smart phone improvement in m-generation. *International Journal of Fuzzy Systems*, 14(2): 204–214.

Hu, S.K., Lu, M.T., and Tzeng, G.H. (2014). Exploring smart phone improvements based on a hybrid MCDM model. *Expert Systems with Applications*, 41(9): 4401–4413.

Hu, S.K., Tzeng, G.H., and Lu, M.T. (2015). Improving mobile commerce adoption using a new hybrid fuzzy MADM model. *International Journal of Fuzzy Systems*, 17(3): 399–413.

Hu, Y.C., Chen, R.S., Hsu, Y.T., and Tzeng, G.H. (2002a). Grey self-organizing feature maps. *Neurocomputing*, 48(4): 863–877.

Hu, Y.C., Chen, R.S., and Tzeng, G.H. (2002b). Generating learning sequences for decision makers through data mining and competence set expansion. *IEEE Transactions on Systems, Man and Cybernetics (Part B)*, 32(5): 679–686.

Hu, Y.C., Chen, R.S., and Tzeng, G.H. (2003a). An effective learning algorithm for discovering fuzzy sequential patterns. *International Journal Uncertainty, Fuzziness and Knowledge-Based Systems*, 11(2): 173–193.

Hu, Y.C., Chen, R.S., and Tzeng, G.H. (2003b). Finding fuzzy classification rules using data mining techniques. *Pattern Recognition Letters*, 24(1–3): 509–519.

Hu, Y.C., Chen, R.S., and Tzeng, G.H. (2003c). Discovering fuzzy association rules using fuzzy partition methods. *Knowledge-Based Systems*, 16(3): 137–147.

Hu, Y.C., Chen, R.S., Tzeng, G.H., and Chiu, Y.J. (2003d). Acquisition of compound skills and learning costs for expanding competence sets. *Computers and Mathematics with Applications*, 46(5–6): 831–848.

Hu, Y.C., Chen, R.S., Tzeng, G.H., and Shieh, J.H. (2003e). A fuzzy data mining algorithm for finding sequential patterns. *International Journal of Uncertainty, Fuzziness and Knowledge-Based Systems*, 11(2): 173–193.

Hu, Y.C., Chiu, Y.J., Chen, C.M., and Tzeng, G.H. (2003f). Competence set expansion for obtaining scheduling plans in intelligent transportation security systems. In: Tanino, T., Tanaka, T., and Inuiguchi, M. (eds), *Multi-objective Programming and Goal-Programming: Theory and Applications* (pp. 347–352). Berlin Heidelberg: Springer.

Hu, Y.C., Chiu, Y.J., and Tzeng, G.H. (2003g). Grey theory and competence sets for multiple criteria project scheduling. *Management Review*, 1(2): 257–273 (in Chinese).

Hu, Y.C., Chiu, Y.J., Hu, J.S., Tzeng, G.H., and Huang, Y.S. (2002). Acquisitions of learning costs for expanding competence sets using grey relations and neural networks. *Journal of the Chinese Grey System Association*, 5(2): 75–82 (in Chinese).

Hu, Y.C., Hu, J.S., Chen, R.S., and Tzeng, G.H. (2004a). Assessing weights of product attributes from fuzzy knowledge in a dynamic environment. *European Journal of Operational Research*, 154(1): 125–143.

Hu, Y.C., and Tzeng, G.H. (2002). Mining fuzzy association rule for classification problems. *Computers and Industrial Engineering*, 43(4): 735–750.

Hu, Y.C., and Tzeng, G.H. (2003). Elicitation of classification rules by fuzzy data mining. *Engineering Applications of Artificial Intelligence*, 16(7–8): 709–716.

Hu, Y.C., Tzeng, G.H., and Chen, C.M. (2004b). Deriving two stage learning sequence from knowledge in fuzzy sequential pattern mining. *Information Sciences*, 159(1–2): 69–86.

Huang, C.Y., Chang, S.Y., Yang, Y.H., and Tzeng, G.H. (2010). Next generation passive optical networking technology predictions by using hybrid MCDM methods. *Journal of Advanced Computational Intelligence and Intelligent Informatics*, 15(4): 400–405.

Huang, C.Y., Hung, Y.H., and Tzeng, G.H. (2010). Using hybrid MCDM methods to assess fuel cell technology for the next generation of hybrid power automobiles. *Journal of Advanced Computational Intelligence and Intelligent Informatics*, 15(4): 406–417.

Huang, C.Y., Shyu, J.Z., and Tzeng, G.H. (2006). Reconfiguring the innovation policy portfolio for the SIP Mall industry using novel MCDM models. *International Journal of the Information System for Logistics and Management*, 1(2): 69–87.

Huang, C.Y., Shyu, J.Z., and Tzeng, G.H. (2007). Reconfiguring the innovation policy portfolios for Taiwan's SIP mall industry. *Technovation*, 27(12): 744–765.

Huang, C.Y., and Tzeng, G.H. (2007). Post-merger high technology R&D human resources optimization through the de novo perspective. In: *Advances in Multiple Criteria Decision Making and Human Systems Management: Knowledge and Wisdom* (pp. 47–64). Amsterdam, the Netherlands: IOS Press.

Huang, C.Y., and Tzeng, G.H. (2008). Multiple generation product life cycle predictions using a novel two-stage fuzzy piecewise regression analysis method. *Technological Forecasting and Social Change*, 75(1): 12–31.

Huang, C.Y., Tzeng, G.H., Chan, C.C., and Wu, H.C. (2008). Semiconductor market fluctuation indicators and rules derivations by using the rough set theory. *International Journal of Innovative Computing, Information and Control*, 5(6): 1485–1503.

Huang, C.Y., Tzeng, G.H., Chen, Y.T., and Chen, H. (2012a). Performance evaluation of leading fabless integrated circuit design houses by using a multiple objective programming based data envelopment analysis approach. *International Journal of Innovative Computing, Information and Control*, 8(8): 5899–5916.

Huang, C.Y., Tzeng, G.H., and Ho, W.R. (2011). System on chip design service e-business value maximization through a novel MCDM framework. *Expert Systems with Applications*, 38(7): 7947–7962.

320 ■ *References*

Huang, C.Y., Wang, P.Y., and Tzeng, G.H. (2012b). Evaluating top information technology firms in Standard and Poor's 500 index by using a multiple objective programming based data envelopment analysis. In: *Advanced Research in Applied Artificial Intelligence, Lecture Notes in Computer Science* (pp. 720–730), LNAI 7345. Berlin Heidelberg: Springer.

Huang, C.Y., Wu, M.J., Liu, Y.W., and Tzeng, G.H. (2012c). Using the DEMATEL based network process and structural equation modeling methods for deriving factors influencing the acceptance of smart phone operation systems. In: *Advanced Research in Applied Artificial Intelligence* (pp. 731–741), LNAI 7345. Berlin Heidelberg: Springer.

Huang, J.J., Chen, C.Y., Liu, H.H., and Tzeng, G.H. (2011). A multiobjective programming model for partner selection: Perspectives of objective synergies and resource allocations. *Expert Systems with Applications*, 37(5): 3530–3536.

Huang, J.J., Chen, C.Y., and Tzeng, G.H. (2016). Generalized DEMATEL technique with centrality measurements. *Technological and Economic Development of Economy* (in press).

Huang, J.J., Ong, C.S., and Tzeng, G.H. (2005a). A novel hybrid model for portfolio selection. *Applied Mathematics and Computation*, 169(2): 1195–1210.

Huang, J.J., Ong, C.S., and Tzeng, G.H. (2005b). Building credit scoring models using genetic programming. *Expert Systems with Applications*, 29(1): 41–47.

Huang, J.J., Ong, C.S., and Tzeng, G.H. (2005c). Model identification of ARIMA family using genetic algorithms. *Applied Mathematics and Computation*, 164(3): 885–912.

Huang, J.J., Ong, C.S., and Tzeng, G.H. (2005d). Motivation and resource-allocation for strategic alliance through the De Novo perspective. *Mathematical and Computer Modelling*, 41(6–7): 711–721.

Huang, J.J., Ong, C.S., and Tzeng, G.H. (2006a). Fuzzy principal component regression (FPCR) for fuzzy input and output data. *International Journal of Uncertainty and Knowledge-Based Systems*, 14(1): 87–100.

Huang, J.J., Ong, C.S., and Tzeng, G.H. (2006b). Optimal fuzzy multi-criteria expansion of competence sets using multi-objectives evolutionary algorithms. *Expert Systems with Applications*, 30(4): 739–745.

Huang, J.J., and Tzeng, G.H. (2007). Marketing segmentation using support vector clustering. *Expert Systems with Applications*, 32(2): 313–317.

Huang, J.J., and Tzeng, G.H. (2014). New thinking of multi-objective programming with changeable space–in search of excellence. *Technological and Economic Development of Economy*, 20(2): 254–273.

Huang, J.J., Tzeng, G.H., and Ong, C.S. (2005a). Motivation and resource-allocation for strategic alliances through the De Novo perspective. *Mathematical and Computer Modelling*, 41(6): 711–721.

Huang, J.J., Tzeng, G.H., and Ong C.S. (2005b). Multidimensional data in multidimensional scaling using the analytic network process. *Pattern Recognition Letters*, 26(6): 755–767.

Huang, J.J., Tzeng, G.H., and Ong, C.S. (2006a). A novel algorithm for dynamic factor analysis. *Applied Mathematics and Computation*, 175(2): 1288–1297.

Huang, J.J., Tzeng, G.H., and Ong, C.S. (2006b). A novel algorithm for uncertain portfolio selection. *Applied Mathematics and Computation*, 173(1): 350–359.

Huang, J.J., Tzeng, G.H., and Ong, C.S. (2006c). Choosing best alliance partners and allocating optimal alliance resources using the fuzzy multi-objective dummy programming model. *Journal of Operational Research Society*, 57(10): 1216–1223.

Huang, J.J., Tzeng, G.H., and Ong, C.S. (2006d). Interval multidimensional scaling for group decision using rough set concept. *Expert Systems with Applications*, 31(3): 525–530.

Huang, J.J., Tzeng, G.H., and Ong, C.S. (2006e). Two-stage genetic programming (2SGP) for the credit scoring model. *Applied Mathematics and Computation*, 174(2): 1039–1053.

Huang, K.W., Huang, J.H., and Tzeng, G.H. (2016). New hybrid MADM model for improving competence sets: Enhancing a company's core competitiveness. *Sustainability*, 8(2): 175–201.

Hung, C., and Chen, J.H. (2009). A selective ensemble based on expected probabilities for bankruptcy prediction. *Expert Systems with Applications*, 36(3): 5297–5303.

Hung, W., Huang, S.T., Lu, C.C., and Liu, N. (2015). Trading behavior and stock returns in Japan. *The Quarterly Review of Economics and Finance*, 58: 200–212.

Hung, Y.H., Chou, S.C.T., and Tzeng, G.H. (2011). Knowledge management adoption and assessment for SMEs by a novel MCDM approach. *Decision Support Systems*, 51(2): 270–291.

Hung, Y.H., Huang, T.L., Hsieh, J.C., Tsuei, H.J., Cheng, C.C., and Tzeng, G.H. (2012). Online reputation management for improving marketing by using a hybrid MCDM model. *Knowledge-Based Systems*, 35: 87–93.

Hwang, C.L., and Yoon, K. (1981). *Multiple Attribute Decision Making: Methods and Application*. Lecture Notes in Economics and Mathematical Systems, vol. 186. New York: Springer.

Iacovou, C.L., Benbasat, I., and Dexter, A.S. (1995). Electronic data interchange and small organizations: Adoption and impact of technology. *MIS Quarterly*, 19(4): 465–485.

IEK. Online: http://ieknet.iek.org.tw (accessed in 2014).

Ignizio, J.P. (1976). *Goal Programming and Extensions*. Michigan: Lexington Books.

Ijiri, Y. (1965). *Management Goals and Accounting for Control*, vol. 3., Amsterdam Chicago: North-Holland.

Ishii, H., Shiode, S., Hark Hwang, H., Tzeng, G.H., and Seki, H. (2012). Preface: "Soft computing for management systems." *Computers and Industrial Engineering*, 62(3): 687.

Jang, J.S. (1993). ANFIS: Adaptive-network-based fuzzy inference system. *IEEE Transactions on Systems, Man, and Cybernetics*, 23(3): 665–685.

Jeng, J.F., and Tzeng, G.H. (2012). Social influence on the use of clinical decision support systems: Revisiting the unified theory of acceptance and use of technology by the fuzzy DEMATEL technique. *Computers and Industrial Engineering*, 62(3): 819–828.

Jeyaraj, A., Rottman, J., and Lacity, M.J. (2006). A review of the predictors, linkages, and biases in IT innovation adoption research. *Journal of Information Technology*, 21(1): 1–23.

Kahneman, D., Knetsch, J.L., and Thaler, R.H. (1991). Anomalies: The endowment effect, loss aversion, and status quo bias. *Journal of Economic Perspectives*, 5(1): 193–206.

Kahneman, D., and Tversky, A. (1979). Prospect theory: An analysis of decision under risk. *Econometrica: Journal of the Econometric Society*, 47(2): 263–291.

Kahneman, D., and Tversky, A. (1984). Choices, values, and frames. *American Psychologist*, 39(4): 341–350.

Kahneman, D., and Tversky, A. (2000). *Choices, Values, and Frames*. Cambridge, UK: Cambridge University Press.

322 ■ *References*

Kahraman, C., Ertay, T., and Büyüközkan, G. (2006). A fuzzy optimization model for QFD planning process using analytic network approach. *European Journal of Operational Research*, 171(22): 390–411.

Kao, C., and Liu, S.T. (2014). Multi-period efficiency measurement in data envelopment analysis: The case of Taiwanese commercial banks. *Omega*, 47: 90–98.

Keeney, R.L., and Raiffa, H. (1972). A critique of formal analysis in public decision making. In: Drake, A.W., Keeney, R.L., Morse, P.M. (eds), *Analysis of Public Systems* (pp. 64–75). Cambridge, MA: MIT Press.

Keeney, R.L., and Raiffa, H. (1976). *Decision Analysis with Multiple Conflicting Objectives.* New York: John Wiley.

Keeney, R.L., and Raiffa, H. (1993). *Decisions with Multiple Objectives: Preferences and Value Trade-Offs*, Second Edition. Cambridge, UK and New York: Cambridge University Press.

Kim, H., Ida, K., and Gen, M. (1993). A De Novo approach for bicriteria 0–1 linear programming with interval coefficients under GUB structure. *Computers and Industrial Engineering*, 25(1): 17–20.

Ko, Y.C., Fujita, H., and Tzeng, G.H. (2013a). An extended fuzzy measure on competitiveness correlation based on WCY 2011. *Knowledge-Based Systems*, 37: 86–93.

Ko, Y.C., Fujita, H., and Tzeng, G.H. (2013b). A fuzzy integral fusion approach in analyzing competitiveness patterns from WCY2010. *Knowledge-Based Systems*, 49: 1–9.

Ko, Y.C., Fujita, H., and Tzeng, G.H. (2014). A simple utility function with the rules-verified weights for analyzing the top competitiveness of WCY 2012. *Knowledge-Based Systems*, 58: 58–65.

Kogut, B. (1988). Joint ventures: Theoretical and empirical perspectives. *Strategic Management Journal*, 9(4): 319–332.

Köksalan, M.M., Wallenius, J., and Zionts, S. (2011). *Multiple Criteria Decision Making: From Early History to the 21st Century*. Singapore: World Scientific.

Koopmans, T.C. (1951). Efficient allocation of resources. *Econometrica: Journal of the Econometric Society*: 455–465.

Kosmidou, K., and Zopounidis, C. (2008). Predicting US commercial bank failures via a multicriteria approach. *International Journal of Risk Assessment and Management*, 9(1): 26–43.

Kuan, M.J., Tzeng, G.H., and Hsiang, C.C. (2012). Exploring the quality assessment system for new product development process by combining DANP with MCDM model. *International Journal of Innovative Computing, Information and Control*, 8(8): 5745–5762.

Kuhn, H.W., and Tucker, A.W. (1951). Nonlinear programming. In: *Proceedings of the Berkeley Symposium of Mathematical Statistics and Probability* (pp. 481–491). Berkeley, CA: University of California Press.

Kuo, M.S. (2011). A novel interval-valued fuzzy MCDM method for improving airlines' service quality in Chinese cross-strait airlines. *Transportation Research Part E: Logistics and Transportation Review*, 47(6): 1177–1193.

Kuo, M.S., Tzeng, G.H., and Huang, W.C. (2007). Group decision making based on concepts of ideal and anti-ideal points in fuzzy environment. *Mathematical and Computer Modeling*, 45(3–4): 324–339.

Kuo, R.J., Wang, Y.C., and Tien, F.C. (2010). Integration of artificial neural network and MADA methods for green supplier selection. *Journal of Cleaner Production*, 18(12): 1161–1170.

La Porta, R., Lakonishok, J., Shleifer, A., and Vishny, R. (1997). Good news for value stocks: Further evidence on market efficiency. *The Journal of Finance*, 52(2): 859–874.

Lai, Y.J., Liu, T.Y., and Hwang, C.L. (1994). TOPSIS for MODM. *European Journal of Operational Research*, 76(3): 486–500.

Lakonishok, J., Shleifer, A., and Vishny, R.W. (1994). Contrarian investment, extrapolation, and risk. *The Journal of Finance*, 49(5): 1541–1578.

Lam, M. (2004). Neural network techniques for financial performance prediction: Integrating fundamental and technical analysis. *Decision Support Systems*, 37(4): 567–581.

Larbani, M., Huang, C.Y., and Tzeng, G.H. (2011). A novel method for fuzzy measure identification. *International Journal of Fuzzy Systems*, 13(1): 24–34.

LaValle, S., Hopkins, M.S., Lesser, E., Shockley, R. (2010). Analytics: The new path to value. IBM Institute for Business Value.

LaValle, S., Lesser, E., Shockley, R., Hopkins, M.S., and Kruschwitz, N. (2011). Big data, analytics and the path from insights to value. *MIT Sloan Management Review*, 52: 21–31.

Lee, A.H.I., Kang, H.Y., Hsu, C.F., and Hung, H.C. (2009). A green supplier selection model for high-tech industry. *Expert Systems with Applications*, 36(4): 7917–7927.

Lee, C., Liu, L.C., and Tzeng, G.H. (2001). Hierarchical fuzzy integral evaluation approach for vocational education performance: Case of junior colleges in Taiwan. *International Journal of Fuzzy Systems*, 3(3): 476–485.

Lee, C.C., Tzeng, G.H., and Chiang, C. (2011). Determining key service quality measurement indicators in a travel website using a fuzzy analytic hierarchy process. *International Journal of Electronic Business Management*, 9(4): 322–333.

Lee, C.F., Tzeng, G.H., and Wang, S.Y. (2005a). A fuzzy set approach to generalize CRR model: An empirical analysis of S&P 500 index option. *Review of Quantitative Finance and Accounting*, 25(3): 255–275.

Lee, C.F., Tzeng, G.H., and Wang, S.Y. (2005b). A new application of fuzzy set theory to the Black–Scholes option pricing model. *Expert Systems with Applications*, 29(2): 330–342.

Lee, E.S., and Li, R.J. (1993). Fuzzy multiple objective programming and compromise programming with Pareto optimum. *Fuzzy Sets and Systems*, 53(3): 275–288.

Lee, H.S., Tzeng, G.H., Yeih, W., Wang, Y.J., and Yang, S.C. (2013). Revised DEMATEL: Resolving the infeasibility of DEMATEL. *Applied Mathematical Modelling*, 37(10): 6746–6757.

Lee, S.M. (1972). *Goal Programming for Decision Analysis*. Philadelphia, PA: Auerbach.

Lee, W.S., Tzeng, G.H., Guan, J.L., Chien, K.T., and Huang, J.M. (2009). Combined MCDM techniques for exploring stock selection based on Gordon model. *Expert Systems with Applications*, 36(3): 6421–6430.

Li, C.W., and Tzeng, G.H. (2009a). Identification of a threshold value for the DEMATEL method using the maximum mean de-entropy algorithm to find critical services provided by a semiconductor intellectual property mall. *Expert Systems with Applications*, 36(6): 9891–9898.

Li, C.W., and Tzeng, G.H. (2009b). Identification of interrelationship of key customers' needs based on structural model for services/capabilities provided by a semiconductor-intellectual-property mall. *Applied Mathematics and Computation*, 215(6): 2001–2010.

324 ■ References

Li, R.J., and Lee, E.S. (1990). Multicriteria De Novo programming with fuzzy parameters. *Computers and Mathematics with Applications*, 19(1): 13–20.

Li, R.J., and Lee, E.S. (1993). De Novo programming with fuzzy coefficients and multiple fuzzy goals. *Journal of Mathematical Analysis and Applications*, 172(1): 212–220.

Liao, Z., and Rittscher, J. (2007). A multi-objective supplier selection model under stochastic demand conditions. *International Journal of Production Economics*, 105(1): 150–159.

Lin, C.L., Chen, C.W., and Tzeng, G.H. (2010a). Planning the development strategy for the mobile communication package based on consumers' choice preferences. *Expert Systems with Applications*, 37(7): 4749–4760.

Lin, C.L., Hsieh, M.S., and Tzeng, G.H. (2010b). Evaluating vehicle telematics system by using a novel MCDM techniques with dependence and feedback. *Expert Systems with Applications*, 37(10): 6723–6736.

Lin, C.L., Shih, Y.H., Tzeng, G.H., and Yu, H.C. (2016). A service selection model for digital music service platforms using a hybrid MCDM approach. *Applied Soft Computing* (in press).

Lin, C.L., and Tzeng, G.H. (2010). A value-created system of science (technology) park by using DEMATEL. *Expert Systems with Applications*, 36(6): 9683–9697.

Lin, C.M., Huang, J.J., Gen, M., and Tzeng, G.H. (2006). Recurrent neural network for dynamic portfolio selection. *Applied Mathematics and Computation*, 175(2): 1139–1146.

Lin, C.S., Tzeng, G.H., and Chin, Y.C. (2011). Combined rough set theory and flow network graph to predict customer churn in credit card accounts. *Expert Systems with Applications*, 38(1): 8–15.

Lin, C.S., Tzeng, G.H., Chin, Y.C., and Chang, C.C. (2010a). Recommendation sources on the intention to use e-books in academic digital libraries. *The Electronic Library*, 28(6): 844–857.

Lin, W.Y., Hu, Y.H., Tsai, C.F. (2012). Machine learning in financial crisis prediction: A survey. *IEEE Transactions on Systems, Man, and Cybernetics, Part C: Applications and Reviews*, 42(4): 421–436.

Lin, Y.T., Lin, C.L., Yu, H.C., and Tzeng, G.H. (2010b). A novel hybrid MCDM approach for outsourcing vendor selection: A case study for a semiconductor company in Taiwan. *Expert Systems with Applications*, 37(7): 4796–4804.

Lin, Y.T., Lin, C.L., Yu, H.C., and Tzeng, G.H. (2011). Utilization of interpretive structural modeling method in the analysis of interrelationship of vendor performance factors. *International Journal of Business Performance Management*, 12(3): 260–275.

Liou, J.J.H. (2013). New concepts and trends of MCDM for tomorrow: In honor of Professor Gwo-Hshiung Tzeng on the occasion of his 70th birthday. *Technological and Economic Development of Economy*, 19(2): 367–375.

Liou, J.J.H., Chuang, Y.C., and Tzeng, G.H. (2014). A fuzzy integral-based model for supplier evaluation and improvement. *Information Sciences*, 266: 199–217.

Liou, J.J., and Chuang, Y.T. (2010). Developing a hybrid multi-criteria model for selection of outsourcing providers. *Expert Systems with Applications*, 37(5): 3755–3761.

Liou, J.J.H., Tamosaitiene, J.T., Zavadskas, E., and Tzeng, G.H. (2016). A new hybrid COPRAS-G MADM model for improving and selecting suppliers in green supply chain management. *International Journal of Production Research*, 54(1): 114–134.

Liou, J.J.H., Tsai, C.Y., Lin, R.H., and Tzeng, G.H. (2011). A modified VIKOR multiple-criteria decision method for improving domestic airlines service quality. *Journal of Air Transport Management*, 17(2): 57–61.

Liou, J.J.H., and Tzeng, G.H. (2007). A non-additive model for evaluating airline service quality. *Journal of Air Transport Management*, 13(3): 131–138.

Liou, J.J.H., and Tzeng, G.H. (2010). A dominance-based rough set approach to customer behavior in the airline market. *Information Sciences*, 180(11): 2230–2238.

Liou, J.J.H., and Tzeng, G.H. (2012). Comments on "Multiple criteria decision making (MCDM) methods in economics: An overview." *Technological and Economic Development of Economy*, 18(4): 672–695.

Liou, J.J.H., Tzeng, G.H., and Chang, H.C. (2007). Airline safety measurement using a hybrid model. *Journal of Air Transport Management*, 13(4): 243–249.

Liou, J.J.H., Tzeng, G.H., Hsu, C.C., and Yeh, W.C. (2012). Reply to comment on using a modified grey relation method for improving airline service quality. *Tourism Management*, 33(3): 719–720.

Liou, J.J.H., Tzeng, G.H., Tsai, C.Y., and Hsu, C.C. (2011). A hybrid ANP model in fuzzy environments for strategic alliance partner selection in the airline industry. *Applied Soft Computing*, 11(4): 3515–3524.

Liou, J.J.H., Yen, L., and Tzeng, G.H. (2008). Building an effective safety management system for airlines. *Journal of Air Transport Management*, 14(1): 20–26.

Liou, J.J.H., Yen, L., and Tzeng, G.H. (2010). Using decision rules to achieve mass customization of airline services. *European Journal of Operational Research*, 205(3): 680–686.

Liu, B.C., Tzeng, G.H., and Hsieh C.T. (1992). Energy planning and environment-quality management: A decision support system approach. *Energy Economics*, 14(4): 302–307.

Liu, C.H., Tzeng, G.H., and Lee, M.H. (2012). Improving tourism policy implementation: The use of hybrid MCDM models. *Tourism Management*, 33(2): 239–488.

Liu, C.H., Tzeng, G.H., and Lee, M.H. (2013a). Strategies for improving cruise product sales in the travel agency: Using hybrid MCDM models. *The Service Industry Journal*, 33(5): 542–563.

Liu, C.H., Tzeng, G.H., Lee, M.H., and Lee, P.Y. (2013b). Improving metro–airport connection service for tourism development: Using hybrid MCDM models. *Tourism Management Perspectives*, 6: 95–107.

Liu, H.T. (2011). Product design and selection using fuzzy QFD and fuzzy MCDM approaches. *Applied Mathematical Modelling*, 35(1): 482–496.

Liu, L.C., Lee, C., and Tzeng, G.H. (2001). Hierarchical fuzzy integral evaluation approach for vocational education performance: Case of junior colleges in Taiwan. *International Journal of Fuzzy Sets*, 3(3): 476–485.

Liu, L.C., Lee, C., and Tzeng, G.H. (2003). Using DEA of REM and EAM for efficiency assessment of technology institutes upgraded from junior colleges: The case in Taiwan. In Tanino, T., Tanaka, T., and Inuiguchi, M. (eds), *Multi-objective Programming and Goal-Programming: Theory and Applications* (pp. 361–366). New York: Springer-Verlag.

Liu, L.C., Lee, C., and Tzeng, G.H. (2004a). DEA approach of the current-period and cross-period efficiency for evaluating the vocational education. *International Journal of Information Technology and Decision Making*, 3(2): 353–374.

Liu, Y.H., Tzeng, G.H., and Park, D.H. (2004b). Set covering problem and the reliability of the covers. *International Journal of Reliability and Applications*, 5(4): 145–154.

Loban, S.R. (1997). A framework for computer-assisted travel counseling. *Annals of Tourism Research*, 24(4): 813–834.

326 ■ References

Lu, I.Y., Kuo, T., Lin, T.S., Tzeng, G.H., and Huang, S.L. (2016). Multicriteria decision analysis to develop effective sustainable development strategies for enhancing competitive advantages: Case of the TFT-LCD industry in Taiwan. *Sustainability*, 8(7): 646–677.

Lu, M.T., Lin, S.W., and Tzeng, G.H. (2013). Improving RFID adoption in Taiwan's healthcare industry based on a DEMATEL technique with a hybrid MCDM model. *Decision Support Systems*, 56: 259–269.

Lu, M.T., Tzeng, G.H., Cheng, H., and Hsu, C.C. (2015a). Exploring mobile banking services for user behavior in intention adoption: Using new hybrid MADM model. *Service Business*, 9(3): 541–565.

Lu, M.T., Tzeng, G.H., and Hu, S.K. (2015b). Evaluating the implementation of business-to-business m-commerce by SMEs based on a new hybrid MADM model. *Management Decision*, 53(2): 290–317.

Lu, M.T., Tzeng, G.H., and Tang, L.L. (2013). Environmental strategic orientations for improving green innovation performance in the electronics industry: Using fuzzy hybrid MCDM model. *International Journal of Fuzzy Systems*, 15(3): 297–316.

Lu, S.T., Hsieh, T.Y., and Tzeng, G.H. (2004). Fuzzy MCDM approach for planning and design tenders selection in public office buildings. *International Journal of Project Management*, 22(7): 574–584.

Luo, S.T., Cheng, B.W., and Hsieh, C.H. (2009). Prediction model building with clustering-launched classification and support vector machines in credit scoring. *Expert Systems with Applications*, 36(4): 7562–7566.

Malakooti, B., and Zhou, Y.Q. (1994). Feed-forward artificial neural networks for solving discrete multiple criteria decision making problems. *Management Science*, 40(11): 1542–1561.

Mardani, A., Jusoh, A., and Zavadskas, E.K. (2015). Fuzzy multiple criteria decision-making techniques and applications: Two decades review from 1994 to 2014. *Expert Systems with Applications*, 42(8): 4126–4148.

Mareschal, B., and Brans, J.P. (1991). BANKADVISER: An industrial evaluation system. *European Journal of Operational Research*, 54(3): 318–324.

Mareschal, B., and Mertens, D. (1992). BANKS: A multicriteria, PROMETHEE-based, decision support system for the evaluation of the international banking sector. *Journal of Decision Systems*, 1(2–3): 175–189.

Markowitz, H. (1952). Portfolio selection. *Journal of Finance*, 7(1): 77–91.

Markowitz, H. (1959). *Portfolio Selection: Efficient Diversification of Investments*. New York: John Wiley.

Markowitz, H. (1987). *Mean-Variance Analysis in Portfolio Choice and Capital Market*. New York: Basil Blackwell.

Menkhoff, L. (2010). The use of technical analysis by fund managers: International evidence. *Journal of Banking and Finance*, 34(11): 2573–2586.

Menkhoff, L., and Taylor, M.P. (2007). The obstinate passion of foreign exchange professionals: technical analysis. *Journal of Economic Literature*, 45(4): 936–972.

Michalewicz, Z., and Schoenauer, M. (1996). Evolutionary algorithms for constrained parameter optimization problems. *Evolutionary Computation*, 4(1): 1–32.

Mirchandani, A.A., and Motwani, J. (2001). Understanding small business electronic commerce adoption: An empirical analysis. *Journal of Computer Information Systems*, 41(3): 70–73.

Moghadam, M.R.S., Afsar, A., and Sohrabi, B. (2008). Inventory lot-sizing with supplier selection using hybrid intelligent algorithm. *Applied Soft Computing*, 8(4): 1523–1529.

Mohanram, P.S. (2005). Separating winners from losers among low book-to-market stocks using financial statement analysis. *Review of Accounting Studies*, 10(2–3) (2005): 133–170.

Mon, D.L., Tzeng, G.H., and Lu, H.C. (1995). Grey decision making in weapon system evaluation. *Journal of Chung Cheng Institute of Technology*, 24(1): 73–84.

Mulpuru, S., Sehgal, V., Evans, P.F., and Roberge, D. (2011). US online retail forecast, 2010 to 2015. http://www.forrester.com/rb/Research/us_online_retail_forecast%2C_2010_to_2015/q/id/58596/t/2.

Mulpuru, S., Sehgal, V., Evans, P.F., Hoar, A., and Roberge, D. (2012). US online retail forecast, 2011 to 2016. Reported by InternetRetailing.net, http://www.forrester.com/US+Online+Retail+Forecast+2011+To+2016/fulltext/-/E-RES60672?docid=60672.

Narayanasamy, K., Rasiah, D., and Tan, T.M. (2011). The adoption and concerns of e-finance in Malaysia. *Electronic Commerce Research*, 11(4): 383–400.

Niemira, M.P., and Saaty, T.L. (2004). An analytic network process model for financial-crisis forecasting. *International Journal of Forecasting*, 20(4): 573–587.

Ong, C.S., Huang, J.J., and Tzeng, G.H. (2005a). Building credit scoring models using genetic programming. *Expert Systems with Applications*, 29(1): 41–47.

Ong, C.S., Huang, J.J., and Tzeng, G.H. (2005b). Motivation and resource allocation for strategic alliance through De Novo perspective. *Mathematical and Computer Modeling*, 41(6–7): 711–721.

Opricovic S. (1986). Višekriterijumska optimizacija. Faculty of Civil Engineering, University of Zagreb, Serbia.

Opricovic, S. (1998). Multicriteria optimization of civil engineering systems. Faculty of Civil Engineering, University of Belgrade, Serbia.

Opricovic, S., and Tzeng, G.H. (2002). Multicriteria planning of post-earthquake sustainable reconstruction. *Computer-Aided Civil and Infrastructure Engineering*, 17(3): 211–220.

Opricovic, S., and Tzeng, G.H. (2003a). Comparing DEA and MCDM method. In Tanino, T., Tanaka, T., and Inuiguchi, M. (eds), *Multi-objective Programming and Goal-Programming: Theory and Applications* (pp. 227–232). Berlin Heidelberg: Springer-Verlag.

Opricovic, S., and Tzeng, G.H. (2003b). Defuzzification within a fuzzy multicriteria decision model. *International Journal of Uncertainty, Fuzziness and Knowledge-Based Systems*, 11(5): 635–652.

Opricovic, S., and Tzeng, G.H. (2003c). Fuzzy multicriteria model for post-earthquake land-use planning. *Natural Hazards Review*, 4(2): 59–64.

Opricovic, S., and Tzeng, G.H. (2003d). Multicriteria expansion of a competence set using genetic algorithm. In Tanino, T., Tanaka, T., and Inuiguchi, M. (eds), *Multi-objective Programming and Goal-Programming: Theory and Applications* (pp. 221–226). Berlin Heidelberg: Springer-Verlag.

Opricovic, S., and Tzeng, G.H. (2004). Compromise solution by MCDM methods: A comparative analysis of VIKOR and TOPSIS. *European Journal of Operational Research*, 156(2): 445–455.

Opricovic, S., and Tzeng, G.H. (2007). Extended VIKOR method in comparison with outranking methods. *European Journal of Operational Research*, 178(2): 514–529.

Ou Yang, Y.P., Shieh, H.M., Leu, J.D., and Tzeng, G.H. (2008). A novel hybrid MCDM model combined with DEMATEL and ANP with applications. *International Journal of Operations Research*, 5(3): 160–168.

328 ■ *References*

Ou Yang, Y.P., Shieh, H.M., and Tzeng, G.H. (2009). A VIKOR-based multiple criteria decision method for improving information security risk. *International Journal of Information Technology and Decision Making*, 8(2): 267–287.

Ou Yang, Y.P., Shieh, H.M., and Tzeng, G.H. (2013). A VIKOR technique based on DEMATEL and ANP for information security risk control assessment. *Information Sciences*, 232: 482–500.

Ou Yang, Y.P., Shieh, H.M., Tzeng, G.H., Yen, L., and Chan, C.C. (2011). Combined rough sets with flow graph and formal concept analysis for business aviation decision-making. *Journal of Intelligent Information Systems*, 36(3): 347–366.

Ozaki, T., Lo, M.C., Kinoshita, E., and Tzeng, G.H. (2011). Decision-making for the best selection of suppliers by using minor ANP. *Journal of Intelligent Manufacturing*, 23(6): 2171–2178.

Ozaki, T., Miwa, K., Itoh, A., Sugiura, S., Kinoshita, E., and Tzeng, G.H. (2013). Dissolution of dilemma by newly defining criteria matrix in ANP. *Journal of the Operations Research Society of Japan*, 56(2): 93–110.

Ozaki, T., Sugiura, S., Kinoshita, E., and Tzeng, G.H. (2010). Dissolution of the dilemma problems by defining the criteria matrix in ANP. *Nagoya Gakuin Daigaku*, 23: 31–50.

Pan, M.J., and Jang, W.Y. (2008). Determinants of the adoption of enterprise resource planning within the technology–organization–environment framework: Taiwan's communications industry. *Journal of Computer information Systems*, 48(3): 94–102.

Papazafeiropoulou, A. (2004). Inter-country analysis of electronic commerce adoption in South Eastern Europe: Policy recommendations for the region. *Journal of Global Information Technology Management*, 7(2): 54–69.

Pareto, V. (1906). *Manuale di economia politica*, vol. 13. Pordenone, Italy: Societa Editrice.

Park, C.H., and Irwin, S.H. (2007). What do we know about the profitability of technical analysis? *Journal of Economic Surveys*, 21(4): 786–826.

Pawlak, Z. (1982). Rough sets. *International Journal of Computer and Information Science*, 11(5): 341–356.

Pawlak, Z. (1984). Rough classification. *International Journal of Man–Machine Studies*, 20(5): 469–483.

Pawlak, Z. (2002a). Rough sets, decision algorithms and Bayes' theorem. *European Journal of Operational Research*, 136(1): 181–189.

Pawlak, Z. (2002b). Rough set theory and its applications. *Journal of Telecommunications and Information Technology*, 3(1): 7–10.

Pawlak, Z., and Slowinski, R. (1994). Decision analysis using rough sets. *International Transactions in Operational Research*, 1(1): 107–114.

Pedrycz, W. (2013). *Granular Computing: Analysis and Design of Intelligent Systems*. Boca Raton, FL: CRC Press.

Pedrycz, W., Skowron, A., and Kreinovich, V. (eds) (2008). *Handbook of Granular Computing*. New York: John Wiley.

Peng, K.H., and Tzeng, G.H. (2012). Strategies for promoting tourism competitiveness using a hybrid MCDM model. In: Watada J., Watanabe T., Phillips-Wren G., Howlett R., Jain L. (eds) *Intelligent Decision Technologies* (pp. 107– 115), Smart Innovation, Systems and Technologies, vol 16. Berlin, Germany: Springer.

Peng, K.H., and Tzeng, G.H. (2013). A hybrid dynamic MADM model for problem-improvement in economics and business. *Technological and Economic Development of Economy*, 19(4): 638–660.

Peng, K.H., and Tzeng, G.H. (2017). Exploring the heritage tourism performance improvement for making the sustainable development strategies using hybrid modified MADM model. *Current Issues in Tourism*, DOI: 10.1080/13683500.2017.1306030

Penman, S.H. (2007). *Financial Statement Analysis and Security Valuation*. New York: McGraw-Hill.

Penman, S.H., and Zhang, X.J. (2002). Accounting conservatism, the quality of earnings, and stock returns. *The Accounting Review*, 77(2): 237–264.

Pillania, R.K. (2008). Creation and categorization of knowledge in automotive components SMEs in India. *Management Decision*, 46(10): 1452–1464.

Piotroski, J.D. (2000). Value investing: The use of historical financial statement information to separate winners from losers. *Journal of Accounting Research*, 38 (Supplement): 1–41.

Piotroski, J.D., and So, E.C. (2012). Identifying expectation errors in value/glamour strategies: A fundamental analysis approach. *Review of Financial Studies*, 25(9): 2841–2875.

Pooters, I. (2010). Full user data acquisition from Symbian smart phones. *Digital Investigation*, 6(3–4): 125–135.

Porter, M.E. (2001). Strategy and the Internet. *Harvard Business Review*, 79(3): 62–78.

Pourahmad, A., Hosseini, A., Banaitis, Z.A., Nasiri, H., Banaitienė, N., and Tzeng, G.H. (2015). Combination of fuzzy-AHP and DEMATEL-ANP with GIS in a new hybrid MCDM model used for the selection of the best space for leisure in a blighted urban site. *Technological and Economic Development of Economy*, 21(5): 773–796.

Precup, R.E., and Hellendoorn, H. (2011). A survey on industrial applications of fuzzy control. *Computers in Industry*, 62(3): 213–226.

Ravi, V., Kurniawan, H., Thai, P.N.K., and Kumar, P.R. (2008). Soft computing system for bank performance prediction. *Applied Soft Computing*, 8(1): 305–315.

Ravi, V., and Pramodh, C. (2008). Threshold accepting trained principal component neural network and feature subset selection: Application to bankruptcy prediction in banks. *Applied Soft Computing*, 8(4): 1539–1548.

Rene, S.G. (2005). Information security management best practice based on ISO/IEC 17799. *Information Management*, 39(4): 60–66.

Rogers, E.M. (1995). *The Diffusion of Innovations*. New York: Free Press.

Romero, C. (2004). A general structure of achievement function for a goal programming model. *European Journal of Operational Research*, 153(3): 675–686.

Roy, B. (1968). Classement et choix en présence de points de vue multiples. *Revue française d'automatique: D'informatique et de recherche opérationnelle*, 8(1): 57–75.

Roy, B. (1971). Problems and methods with multiple objective functions. *Mathematical Programming*, 1(1): 239–266.

Roy, B. (1976). From optimization to multicriteria decision aid: Three main operational attitudes. In: *Multiple Criteria Decision Making* (pp. 1–34). Berlin Heidelberg: Springer.

Roy, B. (1981). Multicriteria analysis: Survey and new directions. *European Journal of Operational Research*, 8(3): 207–218.

Rust, R.T., Lemon, K.N., and Zeithaml, V.A. (2004). Return on marketing: Using-customer equity to focus marketing strategy. *Journal of Marketing*, 68(1): 109–127.

Rust, R.T., Zeithaml, V.A., and Lemon, K.N. (2000). *Driving Customer Equity: How Customer Lifetime Value Is Reshaping Corporate Strategy*. New York: Free Press.

Saaty, T.L. (1972). An eigenvalue allocation model for prioritization and planning. Energy Management and Policy Center, University of Pennsylvania.

330 ■ References

Saaty, T.L. (1977). A scaling method for priorities in hierarchical structures. *Journal of Mathematical Psychology*, 15(3): 234–281.

Saaty, T. L. (1980). *The Analytic Hierarchy Process*. New York: McGraw-Hill.

Saaty, T. L. (1986a). *Decision Making with Dependence and Feedback: The Analytic Network Process*. Pittsburgh, PA: RWS Publications.

Saaty, T. L. (1986b). Absolute and relative measurement with the AHP. The most livable cities in the United States. *Socio-Economic Planning Sciences*, 20(6): 327–331.

Saaty, T.L. (1988). What is the analytic hierarchy process? In: *Mathematical Models for Decision Support* (pp. 109–121). Berlin Heidelberg: Springer.

Saaty, T.L. (1992). A natural way to make momentous decision. *Journal of Science and Industrial Research*, 61: 561–571.

Saaty, T.L. (1996). *Decision Making with Dependence and Feedback: The Analytic Network Process*. Pittsburgh, PA: RWS.

Saaty, T. L. (1999). Fundamentals of the analytic network process. *International Symposium on the Analytic Hierarchy Process*, Kobe, Japan, August 12–14.

Saaty, T. L. (2003). Decision-making with the AHP: Why is the principal eigenvector necessary? *European Journal of Operational Research*, 145(1): 85–91.

Saaty, T.L. (2004). Decision making: The analytic hierarchy and network processes (AHP/ANP). *Journal of Systems Science and Systems Engineering*, 13(1): 1–35.

Sabherwal, R., Jeyaraj, A., and Chowa, C. (2006). Information system success: Individual and organizational determinants. *Management Science*, 52(12): 1849–1864.

Sakawa, M., and Yano, H. (1985). Interactive fuzzy decision-making for multi-objective nonlinear programming using reference membership intervals. *International Journal of Man–Machine Studies*, 23(4): 407–421.

Sakawa, M., and Yumine, T. (1983). Interactive fuzzy decision making for multiobjective linear fractional-programming problems. *Large Scale Systems in Information and Decision Technologies*, 5(2): 105–113.

Sakawa, M., Yumine, T., and Nango, Y. (1984). Interactive fuzzy decision making for multiobjective nonlinear programming problems. *Electronics and Communications in Japan (Part I: Communications)*, 67(4): 31–38.

Sevkli, M., Lenny Koh, S.C., Zaim, S., Demirbag, M., and Tatoglu, E. (2007). An application of data envelopment analytic hierarchy process for supplier selection: A case study of BEKO in Turkey. *International Journal of Production Research*, 45(9): 1973–2003.

Sharma, R., Reynolds, P., Scheepers, R., and Seddon, P.B. (2010). Business analytics and competitive advantage: A review and a research agenda. In: Respicio, A., Adam, F., Phillips, G., Teixeira, C., and Telhada, J. (eds), *Bridging the Sociotechnical Gap in Decision Support Systems* (pp. 187–198). Washington, DC: IOS Press.

Shaw, K., Shankar, R., Yadav, S.S., and Thakur, L.S. (2012). Supplier selection using fuzzy AHP and fuzzy multi-objective linear programming for developing low carbon supply chain. *Expert Systems with Applications*, 39(9): 8182–8192.

Shee, D.Y., Tang, T., and Tzeng, G.H. (2000). Modeling the supply–demand interaction in electronic commerce: A bi-level programming approach. *Journal of Electronic Commerce Research*, 1(2): 79–93.

Shee, D.Y., and Tzeng, G.H. (2002). The key dimensions of criteria for the evaluation of ISPs: An exploratory study. *The Journal of Computer Information System*, 42(4): 112–121.

References ▪ 331

Shee, D.Y., Tzeng, G.H., and Tang, T. (2003). AHP, fuzzy measure and fuzzy integral approaches for the appraisal of information service providers in Taiwan. *Journal of Global Information Technology Management*, 6(1): 8–30.

Shen, K.Y. (2011). Implementing value investing strategy by artificial neural network. *International Journal of Business and Information Technology*, 1(1): 12–22.

Shen, K.Y. (2013). Implementing value investing strategy through an integrated fuzzy-ANN model. *Journal of Theoretical and Applied Information Technology*, 51(1): 150–157.

Shen, K.Y., Hu, S.K., and Tzeng, G.H. (2017). Financial modeling and improvement planning for the life insurance industry by using a rough knowledge based hybrid MCDM model. *Information Sciences*, 375: 296–313.

Shen, K.Y., and Tzeng, G.H. (2014a). DRSA-based neuro-fuzzy inference systems for the financial performance prediction of commercial banks. *International Journal Fuzzy Systems*, 16(2): 173–183.

Shen, K.Y., and Tzeng, G.H. (2014b). Decision rules-based probabilistic MCDM evaluation method: An empirical case from semiconductor industry. In: Kryszkiewicz, M., Cornelis, C., Ciucci D., Medina-Moreno, J., Motoda, H., and Raś Z.W. (eds), *International Conference on Rough Sets and Intelligent Systems Paradigms* (pp. 179–190). Springer.

Shen, K.Y., and Tzeng, G.H. (2015a). A decision rule-based soft computing model for supporting financial performance improvement of the banking industry. *Soft Computing*, 19(4): 859–874.

Shen, K.Y., and Tzeng, G.H. (2015b). Knowledge supported refinements for rough granular computing: A case of life insurance industry. In: Yao, Y., Hu, Q., Yu, H., and Grzymala-Busse, J. (eds), *Rough Sets, Fuzzy Sets, Data Mining, and Granular Computing* (pp. 233–244). Springer.

Shen, K.Y., and Tzeng, G.H. (2015c). Fuzzy inference-enhanced VC-DRSA model for technical analysis: Investment decision aid. *International Journal of Fuzzy Systems*, 17(3): 375–389.

Shen, K.Y., and Tzeng, G.H. (2015d). Combined soft computing model for value stock selection based on fundamental analysis. *Applied Soft Computing*, 37: 142–155.

Shen, K.Y., and Tzeng, G.H. (2015e). Combining DRSA decision-rules with FCA-based DANP evaluation for financial performance improvements. *Technological and Economic Development of Economy* (in press). DOI:10.3846/20294913.2015.1071295.

Shen, K.Y., and Tzeng, G.H. (2015f). A new approach and insightful financial diagnoses for the IT industry based on a hybrid MADM model. *Knowledge-Based Systems*, 85: 112–130.

Shen, K.Y., and Tzeng, G.H. (2015g). An emerging trend in multiple rule-based decision-making by rough machine learning. Reported in: *2015 International Conference on Fuzzy Theory and Its Applications (iFUZZY 2015)*, Yilan, Taiwan.

Shen, K.Y., and Tzeng, G.H. (2016). Contextual improvement planning by fuzzy-rough machine learning: A novel bipolar approach for business analytics. *International Journal of Fuzzy Systems* (in press). DOI:10.1007/s40815-016-0215-8.

Shen, K.Y., Yan, M.R., and Tzeng, G.H. (2014). Combining VIKOR–DANP model for glamor stock selection and stock performance improvement. *Knowledge-Based Systems*, 58: 86–97.

Shen, Y.C., Grace, T.R., Lin, G.T.R., and Tzeng, G.H. (2012). A novel multi-criteria decision-making combining decision making trial and evaluation laboratory technique for technology evaluation. *Foresight*, 14(2): 139–153.

332 ■ References

Shi, Y. (1995). Studies on optimum-path ratios in De Novo programming problems. *Computers and Mathematics with Applications*, 29(1): 43–50.

Shiller, R.J. (2003). From efficient markets theory to behavioral finance. *The Journal of Economic Perspectives*, 17(1): 83–104.

Shuai, J.J., and Li, H.L. (2005). Using rough set and worst practice DEA in business failure prediction. In: Ślęzak, D., Yao, J., Peters, J.F., Ziarko, W., and Hu, X. (eds), *Rough Sets, Fuzzy Sets, Data Mining, and Granular Computing* (pp. 503–510). Heidelberg, Germany: Springer.

Shyng, J.Y., Shieh, H.M., and Tzeng, G.H. (2010a). An integration method combining rough set theory with formal concept analysis for personal investment portfolios. *Knowledge-Based Systems*, 23(6): 586–597.

Shyng, J.Y., Shieh, H.M., Tzeng, G.H., and Hsieh, S.H. (2010b). Using FSBT technique with rough set theory for personal investment portfolio analysis. *European Journal of Operational Research*, 201(2): 601–607.

Shyng, J.Y., Shieh, H.M., and Tzeng, G.H. (2011). Compactness rate as a rule selection index based on rough set theory to improve data analysis for personal investment portfolios. *Applied Soft Computing*, 11(4): 3671–3679.

Shyng, J.Y., Tzeng, G.H., and Wang, F K. (2007). Rough set theory in analyzing the attributes of combination values for insurance market. *Expert Systems with Applications*, 32(1): 56–64.

Siddiqi, K.O. (2011). Interrelations between service quality attributes, customer satisfaction and customer loyalty in the retail banking sector in Bangladesh. *International Journal of Business and Management*, 6: 12–36.

Sila, I. (2013). Factors affecting the adoption of B2B e-commerce technologies. *Electronic Commerce Research*, 13(2): 199–236.

Simon, H.A. (1955). A behavioral model of rational choice. *The Quarterly Journal of Economics*, 69(1): 99–118.

Simon, H.A. (1956). Rational choice and the structure of the environment. *Psychological Review*, 63(2): 129–138.

Simon, H.A. (1959). Theories of decision-making in economics and behavioral science. *The American Economic Review*, 49(3): 253–283.

Simon, H.A. (1972). Theories of bounded rationality. *Decision and Organization*, 1(1): 161–176.

Simon, H.A. (1977). *The New Science of Management Decision*. New York: Prentice Hall PTR.

Simon, H.A. (1982). *Models of Bounded Rationality: Empirically Grounded Economic Reason*, vol. 3. Cambridge, MA: MIT Press.

Słowiński, R. (2008). Dominance-based rough set approach to reasoning about ordinal data: A tutorial. In: *International Conference on Rough Sets and Knowledge Technology* (pp. 21–22). Berlin Heidelberg: Springer.

Słowiński, R., Greco, S., and Matarazzo, B. (2005). Rough set based decision support. In: *Search Methodologies* (pp. 475–527). New York: Springer.

Soliman, K.S., Chen, L.D., and Flolick, M.N. (2003). ASPs: Do they work? *Information Systems Management*, 20(4): 50–57.

Soliman, K.S., and Janz, B.D. (2004). An exploratory study to identify the critical factors affecting the decision to establish Internet-based interorganizational information systems. *Information and Management*, 41(6): 697–706.

Sousa, J.M., and Kaymak, U. (2002). *Fuzzy Decision Making in Modeling and Control*, vol. 27. Singapore: World Scientific.

References ■ 333

Spronk, J., Steuer, E., and Zopounidis, C. (2005). Multicriteria decision aid/analysis in finance. In: Figueira, J., Greco, S., and Ehrgott, M. (eds), *Multiple Criteria Decision Analysis: State of the Art Surveys* (pp. 799–857). New York: Springer Science & Business.

Steuer, R.E., and Na, P. (2003). Multiple criteria decision making combined with finance: A categorized bibliographic study. *European Journal of Operational Research*, 150(3): 496–515.

Streimikiene, D., Balezentis, T., Krisciukaitienė, I., and Balezentis, A. (2012). Prioritizing sustainable electricity production technologies: MCDM approach. *Renewable and Sustainable Energy Reviews*, 16(5): 3302–3311.

Su, C.H., Tzeng, G.H., and Hu, S.K. (2015). Cloud e-learning service strategies for improving e-learning innovation performance in a fuzzy environment by using a new hybrid fuzzy multiple attribute decision-making model. *Interactive Learning Environments* (in press). DOI:10.1080/10494820.2015.1057742.

Sugeno, M. (1974). *Theory of Fuzzy Integrals and Its Applications*. PhD dissertation, Tokyo Institute of Technology, Japan.

Sugeno, M., Fujimoto, K., and Murofushi, T. (1995). A hierarchical decomposition of choquet integral model. *International Journal of Uncertainty, Fuzziness and Knowledge-Based Systems*, 3(1): 213–222.

Sugeno, M., Narukawa, Y., and Murofushi, T. (1998). Choquet integral and fuzzy measures on locally compact space. *Fuzzy Sets and Systems*, 99(2): 205–211.

Szeląg, M., Greco, S., and Słowiński, R. (2013). Rule-based approach to multicriteria ranking. In: *Multicriteria Decision Aid and Artificial Intelligence* (pp. 127–160). New York: John Wiley.

Szymanski, D.M., and Hise, R.T. (2000). E-satisfaction: An initial examination. *Journal of Retailing*, 76: 309–322.

Taha, Z., and Rostam, S. (2011). A fuzzy AHP–ANN-based decision support system for machine tool selection in a flexible manufacturing cell. *The International Journal of Advanced Manufacturing Technology*, 57(5–8): 719–733.

Tamura, M., Nagata, H., and Akazawa, K. (2002). Extraction and systems analysis of factors that prevent safety and security by structural models. *Proceedings of the 41st SICE Annual Conference* (pp. 1752–1759), Osaka, Japan.

Tang, M.T., Tzeng, G.H., and Wang, S.W. (1999). A hierarchy fuzzy MCDM method for studying electronic marketing strategies in the information service industry. *Journal of International Information Management*, 8(1): 1–22.

Tang, T.I., Shee, D.Y., and Tzeng, G.H. (2002). An MCDM framework for assessing ISPs: The fuzzy synthesis decisions of additive and non-additive measurements. *Journal of Information Management*, 8(2): 175–192.

Tang, T.I., and Tzeng, G.H. (1998). Fuzzy MCDM model for pricing strategies in the internet environment. *Journal of Commercial Modernization*, 1(1): 19–34.

Tay, E.H., and Shen, L. (2002). Economic and financial prediction using rough sets model. *European Journal of Operational Research*, 141(3): 641–659.

Taylor, N. (2014). The rise and fall of technical trading rule success. *Journal of Banking & Finance*, 40: 286–302.

Teece, D.J. (2009). *Dynamic Capabilities and Strategic Management: Organizing for Innovation and Growth*. Oxford, UK: Oxford University Press.

Teng, J.Y., and Tzeng, G.H. (1996). Fuzzy multicriteria ranking of urban transportation investment alternatives. *Transportation Planning and Technology*, 20(1): 15–31.

334 ■ References

Tilley, T., Cole, R., Becker, P., and Eklund, P. (2005). A survey of formal concept analysis support for software engineering activities. In: Ganter, B., Stumme, G., and Wille, R. (eds), *Formal Concept Analysis* (pp. 250–271). Heidelberg, Germany: Springer.

Ting, C.W., Huang, J.W., Wang, D.S., and Tzeng, G.H. (2013). To identify or not to identify: A weighted multidimensional scaling in identifying the similarities of e-shopping stores. *African Journal of Business Management*, 7(22): 2206–2218.

Ting, S.C., and Tzeng, G.H. (2003). Ship scheduling and cost analysis for route planning in liner shipping. *Maritime Economics and Logistics*, 5(4): 378–392.

Ting, S.C., and Tzeng, G.H. (2003). Ship scheduling and service network integration for liner shipping companies and strategic alliances. *Journal of Eastern Asia Society for Transportation Studies*, 5: 765–777.

Ting, S.C., and Tzeng, G.H. (2004). An optimal containership slot allocation for liner shipping revenue management. *Maritime Policy and Management*, 31(3): 199–211.

Ting, S.C., and Tzeng, G.H. (2016). Bi-criteria approach to containership slot allocation in liner shipping. *Maritime Economics and Logistics*, 18(2): 141–157.

To, P.L., Liao, C., and Lin, T.H. (2007). Shopping motivations on Internet: A study based on utilitarian and hedonic value. *Technovation*, 27: 774–787.

Tone, K., and Tsutsui, M. (2009). Network DEA: A slacks-based measure approach. *European Journal of Operational Research*, 197(1): 243–252.

Tone, K., and Tsutsui, M. (2014). Dynamic DEA with network structure: A slacks-based measure approach. *Omega*, 42(1): 124–131.

Tornatzky, L.G., and Fleischer, M. (1990). *The Processes of Technological Innovation*. Lexington, MA: Lexington Books.

Tsai, H.C., Chen, C.M., and Tzeng, G.H. (2006). The comparative productivity efficiency for global telecoms. *International Journal of Production Economics*, 103(2): 509–526.

Tsaur, S.H., and Tzeng, G.H. (1996). Multiattribute decision making analysis for customer preference of tourist hotels. *Journal of Travel and Tourism Marketing*, 4(4): 55–69.

Tsaur, S.H., Tzeng, G.H., and Chang, T.Y. (1997a). Travel agency organization buying behavior: An application of logit model. *Journal of Management and Systems*, 4(2): 127–46.

Tsaur S.H., Tzeng, G.H., and Wang, G.C. (1997b). Evaluating tourist risks from fuzzy perspectives. *Annals of Tourism Research*, 24(4): 796–812.

Tseng, F.M., and Tzeng, G.H. (1999). Forecast seasonal time series by comparing five kinds of hybrid grey models. *Journal of the Chinese Fuzzy Systems Association*, 5(2): 45–55.

Tseng, F.M., and Tzeng, G.H. (2002). A fuzzy seasonal ARIMA model for forecasting. *Fuzzy Sets and Systems*, 126(3): 367–376.

Tseng, F.M., Tzeng, G.H., and Yu, H.C. (1999a). A comparison of four kinds of prediction methods: ARIMA, fuzzy time series, fuzzy regression time series, and grey forecasting; An example of the production value forecast of the mechanical industry in Taiwan. *Journal of the Chinese Grey System Association*, 2(2): 83–98 (in Chinese).

Tseng, F.M., Tzeng, G.H., and Yu, H.C. (1999b). Fuzzy seasonal time series for forecasting the production value of mechanical industry in Taiwan. *Technological Forecasting and Social Change*, 60(3): 263–273.

Tseng, F.M., Tzeng, G.H., Yuan, B.J.C., and Yu, H.C. (2001a). Fuzzy ARIMA model for forecasting the foreign exchange market. *Fuzzy Sets and Systems*, 118(1): 9–19.

Tseng, F.M., Yu, H.C., and Tzeng, G.H. (2001b). Applied hybrid grey model to forecast seasonal time series. *Technological Forecasting and Social Change*, 67(2): 291–302.

Tseng, M.L. (2011). Using hybrid MCDM to evaluate the service quality expectation in linguistic preference. *Applied Soft Computing*, 11(8): 4551–4562.

Tseng, M.L., and Chiu, A.S.F. (2013). Evaluating firm's green supply chain management in linguistic preferences. *Journal of Cleaner Production*, 40: 22–31.

Tseng, Y.H., Durbin, P., and Tzeng, G.H. (2001c). Using a fuzzy piecewise regression to predict the nonlinear time-series of turbulent flows with automatic change-point detection flow. *Turbulence and Combustion*, 67(2): 81–106.

Tsui, C.W., Tzeng, G.H., and Wen, U.P. (2015). A hybrid MCDM approach for improving the performance of green suppliers in the TFT-LCD industry. *International Journal of Production Research*, 53(21): 6436–6454.

Tversky, A., and Kahneman, D. (1986). Rational choice and the framing of decisions. *Journal of Business*, 59(4): 251–278.

Tversky, A., and Kahneman, D. (1992). Advances in prospect theory: Cumulative representation of uncertainty. *Journal of Risk and Uncertainty*, 5(4): 297–323.

Tzeng, G.H., Chang, C.Y., and Lo, M.C. (2005a). MADM approach for effecting information quality of knowledge management. *International Journal of Information Systems for Logistics and Management*, 1(1): 55–67.

Tzeng, G.H., Chang, C.Y., and Lo, M.C. (2005b). The simulation and forecast model for human resources of semiconductor wafer fab operation. *International Journal of Industrial Engineering and Management Systems*, 4(1): 47–53.

Tzeng, G.H., Chang, J.R., Lin, J.D., and Hung, C.T. (2002). Non-additive grey relation model for the evaluation of flexible pavement condition. *International Journal of Fuzzy Systems*, 4(2): 715–724.

Tzeng, G.H., Chang, S.L., Wang, J.C., Hwang, M.J., Yu, G.C., and Juang, M.C. (1998). Application of fuzzy multiobjective programming to the economic energy environment model. *Journal of Management*, 15(4): 683–707 (in Chinese).

Tzeng, G.H., and Chen, C.H. (1993). Multiobjective decision making for traffic assignment. *IEEE Transactions on Engineering Management*, 40(2): 180–187.

Tzeng, G.H., and Chen, J.J. (1997). Developing Taipei automobile driving cycles for emissions. *Energy and Environment*, 8(3): 227–238.

Tzeng, G.H., and Chen, J.J. (1998). Developing Taipei motorcycle driving cycle for emissions and fuel economy. *Transportation Research (Part D)*, 3D(1): 19–27.

Tzeng, G.H., Chen, J.J., and Teng, J.D. (1997). Evaluation and selection of suitable battery for electrics motorcycle in Taiwan: Application of fuzzy multiple attribute decision making. *Journal of the Chinese Institute of Industrial Engineers*, 14(3): 319–331.

Tzeng, G.H., Chen, J.J., and Yen, Y.K. (1996). The strategic model of multicriteria decision making for managing the quality of the environment in metropolitan Taipei. *Asian Journal of Environmental Management*, 4(1): 41–52.

Tzeng, G.H., Chen, T.Y., and Wang, J.C. (1998). A weight-assessing method with habitual domains. *European Journal of Operational Research*, 110(2): 342–367.

Tzeng, G.H., Chen, W.H., Yu, R., and Shih, M.L. (2010). Fuzzy decision maps: A generalization of the DEMATEL methods. *Soft Computing*, 14(11): 1141–1150.

Tzeng, G.H., and Chen, Y.W. (1998). Implementing an effective schedule for reconstructing post-earthquake road-network based on asymmetric traffic assignment: An application of genetic algorithm. *International Journals of Operations and Quantitative Management*, 4(3): 229–246.

Tzeng, G.H., and Chen, Y.W. (1999). The optimal location of airport fire stations: A fuzzy multi-objective programming through revised genetic algorithm. *Transportation Planning and Technology*, 23(1): 37–55.

336 ■ *References*

Tzeng, G.H., Chen, Y.W., and Lin C.Y. (2000). Fuzzy multi-objective reconstruction plan for post-earthquake road-network by genetic algorithm. In: *Research and Practice in Multiple Criteria Decision Making* (pp. 510–528). Lecture Notes in Economics and Mathematical Systems, vol. 487, Part III. Heidelberg, Germany: Springer.

Tzeng, G.H., Cheng, H.J., and Huang, T.D. (2007a). Multi-objective optimal planning for designing relief delivery systems. *Transportation Research Part E: Logistics and Transportation Review*, 43(6): 673–686.

Tzeng, G.H., Chiang, C.H., and Li, C.W. (2007b). Evaluating intertwined effects in e-learning programs: A novel hybrid MCDM model based on factor analysis and DEMATEL. *Expert Systems with Applications*, 32(4): 1028–1044.

Tzeng, G.H., and Chiang, C.I. (1998). Applying possibility regression to grey model. *Journal of the Chinese Grey System Association*, 1(1): 19–31.

Tzeng, G.H., Chiang, C.I., and Hwang, M.J. (1996). Multiobjective programming approach to the allocation of air pollution monitoring station. *Journal of the Chinese Institute of Environment Engineering*, 6(1): 99–105.

Tzeng, G.H., Chiou, Y.C., and Sheu, S.K. (1997). A comparison of genetic and stepwise algorithms for optimal sites of mainline barrier-type toll station. *Journal of the Chinese Institute of Civil and Hydraulic Engineering*, 9(1): 171–178.

Tzeng, G.H., Feng, G.M., and Kang, C.C. (2001). The fuzzy set theory and DEA model for forecasting production efficiency: Case study for Taipei city bus company. *Journal of Advance Computational Intelligences*, 5(3): 128–138.

Tzeng, G.H., and Hu, Y.C. (1996). The section of bus system operation and service performance indicators: Application of grey relation analysis. *Journal of the Chinese Fuzzy Systems Association*, 2(1): 73–82.

Tzeng, G.H., and Huang, C.Y. (2012). Combined DEMATEL technique with hybrid MCDM methods for creating the aspired intelligent global manufacturing & logistics systems. *Annals of Operations Research*, 197(1): 159–190.

Tzeng, G.H., and Huang, J.J. (2011). *Multiple Attribute Decision Making: Methods and Applications*. Boca Raton, FL: CRC Press.

Tzeng, G.H., and Huang, J.J. (2014). *Fuzzy Multiple Objective Decision Making*. Boca Raton, FL: CRC Press.

Tzeng, G.H., Huang, K.W., Lin, C.W., and Yuan, B.J. (2014). New idea of multi-objective programming with changeable spaces for improving the unmanned factory planning. In: *Portland International Center for Management of Engineering and Technology (PICMET '14), "Infrastructure and Service Integration"* (pp. 564–570). Kanazawa, Japan: IEEE.

Tzeng, G.H., and Huang, W.C. (1997). The spatial and temporal bi-criteria parallel savings based heuristic algorithm for vehicle routing problem with time windows. *Transportation Planning and Technology*, 20(2): 163–181.

Tzeng, G.H., Hwang, M.J., and Liu, Y.H. (1997). Dynamic optimal expansion for fuzzy competence sets of multi-stage multi-objective planning. *Pan-Pacific Management Review*, 1(1): 55–70.

Tzeng, G.H., Hwang, M.J., and Yeh, W.C. (1998). Multiobjective planning for integrated land use and outside-allied transportation systems in recreation areas. *Journal of Management*, 15(1): 133–161 (in Chinese).

Tzeng, G.H., Jen, W., and Hu, K.C. (2002). Fuzzy factor analysis for selecting service quality factors: A case of the service quality of city bus service. *International Journal of Fuzzy Systems*, 4(4): 911–921.

Tzeng, G.H., and Kuo, J.S. (1996). Fuzzy multiobjective-double sampling plans with genetic algorithms based on Bayesian model. *Journal of the Chinese Fuzzy Systems Association*, 2(2): 57–74.

Tzeng, G.H., and Lee, M.Y. (2001). Intellectual capital in the information industry. In Chang, C.Y., and Yu, P.L. (eds), *Made by Taiwan: Booming in the Information Technology Era* (pp. 298–344). Singapore: World Science.

Tzeng, G.H., and Lin, C.W. (2000). Evaluation of new substitute fuel modes of buses for suitable urban public transportation. *Transportation Planning Journal*, 29(3): 665–692 (in Chinese).

Tzeng, G.H., Lin, C.W., and Opricovic, S. (2005a). Multi-criteria analysis of alternative-fuel buses for public transportation. *Energy Policy*, 33(11): 1373–1383.

Tzeng, G.H., Ou Yang, Y.P., Lin, C.T., and Chen, C.B. (2005b). Hierarchical MADM with fuzzy integral for evaluating enterprise intranet web sites. *Information Sciences*, 169(3–4): 409–426.

Tzeng, G.H., Shiah, J.Y., and Chiang, C.I. (1998). Application of De Novo programming to land-use plans of theory of Chiao Tung University. *Journal of City and Planning*, 25(1): 93–105.

Tzeng, G.H., Shiau, T.A., and Lin, C.Y. (1992). Application of multicriteria decision making to the evaluation of a new energy system development in Taiwan. *Energy*, 17(10): 983–992.

Tzeng, G.H., Shiau, T.A., and Teng, J.Y. (1994). Multiobjective decision making approach to energy supply mix decisions in Taiwan. *Energy Sources*, 16(3): 301–316.

Tzeng, G.H., Shieh, H.M., and Shiau, T.A. (1989). Route choice behavior in transportation: An application of the multiattribute utility theorem. *Transportation Planning and Technology*, 13(4): 289–301.

Tzeng, G.H., Tang T.I., Hung, Y.M., and Chang, M.L. (2006). Multiple-objective planning for a production and distribution model of the supply chain: Case of a bicycle manufacturer. *Journal of Scientific and Industrial Research*, 65(4): 309–320.

Tzeng, G.H., and Teng, J.Y. (1998). Transportation investment project selection using fuzzy multi-objective programming. *Fuzzy Sets and Systems*, 96(3): 259–280.

Tzeng, G.H., Teng, M.H., Chen, J.J., and Opricovic, S. (2002). Multicriteria selection for a restaurant location in Taipei. *International Journal of Hospitality Management*, 21(2): 171–187.

Tzeng, G.H., Teodorovic, D., and Hwang, M.J. (1996). Fuzzy bi-criteria multi-index transportation problem for coal allocation planning of Taipower. *European Journal of Operational Research*, 95(1): 62–72.

Tzeng, G.H., and Tsaur, S.H. (1993). Application of multicriteria decision making to old vehicle elimination in Taiwan. *Energy and Environment*, 4(2): 268–283.

Tzeng, G.H., and Tsaur S.H. (1994). The multiple criteria evaluation of grey relation model. *The Journal of Grey System*, 6(2): 87–108.

Tzeng, G.H., and Tsaur, S.H. (1997). Application of multiple criteria decision making for network improvement. *Journal of Advanced Transportation*, 31(1): 49–74.

Tzeng, G.H., Tsaur, S.H., Laiw, Y.D., and Serafim, O. (2002). Multicriteria analysis of environmental quality in Taipei: Public preferences and improvement strategies. *Journal of Environmental Management*, 65(2): 109–120.

Tzeng, G.H., Wang, H.F., Wen, U.P., and Yu, P.L. (1994). *Multiple Criteria Decision Making*. New York: Springer.

338 ■ References

Tzeng, G.H., Wang, J.C., and Hwang, M.J. (1996). Using genetic algorithms and the template path concept to solve the traveling salesman problem. *Transportation Planning Journal*, 25(3): 493–516.

Tseng, F.M., Yu, H.C., and Tzeng, G.H. (2002). Combing neural network with seasonal time series ARIMA model. *Technological Forecasting and Social Chang*, 69(1): 71–87.

Verikas, A., Kalsyte, Z., Bacauskiene, M., and Gelzinis, A. (2010). Hybrid and ensemble-based soft computing techniques in bankruptcy prediction: A survey. *Soft Computing*, 14(9): 995–1010.

Von Neumann, J., and Morgenstern, O. (1947). *Theory of Games and Economic Behavior*, 2nd edn. Princeton, NJ: Princeton University Press.

Wang, B., Shyu, J.Z., and Tzeng, G.H. (2007). Appraisal model for admitting new tenants to the incubation center at ITRT. *International Journal of Innovative Computing Information and Control*, 3(1): 119–130.

Wang, C.H., Chin, Y.C., and Tzeng, G.H. (2010). Mining the R&D innovation performance processes for high-tech firms based on rough set theory. *Technovation*, 30(7–8): 447–458.

Wang, F.K., Hsu, C.H., and Tzeng, G.H. (2014). Applying a hybrid MCDM model for six sigma project selection. *Mathematical Problems in Engineering*, Article ID 730934, https://www.hindawi.com/journals/mpe/2014/730934/abs/.

Wang, J.L., and Chan, S.H. (2006). Stock market trading rule discovery using two-layer bias decision tree. *Expert Systems with Applications*, 30(4): 605–611.

Wang, T.C., and Lee, H.D. (2009). Developing a fuzzy TOPSIS approach based on subjective weights and objective weights. *Expert Systems with Applications*, 36(5): 8980–8985.

Wang, Y., Wang, S., and Lai, K.K. (2005). A new fuzzy support vector machine to evaluate credit risk. *IEEE Transactions on Fuzzy Systems*, 13(6): 820–831.

Wang, Y.L., and Tzeng, G.H. (2012). Brand marketing for creating brand value based on a MCDM model combining DEMATEL with ANP and VIKOR methods. *Expert Systems with Applications*, 39(5): 5600–5615.

Warfield, J.N. (ed.) (1976). *Societal Systems, Planning, Policy and Complexity*. New York: John Wiley.

Wassmer, U. (2008). Alliance portfolios: A review and research agenda. *Journal of Management*, 36(1): 141–171.

Weber, C.A., Current, J.R., and Benton, W.C. (1991). Vendor selection criteria and methods. *European Journal of Operational Research*, 50(1): 2–18.

Wei, L.Y., Chen, T.L., and Ho, T.H. (2011). A hybrid model based on adaptive-network-based fuzzy inference system to forecast Taiwan stock market. *Expert Systems with Applications*, 38(11): 13625–13631.

Wille, R. (2005). Formal concept analysis as mathematical theory of concepts and concept hierarchies. In: *Formal Concept Analysis* (pp. 1–33). Heidelberg, Germany: Springer.

Williamson, O.E. (1991). Comparative economic organization: The analysis of discrete structural alternatives. *Administrative Science Quarterly*, 36(2): 269–296.

Wong, Y.K., and Hsu, C.J. (2008). A confidence-based framework for business to consumer (B2C) mobile commerce adoption. *Personal and Ubiquitous Computing*, 12: 77–84.

Wu, C.H., Tzeng, G.H., Goo, Y.J., and Fang, W.C. (2007). A real-valued genetic algorithm to optimize the parameters of support vector machine for predicting bankruptcy. *Expert Systems with Applications*, 32(2): 397–408.

Wu, C.H., Tzeng, G.H., and Lin, R.H. (2009). A novel hybrid genetic algorithm for kernel function and parameter optimization in support vector regression. *Expert Systems with Applications*, 36(3): 4725–4735.

Wu, C.H., Tzeng, Y.L., Kuo, B.C., and Tzeng, G.H. (2014). Affective computing techniques for developing a human affective norm recognition system for u-learning systems. *International Journal of Mobile Learning and Organization*, 8(1): 50–66.

Wu, D., Wu, D.D., Zhang, Y., and Olson, D.L. (2013). Supply chain outsourcing risk using an integrated stochastic-fuzzy optimization approach. *Information Sciences*, 235: 242–258.

Wu, H.Y., Chen, J.K., and Chen, I.S. (2012). Ways to promote valuable innovation: Intellectual capital assessment for higher education system. *Quality and Quantity*, 46(5): 1377–1391.

Wu, H.Y., Tzeng, G.H., and Chen, Y.H. (2009). A fuzzy MCDM approach for evaluating banking performance based on balanced scorecard. *Expert Systems with Applications*, 36(6): 10135–10147.

Xidonas, P., Mavrotas, G., and Psarras, J. (2009). A multicriteria methodology for equity selection using financial analysis. *Computers and Operations Research*, 36(12): 3187–3203.

Yang, C., Fu, G.L., and Tzeng, G.H. (2007). A multicriteria analysis of the strategies to open Taiwan's mobile virtual network operators services. *International Journal of Information Technology and Decision Making*, 6(1): 85–112.

Yang, C.A., Fu, G.L., and Tzeng, G.H. (2005). Creating a win–win in the telecommunications industry: The relationship between MVNOs and MNOs in Taiwan. *Canadian Journal of Administration Sciences*, 22(4): 316–328.

Yang, J.L., Chiu, H.N., and Tzeng, G.H. (2008). Vendor selection by integrated fuzzy MCDM techniques with independent and interdependent relationships. *Information Sciences*, 178(21): 4166–4183.

Yang, J.L., and Tzeng, G.H. (2011). An integrated MCDM technique combined with DEMATEL for a novel cluster-weighted with ANP method original research. *Expert Systems with Applications*, 38(3): 1417–1424.

Yeh, C.H., Deng, H., and Chang, Y.H. (2000). Fuzzy multicriteria analysis for performance evaluation of bus companies. *European Journal of Operational Research*, 126(3): 459–473.

Yu, J.R., and Tzeng, G.H. (2009). A fuzzy multiple objective programming in interval piecewise regression model. *International Journal Uncertainty, Fuzziness and Knowledge-Based Systems*, 17(3): 365–376.

Yu, J.R., Tzeng, G.H., Chiang, C.I., and Sheu, H.J. (2007). Raw material supplier ratings in the semiconductor manufacturing industry through fuzzy multiple objectives programming to DEA. *International Journal of Operations and Quantitative Management*, 13(4): 101–111.

Yu, J.R., Tzeng, G.H., and Li, H.L. (1999). A general piecewise necessity regression analysis based on linear programming. *Fuzzy Sets and Systems*, 105(3): 429–436.

Yu, J.R., Tzeng, G.H., and Li, H.L. (2001). General fuzzy piecewise regression analysis with automatic change-point detection. *Fuzzy Sets and Systems*, 119(2): 247–257.

Yu, J.R., Tzeng, G.H., and Li, H.L. (2005). Interval piecewise regression model with automatic change-point detection by quadratic programming. *International Journal of Uncertainty, Fuzziness and Knowledge-Based Systems*, 13(3): 347–361.

340 ■ *References*

Yu, J.R., Tzeng, Y.C., Tzeng, G.H., Yu, T.Y., and Sheu, H.J. (2004). A fuzzy multiple objective programming approach to DEA with imprecise data. *International Journal of Uncertainty, Fuzziness and Knowledge-Based Systems*, 12(5): 591–600.

Yu, P.L. (1973). A class of solutions for group decision problems. *Management Science*, 19(8): 936–946.

Yu, P.L., and Chianglin, C.Y. (2006). Decision traps and competence dynamics in changeable spaces. *International Journal of Information Technology and Decision Making*, 5(1): 5–18.

Yu, P.L., and Zeleny, M. (1975). The set of all nondominated solutions in linear cases and a multicriteria simplex method. *Journal of Mathematical Analysis and Applications*, 49(2): 430–468.

Yu, R.C., and Tzeng, G.H. (2006). A soft computing method for multi-criteria decision making with dependence and feedback. *Applied Mathematics and Computation*, 180(1): 63–75.

Yuan, B.J.C., Wang, C.P., and Tzeng, G.H. (2005). An emerging approach for strategy evaluation in fuel cell development. *International Journal of Technology Management*, 32(3–4): 302–338.

Zadeh, L.A. (1965). Fuzzy sets. *Information and Control*, 8(3): 338–353.

Zaras, K. (2011). The dominance-based rough set approach (DRSA) applied to bankruptcy prediction modeling for small and medium businesses. In: *Multiple Criteria Decision Making* (pp. 287–295). Katowice, Poland: University of Economics in Katowice.

Zavadskas, E.K., Govindan, K., Antucheviciene, J., and Turskis, Z. (2016). Hybrid multiple criteria decision-making methods: A review of applications for sustainability issues. *Economic Research*, 29(1): 857–887.

Zavadskas, E.K., and Turskis, Z. (2011). Multiple criteria decision making (MCDM) methods in economics: An overview. *Technological and Economic Development of Economy*, 17(2): 397–427.

Zeleny, M. (1972). Linear multiobjective programming. PhD thesis, Graduate School of Management, University of Rochester, UK.

Zeleny, M. (1982). *Multiple Criteria Decision Making*. New York: McGraw-Hill.

Zeleny, M. (1986). Optimal system design with multiple criteria: De Novo programming approach. *Engineering Costs and Production Economics*, 10(2): 89–94.

Zeleny, M. (1990). Optimizing given systems vs. designing optimal systems: The De Novo programming approach. *International Journal of General Systems*, 17(4): 295–307.

Zeleny, M. (1995). Trade-offs-free management via De Novo programming. *International Journal of Operations and Quantitative Management*, 1(1): 3–13.

Zeleny, M. (1998). Multiple criteria decision making: Eight concepts of optimality. *Human Systems Management*, 17(2): 97–107.

Zeleny, M. (2012). *Linear Multiobjective Programming*, vol. 95. Heidelberg New York: Springer-Verlag.

Zhao, H., and Sinha, A.P., and Ge, W. (2009). Effects of feature construction on classification performance: An empirical study in bank failure prediction. *Expert Systems with Applications*, 36(2): 2633–2644.

Zhou, X.S., and Dong, M. (2004). Can fuzzy logic make technical analysis 20/20? *Financial Analysts Journal*, 60(4): 54–75.

Zhu, B.W., Zhang, J.R., Tzeng, G.H., Huang, S.L., and Xiong, L. (2017). Public open space development for elderly people by using the DANP-V Model to establish continuous improvement strategies towards a sustainable and healthy aging society. *Sustainability*, 9(3): 420–449.

Zhu, K., and Kraemer, K.L. (2005). Post-adoption variations in usage and value of e-business by organizations: Cross-country evidence from the retail industry. *Information Systems Research*, 16(1): 61–84.

Zhu, K., Kraemer, K.L., Xu, S., and Dedrick, J. (2004). Information technology payoff in e-business environments: An international perspective on value creation of e-business in the financial services industry. *Journal of Management Information Systems*, 21(1): 17–54.

Zimmermann, H.J. (1978). Fuzzy programming and linear programming with several objective functions. *Fuzzy Sets and Systems*, 1(1): 45–55.

Zimmermann, H.J. (2011). *Fuzzy Set Theory—And Its Applications*. New York: Springer Science & Business Media.

Zopounidis, C., and Doumpos, M. (2000). Building additive utilities for multi-group hierarchical discrimination: The M.H.Dis method. *Optimization Methods and Software*, 14(3): 219–240.

Zopounidis, C., and Doumpos, M. (2013). Multicriteria decision systems for financial problems. *Top*, 21(2): 241–261.

Zopounidis, C., Galariotis, E., Doumpos, M., Sarri, S., and Andriosopoulos, K. (2015). Multiple criteria decision aiding for finance: An updated bibliographic survey. *European Journal of Operational Research*, 247(2): 339–348.

Zopounidis, C., Godefroid, M., and Hurson, C. (1995). Designing a multicriteria decision support system for portfolio selection and management. In: Janssen, J., Skiadas, C., and Zopounidis, C. (eds), *Advances in Stochastic Modelling and Data Analysis* (pp. 261–292). Dordrecht, the Netherlands: Springer.

Zwass, V. (1996). Electronic commerce: Structures and issues. *International Journal of Electronic Commerce*, 1(1): 3–23.

Index

A

Additive-type aggregators, 68
AHP, *see* Analytic hierarchy process (AHP)
Analytic hierarchy process (AHP), 3, 7, 62–63, 202
Analytic network process (ANP), 7, 59, 62–63, 202, 256
ANNs, *see* Artificial neural networks (ANNs)
ANP, *see* Analytic network process (ANP)
Artificial intelligence (AI), 11
Artificial neural networks (ANNs), 12, 240–241
Aspiration level, 24, 27–28, 66–67, 216, 217
Automated factory planning
  case background, 230
  optimization evolution
    conventional Pareto solution, 232
    De Novo programming, 232–233
    new ideas of changeable spaces, 233–235
  overview, 229–230

B

BA, *see* Business analytics (BA)
B2B, *see* Business-to-business (B2B) m-commerce
Biery, K., 110
Big Data era, 26, 93
Bipolar decision model, 104
B/M, *see* Book-to-market-value (B/M)
Book-to-market-value (B/M), 184
Bounded rationality, 25
Brand equity, 153
Business analytics (BA)
  background of, 290–291
  hybrid bipolar MRDM model, 291–298
  overview, 289

semiconductor industry case
  bipolar weighting system, 299–302
  data, 298–299
  fuzzy performance evaluations using modified VIKOR, 302–305
Business-to-business (B2B) m-commerce
  and DEMATEL analysis, 172–174
  and DEMATEL-based ANP (DANP) method, 172–174
  and external environment aspect, 171–172
  modified VIKOR for performance gap aggregation, 174–177
  and organizational environment aspect, 171
  overview, 169–170
  and technological environment aspect, 170–171

C

Cash flow from operations (CFO), 185
Cash flow ROA, 185
CFO, *see* Cash flow from operations (CFO)
Changeable spaces, 25, 73, 78
  MOP with, 83–89
Charnes–Cooper–Rhodes (CCR) model, 10
Choquet, Gustave, 68
Choquet integral, 68, 207
Compromise solution, 3, 8, 67, 74, 76–77, 81–83, 89, 206
Computational intelligence, 11
  and TA, 240–242
Computer-supported linear programming, 11
CORE, 95, 98, 100, 103
Core attribute-based MRDM approach, 100
CRM, *see* Customer relationship management (CRM)
Curse of dimensionality, 8
Customer equity, 153, 166

343

344 ■ *Index*

Customer relationship management (CRM), 177
Customer satisfaction, 120

D

DANP, *see* DEMATEL-based ANP (DANP) method
Data-centric analytics, 29
Data-centric problems, and MRDM, 93–106
  core attribute-based, 100
  hybrid bipolar, 100–105
    dominance-based rough set approach, 101–104
    evaluations for aggregated bipolar decision model, 104
    reference point-based, 96–99
    variable-consistency dominance-based rough set approach, 94–96
Data envelopment analysis (DEA), 10, 256
Data fusion, 250
Data preprocessing, 242–243
  numerical experiment, 244–249
DCs, *see* Decision classes (DCs)
Decision classes (DCs), 16, 17
Decision makers (DMs), 2, 3, 7, 24, 57, 64, 89, 99
Decision-making trial and evaluation laboratory (DEMATEL) analysis, 26–27
  analysis, 156–161
  for assessing information risk, 109–118
    calculating DANP, 114–117
    with INRM, 113–114
    research framework, 110–113
  and business-to-business (B2B) m-commerce, 172–174
  concepts, 33–45
    background and basic notions, 34
    infeasibility, 37–38
    numerical examples, 39–45
    operational steps, 34–37
    revised, 38–39
  and G-score model, 186–192
  and nonadditive hybrid MADM model, 208
  technique for forming INRM, 49–59
    constructing influential network relations map, 50–54
    and DNP, 54–56
    methodology for assessing real-world problems, 49–50

    problem solving for ranking/selection decision, 57–58
Decision-making units (DMUs), 10, 204
Decision rules, 103
Decision space, 27, 236–237
DEMATEL, *see* Decision-making trial and evaluation laboratory (DEMATEL) analysis
DEMATEL-based ANP (DANP) method, 64, 70, 109
  and business-to-business (B2B) m-commerce, 172–174
  e-store business evaluation, 122–126
  and G-score model, 186–192
  information risk assessment, 114–117
  INRM and, 54–58
  and nonadditive hybrid MADM model, 208
  smartphone improvement, 156–161
  thin film transistor liquid crystal display (TFT-LCD) industry, 136–137
De Novo programming, 3, 78–83
  numerical case, 224–226
  optimization, 232–233
  overview, 219–220
  and resource allocation, 221–224
  resource-dependent theory, 220–221
  strategic behavior and organizational learning, 221
  transaction cost theory, 220
DFG, *see* Directional flow graph (DFG)
Directional flow graph (DFG), 100
Directional implication relationship map (DIRM), 278, 287
DMs, *see* Decision makers (DMs)
DMU, *see* Decision-making units (DMUs)
Dominance-based rough set approach (DRSA), 16, 18
Dominance relation, 18
DRSA, *see* Dominance-based rough set approach (DRSA)
Dubois, D., 68

E

Earning-to-price (E/P) ratio, 184
Efficient market hypothesis (EMH), 184
EMH, *see* Efficient market hypothesis (EMH)
E/P, *see* Earning-to-price (E/P) ratio
Equipment manufacturing (OEM), 230
E-store business evaluation, and improvement, 119–132

DANP method, 122–126
performance measures and modified
VIKOR, 126
research framework, 120–122
External environment, and B2B m-commerce,
171–172

F

FA, *see* Fundamental analysis (FA)
FAHP, *see* Fuzzy AHP (FAHP)
FANP, *see* Fuzzy ANP (FANP)
FCA, *see* Formal concept analysis (FCA)-based
DANP model
FCF, *see* Free cash flow (FCF)
Financial performance (FP)
case of five commercial banks, 259–264
and modified VIKOR method, 264–268
and MRDM, 257–259
overview, 255
research background, 255–257
FIS, *see* Fuzzy inference systems (FIS)
Formal concept analysis (FCA)-based
DANP model
case study of semiconductor companies,
280–284
and hybrid MRDM model, 278–279
improvement planning, 286–288
MCDM and soft computing methods,
275–278
overview, 273–275
for ranking improvement plans, 284–286
Forrester research, 119
FP, *see* Financial performance (FP)
Free cash flow (FCF), 184
FSAW, *see* Fuzzy simple additive
weighting (FSAW)
Fundamental analysis (FA), 276
Fuzzy AHP (FAHP), 63
Fuzzy ANP (FANP), 63
Fuzzy inference systems (FIS), 241, 249–250
Fuzzy integrals, 68–71
and nonadditive hybrid MADM model,
206–216
Fuzzy measure, 68
Fuzzy set theory, 7, 13–14
Fuzzy simple additive weighting (FSAW), 264

G

Glamour stock selection (G-score) model
DEMATEL analysis and DANP, 186–192

examination of stock returns, 195–198
research framework, 184–186
Goal programming (GP), 74
GP, *see* Goal programming (GP)
Granular computing, 105, 292
Granules of knowledge, 14, 94
G-score, *see* Glamour stock selection
(G-score) model

H

Hazardous substance management (HSM),
203
HMCDM, *see* Hybrid MCDM (HMCDM)
Holding period return (HPR), 195
HPR, *see* Holding period return (HPR)
HSM, *see* Hazardous substance
management (HSM)
Hybrid bipolar MRDM approach, 100–105,
291–298
dominance-based rough set approach,
101–104
evaluations for aggregated bipolar decision
model, 104
Hybrid MADM model, 136
e-store business evaluation and
improvement using, 119–132
DANP for finding influential weights,
122–126
performance measures and
modified VIKOR, 126
research framework, 120–122
smartphone improvements based
on, 151–167
DANP influential weights for criteria,
156–161
DEMATEL analysis, 156–161
kinds of analytics, 165–166
modified VIKOR for performance gap
aggregation, 161–165
research framework, 152–155
vendors in Taiwan, 152
Hybrid MCDM (HMCDM), 1, 26–33
framework, 28–33
need of, 26–28

I

IDSS, *see* Intelligent decision support
system (IDSS)
Improvement planning, 286–288, 291,
306–307

346 ■ Index

Influential network relationships map (INRM), 37, 174
  constructing, 50–54
  DEMATEL analysis with, 113–114
  problem solving for ranking/selection decision by, and DANP, 57–58
Initial direct-relation matrix, 35
INRM, *see* Influential network relationships map (INRM)
Intelligent decision support system (IDSS), 242–243
ISO/IEC 17799, 110

L

Lambda-max method, 62
Lexicographic GP (LGP), 74, 76
LGP, *see* Lexicographic GP (LGP)
Linear programming, *see* Multi-objective programming (MOP)
Location-based services, 155

M

Machine learning, 10–15, 240, 241–242, 256, 257
MADM, *see* Multiple attribute decision making (MADM) model
Mass customization, 222
Mathematical programming models, and MADM, 203
MAUT, *see* Multiattribute utility theory (MAUT)
MCDM, *see* Multiple criteria decision making (MCDM)
Memory, 155
MGP, *see* Min-max GP (MGP)
Min-max GP (MGP), 74, 76–78
Mobile convenience, 155, 165
Mobile multimedia services, 155
Mobile wallet services, 155, 165–166
Modified VIKOR for performance gap aggregation, 27, 64–68, 126
  and B2B m-commerce, 174–177
  and fuzzy performance evaluations using, 302–305
  modified, for performance gap aggregation, 161–165
  multiple attribute decision making (MADM) model, 192–195

and nonadditive hybrid MADM model, 205–206
MODM, *see* Multiple objective decision making (MODM)
MOP, *see* Multiobjective programming (MOP)
Morgenstern, O., 5
MRDM, *see* Multiple rules-based decision making (MRDM)
Multiattribute utility theory (MAUT), 2–3, 5, 27, 68, 277
Multigraded dominance, 18
Multiobjective programming (MOP), 10, 74, 219, 222, 231
  with changeable parameters, 83–89
Multiple attribute decision making (MADM) model, 2, 3, 23–24, 49–50, 109, 135, 229
  and glamour stock selection (G-score) model
    DEMATEL analysis and DANP, 186–192
    examination of stock returns, 195–198
    research framework, 184–186
  history, 5–7
  modified VIKOR for performance gap aggregation, 192–195
  overview, 183
  research background and investment strategy, 184
Multiple criteria decision making (MCDM), 10, 16, 19, 201, 255
  concepts and trends in, 23–33
    framework of hybrid models, 28–33
    need of hybrid approaches, 26–28
    problem solving in traditional, 25–26
  overview, 1–3
  and soft computing methods, 275–278
  statistics *vs.*, 4–5
Multiple objective decision making (MODM), 2, 3, 25, 203, 229
  with De Novo, 73–91
    concepts and trends, 74–78
    description, 73
    MOP with changeable parameters, 83–89
    overview, 89–91
    programming, 78–83
  history, 8–10
Multiple regression model, 4
Multiple rules-based decision making (MRDM), 16, 23, 239
  and FCA-based DANP model, 278–279

and financial performance (FP), 257–259
for solving data-centric problems, 93–106
    core attribute-based, 100
    hybrid bipolar, 100–105
    reference point-based, 96–99
    variable-consistency dominance-based
        rough set approach, 94–96

N

National Information and Communication
        Security Taskforce (NICST), 110
Net flow score (NFS), 96
New hybrid modified MADM, 63–68
    DEMATEL-based ANP (DANP), 64
    modified VIKOR for measuring
        performance gaps, 64–68
NFS, *see* Net flow score (NFS)
NICST, *see* National Information and
        Communication Security
        Taskforce (NICST)
Nonadditive hybrid MADM model
    DEMATEL and DANP, 208
    description, 203
    and fuzzy integrals, 206–216
    improving aspired levels, 216–217
    modified VIKOR for performance gap
        aggregation, 205–206
    overview, 201–202
    research background and literature review,
        202–204
        combined and integrated approaches,
            203–204
        mathematical programming
            models, 203
        multiple objective decision making
            (MODM), 203
Nonadditive-type aggregators, 24–25, 27,
        68–71
Noninferior nondominated solutions/effective
        solutions, *see* Pareto optimization

O

OECD, *see* Organisation for Economic
        Co-operation and Development
        (OECD)
OEM, *see* Equipment manufacturing (OEM)
Operating system, 155
Optimization
    De Novo programming, 232–233
    new ideas, 233–235

Pareto solution, 232
Organisation for Economic Co-operation and
        Development (OECD), 171
Organizational environment, and B2B
        m-commerce, 171
Organizational learning, and strategic
        behavior, 221
Outranking methods, 7

P

Pairwise comparison table (PCT), 18, 96, 97
Pareto optimization, 3, 74, 232
PCT, *see* Pairwise comparison table (PCT)
PDCA, *see* Plan, do, check, and act (PDCA)
Plan, do, check, and act (PDCA), 117–118
Prade, H., 68
Preference ranking organization method
        for enrichment evaluation
        (PROMETHEE), 135, 137–147
Preferential independence, 5
Processor, 155
Product function, 155
PROMETHEE, *see* Preference ranking
        organization method for enrichment
        evaluation (PROMETHEE)
Prospect theory, 27

R

R&D, *see* Research and development (R&D)
RDEC, *see* Research Development and
        Evaluation Commission (RDEC)
REDUCT, 18, 95, 98, 103, 283
Reference point-based MRDM approach,
        96–99
Remote control services, 155
Research and development (R&D), 186, 236
Research Development and Evaluation
        Commission (RDEC), 110
Resource allocation, and De Novo
        programming, 221–224
Resource-dependent theory, 220–221
Retention equity, 153
Return on assets (ROA), 185, 259
Return on equity (ROE), 184
Risk assessment, 110
Risk control assessment system, 110
Risk monitoring and review, 110
Risk remediation, 110
Risk treatment, 110
ROA, *see* Return on assets (ROA)

348 ■ *Index*

ROE, *see* Return on equity (ROE)
Rough set theory (RST), 14–15
RST, *see* Rough set theory (RST)

S

SAW, *see* Simple additive weighting (SAW) method
SCM, *see* Supply chain management (SCM)
SD, *see* Standard deviation (SD)
Simple additive weighting (SAW) method, 68
Simulated investment performance, and TA, 252–253
Smartphone improvements, based on hybrid MADM model, 151–167
    DANP method, 156–161
    DEMATEL analysis, 156–161
    kinds of analytics, 165–166
    modified VIKOR for performance gap aggregation, 161–165
    research framework, 152–155
    vendors in Taiwan, 152
Soft computing techniques, 12
    fuzzy set theory, 13–14
    and MCDM, 275–278
    rough set theory, 14–15
Standard deviation (SD), 259
Statistics *vs.* MCDM approach, 4–5
Strategic behavior and organizational learning, 221
Sugeno, M., 27, 68
Supermatrix, 55, 59, 62–63, 114, 126, 161, 174, 208, 262
Supply chain management (SCM), 177

T

TDC, *see* Tourism destination competitiveness (TDC)
Technical analysis (TA)
    and computational intelligence, 240–242
    and data preprocessing, 242–243
        numerical experiment, 244–249
    and fuzzy inference systems (FIS), 249–250
    and intelligent decision support system (IDSS), 242–243
    overview, 239
    simulated investment performance, 252–253
    and VC-DRSA, 242–243, 250–252
Technological environment, and B2B m-commerce, 170–171
Technology-organization-environment (TOE), 170

"Theory of Fuzzy Integrals and Its Applications," 27
Thin film transistor liquid crystal display (TFT-LCD), green suppliers in, 133–149
    case study, 134
    DANP method, 136–137
    modified PROMETHEE, 137–147
    research framework and selected criteria, 135–136
    weighted performance gaps, 147–149
TL Company, 134, 135
TOE, *see* Technology-organization-environment (TOE)
Tokyo Institute of Technology, 27
Touch panel, 155
Tourism destination competitiveness (TDC), 50
Trade-offs, 229
Traditional MADM, for ranking and selection, 61–63
Traditional MCDM, 25–26
Transaction cost theory, 220
Tzeng, G.-H., 23

U

UTA, *see* Utility additive (UTA)
UTADIS, *see* Utilités Additives Discriminantes (UTADIS)
Utilités Additives Discriminantes (UTADIS), 256
Utility additive (UTA), 3, 277

V

Value equity, 153
Variable-consistency dominance-based rough set approach (VC-DRSA), 94–96, 239, 242–243, 250–252
VC-DRSA, *see* Variable-consistency dominance-based rough set approach (VC-DRSA)
Vector optimization, 8

W

Weighted GP (WGP), 74, 75–76
WGP, *see* Weighted GP (WGP)

X

XScript, 242